ELEMENTS OF MOLECULAR SYMMETRY

ELEMENTS OF MOLECULAR SYMMETRY

YNGVE ÖHRN
University of Florida
Gainesville, Florida

WILEY-INTERSCIENCE

A John Wiley & Sons, Inc. Publication

New York • Chichester • Weinheim • Brisbane • Singapore • Toronto

This book is printed on acid-free paper.♾

Published simultaneously in Canada.

For ordering and customer service, call 1-800-CALL-WILEY.

Library of Congress Cataloging-in-Publication Data:

Öhrn, Yngve.
Elements of molecular symmetry / Yngve Öhrn.
p. cm.
Includes index.
ISBN 0-471-36323-5 (alk. paper)
1. Molecular theory. 2. Symmetry (Physics) 3. Group theory.
I. Title
QD461.037 2000
541.2′8--dc21 99-39819

10 9 8 7 6 5 4 3 2 1

CONTENTS

PREFACE

Much of the material in this text has existed for a long time as lecture notes. These lectures were presented during the Scandinavian Summer Schools in Quantum Chemistry organized by the Uppsala Quantum Chemistry Group. In these annual schools, which were held from 1958 through 1987, group theory was initially taught by Professors Laurens Jansen and John A. Coleman. The author learned a lot from listening to these two experts. From 1966 on he accepted the responsibility to teach group theory, and this text has grown out of these summer school lectures. Discussions about symmetry and other matters with the author's late friend Jean-Louis Calais have greatly influenced these notes. Some material on elementary molecular orbital theory, space groups, and continuous groups have been added, and the text as presented has formed the basis of a one-semester course for beginning graduate students at the University of Florida.

There are so many texts on group theory and its application to molecular symmetry that it is difficult to justify yet another book on the subject. One can perhaps argue that each text has a different approach and a different combination of applications that might appeal to a different readership. The author has in part followed a rather formal mode of presentation and not shied away from some formal mathematics. However, much material is presented and applied without proofs. There are other texts that take a similar approach, and this is why these notes have remained unpublished for so long. However, with the encouragement from students and colleagues, the author has tried to arrange the notes in a manner that can be enjoyed as a published text.

Much of the notes in their present form were put together for a series of lectures that the author gave at the Université Notre Dame de la Paix, at

Namur, Belgium, when he spent a good part of 1995 as the holder of the International Francqui Chair. The interaction with Jean-Marie André, Joseph Delhalle, Joseph Fripiat, and Benoît Champagne are gratefully acknowledged. The editing and final writing of the book was done while the author spent a summer month at Odense University, Denmark in 1997 and 1998. The hospitality of professor Jens Oddershede is gratefully acknowledged, as are the daily discussions over coffee with my colleague and friend Jack Sabin.

YNGVE ÖHRN

University of Florida
Gainesville, FL

ELEMENTS OF MOLECULAR SYMMETRY

1

INTRODUCTION

The concept of a group pervades mathematics, both pure and applied. In physics and chemistry group theory derives its importance from the fact that it provides the language in which to discuss and describe the symmetry in physical law. The fact that certain conservation laws are linked to symmetries of dynamical equations is perhaps the most fundamental reason for the study of group theory in physics and chemistry. The conservation of total momentum of a physical system is related to *symmetry or invariance* under space translations. Conservation of energy, is similarly linked to the invariance under time translations. The conservation of angular momentum depends on the invariance under space rotations. In this text we deal primarily with molecular systems. When the labels of the electrons and those of identical nuclei that make up molecular systems are permuted the consequences are described by so-called symmetric groups. If such nuclei have positions that are related by rotations of coordinate axes or reflections in coordinate planes, the associated symmetry is described by point groups. Modern group theory is developed in abstract form, but many of its applications still concern symmetry. Pure group theory is seldom as useful as group algebra, and, in particular, representation theory, since those subjects are concerned with how symmetry governs and simplifies the description of physical systems.

This book takes the reader through several topics of group theory and group algebra. The emphasis is on the application of these subjects to the study of molecules. Some parts of the material are presented with detailed proofs, while other parts are simply introduced via applications.

The introduction of group algebra and representation theory follows closely a paper by P. O. Löwdin in *Reviews of Modern Physics* **39**, 259 (1967), which develops the algebra structure without the use of specific carrier spaces. "The Symmetric Group Made Easy" by A. J. Coleman in *Advances in Quantum Chemistry* **4**, 83 (1968) has inspired the chapters on the permutation groups.

A presentation of molecular orbital theory includes discussion of some standard parametrized methods. Chapter 15, on molecular vibrations, applies the group theory results to symmetry modes and a brief discussion of normal modes. Although most of the material applies to molecular symmetry, Chapter 16 discusses introductory space group material. Chapter 17, on continuous groups, is intended to serve as a minimal introduction to the subject. Some of the material on Lie groups follows closely a paper by B. G. Adams, J. Cizek, and J. Paldus, in *Advances in Quantum Chemistry* **19**, 1 (1988).

2

THE GROUP CONCEPT

The idea of a group is a special case of the concept of a *set*, which is, in general, any collection or assembly of objects (elements), abstract or concrete, finite or infinite. Sets are labeled by capital letters and their elements by lowercase letters. One says that an element a belongs to a set A and writes

$$a \in A \tag{2.1}$$

or an element b does not belong to the set,

$$b \notin A. \tag{2.2}$$

Sometimes sets may be specified by a statement, which can be exemplified by the set X of all-black cats: $X = \{x \in \text{cats} | x = \text{black}\}$, where there is a larger set (universe) understood, as the set "cats," and the second part of the statement defines the subset.

Different categories of sets may be distinguished. Two major subdivisions of all possible sets are (1) sets without structure, and (2) sets with structure. Sets of type (2) can be further classified according to kinds of structure relationships:

Sets with Structure

1. Order relations ($a > b$)
2. Neighborhood relations (topology)
3. Algebraic structure (composition law $a \cdot b = c,\ a, b, c \in A$)

Groups are sets with algebraic structure, and the combination law is commonly referred to as *group multiplication.* The concept of a group has four basic postulates or axioms. A set G is a group if it possesses I–IV in the following list.

Group Postulates

I. The *closure relation,* such that for any two elements $a, b \in G$, $ab = c \in G$, where ab means that the element a is combined with the element b using the group multiplication law of G and that the result is an element c also belonging to G.

II. The associative law holds, such that $(ab)c = a(bc)$ for $a, b, c \in G$, which means that first combining a with b and the result with c equals combing a with the result of bc.

III. A unique identity element $e \in G$ exists such that $ea = ae = a$ for all $a \in G$.

IV. Every element $a \in G$ has a unique inverse $a^{-1} \in G$, such that $aa^{-1} = a^{-1}a = e$.

Groups with a finite number of elements are called *finite groups*; groups with an infinite number of elements are called *infinite groups.* The elements of all finite groups and some infinite groups can be labeled by a discrete index

$$G: \{g_1, g_2, \ldots, g_n\} \tag{2.3}$$

and the number of elements $n = |G|$ is called the *order of the group.* When the elements of an infinite group are labeled by one or more indices that can assume continuous values in some domain, the group is called a *continuous group.* In the following some theorems and considerations apply to all groups, which are then stated. However, the development of the basic tools apply principally to finite groups. A special chapter on continuous groups (Chapter 17) discusses some of the details necessary for such groups.

Finite groups of moderate order can be completely specified via a *group multiplication table*, which schematically works as follows for a generic group $G: \{e, a, b, c\}$:

G	e	a	b	c
e	e	a	b	c
a	a	aa	ab	ac
b	b	ba	bb	bc
c	c	ca	cb	cc

This mode of presentation is obviously less useful for groups of large order or for infinite groups. There are other ways to define groups, using so called generators and generating relations that can be used for such cases. This is discussed later.

In general, $ab \neq ba$ for any two elements of a group. Should all elements of a group commute, specifically, $ab = ba$ for all $a, b \in G$, then the group is said to be Abelian (named after the Norwegian mathematician N. H. Abel). One sometimes refers to other groups as being non-Abelian.

Examples of groups are presented in the following paragraphs.

Example 2.1 The group of the four complex numbers $G = \{1, i, -1, -i\}$ with the combination law of ordinary multiplication, obviously has the multiplication table:

G	1	i	-1	$-i$
1	1	i	-1	$-i$
i	i	-1	$-i$	1
-1	-1	$-i$	1	i
$-i$	$-i$	1	i	-1

This table contains only the group elements, showing the closure property. The associative law is obviously valid for the multiplication of numbers. The identity element is readily identified as the number 1, and every element has an inverse. We note that -1 is its own inverse, while $-i$ is the inverse of i (and, of course, vice versa).

Example 2.2 Another example of a finite group is the point group C_{3v}, the elements of which are all geometric operations that move the ammonia molecule (NH_3) in its equilibrium geometry into itself. There are a number of groups of this kind both finite and infinite that are discussed in some detail in Chapter 7. The term *point group* derives from the fact that the geometric operations leave at least one point fixed. The elements of this group C_{3v}: $\{E, C_3, C_3^2, \sigma_1, \sigma_2, \sigma_3\}$ can be understood by referring to Figure 2.1.

The notations are such that C_3 is a rotation of $\frac{1}{3}$ of a full turn, or $2\pi/3$ radians about the (threefold) symmetry axis. The element C_3^2 is the rotation of $4\pi/3$ about the symmetry axis. A full turn, or $C_3^3 = E$, of course, equals the identity element, while the reflections in the three (vertical) symmetry planes, which intersect along the symmetry axis, are labeled $\sigma_1, \sigma_2, \sigma_3$. The combination law is the successive application of these geometric operations. It is important to note that the elements can be applied in two different ways leading to distinct but equivalent interpretations:

- The operations are carried out on the system of Cartesian coordinates in which the molecule is placed. This means that the coordinate axes, symmetry axes, and planes move and the molecule as defined by the nuclear positions stays fixed.

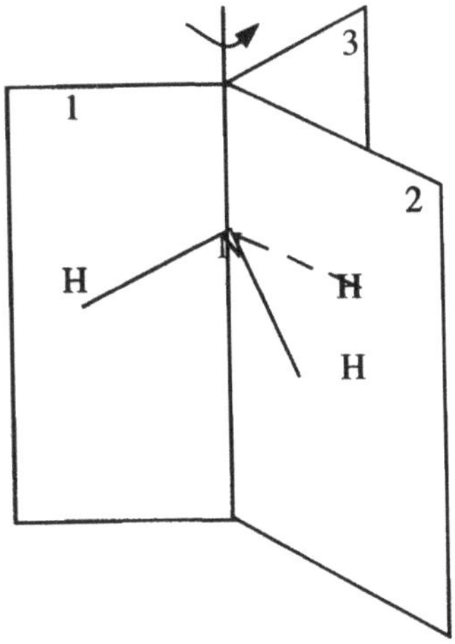

Figure 2.1 Symmetry planes and (rotation) axis of the C_{3v} group.

- The operations are carried out on the molecule. This means that the nuclei move, while the coordinate axes, symmetry axes, and planes remain fixed.

The former mode is often called the "passive" and the latter, the "active" interpretation. Sometimes one is more convenient than the other, and either mode is equally valid, but should never be mixed in a given problem.

Using the active mode and the group multiplication law of successive application, the group multiplication table is as follows:

C_{3v}	E	C_3	C_3^2	σ_1	σ_2	σ_3
E	E	C_3	C_3^2	σ_1	σ_2	σ_3
C_3	C_3	C_3^2	E	σ_3	σ_1	σ_2
C_3^2	C_3^2	E	C_3	σ_2	σ_3	σ_1
σ_1	σ_1	σ_2	σ_3	E	C_3	C_3^2
σ_2	σ_2	σ_3	σ_1	C_3^2	E	C_3
σ_3	σ_3	σ_1	σ_2	C_3	C_3^2	E

To illustrate how successive application works and how this multiplication table was constructed, consider the plane containing the three protons at the corners of an equilateral triangle (viewed from the side of the nitrogen atom in Fig. 2.2) and use the active interpretation.

One should note that the detailed results of the group operations depend on the specific choices made by the user, such as active versus passive mode of application and right-handed (as used here) versus left-handed mode of rotation about symmetry axes.

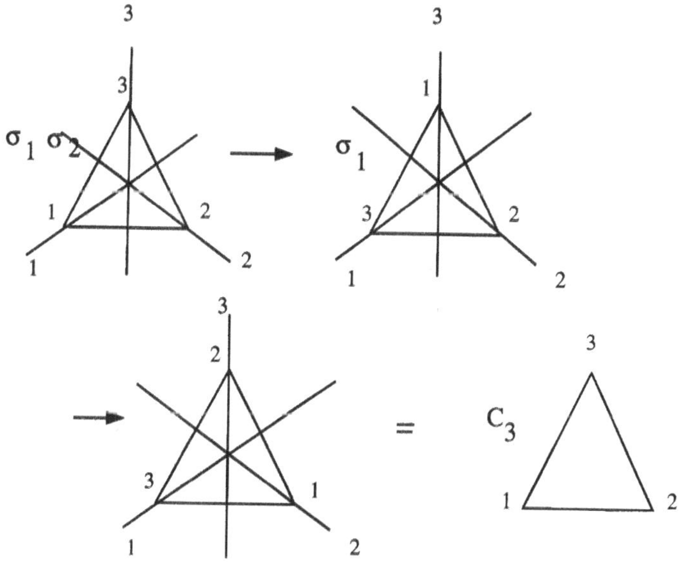

Figure 2.2 Illustration of the action of $\sigma_1\sigma_2 = C_3$.

Example 2.3 The full permutation group S_3, also called the symmetric group S_3, consists of all permutations of three objects. A permutation is an operation that brings one order arrangement of three objects into another. A standard or reference arrangement is one in which each of the three objects is given an order number: 1, 2, and 3. This standard or reference arrangement can then be brought to another arrangement by a permutation; for instance, object 1 and object 2 can change places. The permutation that accomplishes this is given the notation

$$\begin{pmatrix} 123 \\ 213 \end{pmatrix}, \tag{2.4}$$

which can be read column by column "instead of 1, put 2," "instead of 2, put 1," and "instead of 3, put 3." Obviously, in this notation only the columns matter, and not their order. Thus

$$\begin{pmatrix} 123 \\ 213 \end{pmatrix} = \begin{pmatrix} 231 \\ 132 \end{pmatrix} = \begin{pmatrix} 312 \\ 321 \end{pmatrix}, \tag{2.5}$$

and so on.

There are six possible permutations of three objects since $|S_3| = 3!$; thus, the order of S_3 is 6. The six elements are

$$S_3: \left\{\begin{pmatrix} 123 \\ 123 \end{pmatrix}, \begin{pmatrix} 123 \\ 231 \end{pmatrix}, \begin{pmatrix} 123 \\ 312 \end{pmatrix}, \begin{pmatrix} 123 \\ 132 \end{pmatrix}, \begin{pmatrix} 123 \\ 321 \end{pmatrix}, \begin{pmatrix} 123 \\ 213 \end{pmatrix}\right\}, \tag{2.6}$$

and the combination law is successive application:

$$\begin{pmatrix}123\\231\end{pmatrix}\begin{pmatrix}123\\132\end{pmatrix} = \begin{pmatrix}132\\213\end{pmatrix}\begin{pmatrix}123\\132\end{pmatrix} = \begin{pmatrix}123\\213\end{pmatrix}, \tag{2.7}$$

where, in the second step, the freedom to rearrange the columns has been used so that the columns of the second (left) permutation are put in the order that matches its top row with bottom row of the first acting (right) permutation. In this way the bottom row of the first acting permutation and the identical top row of the second acting permutation simply mediate the transfer of the original arrangement, 123, into the final one, 213. From this it is easy to demonstrate that the associative law holds. It also follows that

$$\begin{pmatrix}123\\123\end{pmatrix} \tag{2.8}$$

is the identity, and that an inverse such as

$$\begin{pmatrix}123\\312\end{pmatrix}^{-1} = \begin{pmatrix}312\\123\end{pmatrix} = \begin{pmatrix}123\\231\end{pmatrix} \tag{2.9}$$

is constructed by simply interchanging the top and bottom rows.

A permutation can be expressed as a product of cycles. For example, a particular permutation belonging to the symmetric group S_7 (the 7! permutations of seven objects) such as

$$\begin{pmatrix}1234567\\2135746\end{pmatrix} = (12)(3)(4576). \tag{2.10}$$

The cycle (4576) can be interpreted as meaning "instead of 4, put 5; instead of 5, put 7; instead of 7, put 6; and instead of 6, put 4." Each cycle is, of course, a permutation in its own right, and it becomes intuitively clear that any permutation can be expressed as a product of nonoverlapping cycles. We call a cycle with n entries an n-cycle. Obviously, 1-cycles can be left off.

Using this way of expressing the permutations the elements of S_3 are

$$S_3\colon \{e = (1)(2)(3), (123), (321), (23), (13), (12)\} \tag{2.11}$$

and using the multiplication law of successive application as described above, we obtain the following multiplication table:

S_3	e	(123)	(321)	(23)	(13)	(12)
e	e	(123)	(321)	(23)	(13)	(12)
(123)	(123)	(321)	e	(12)	(23)	(13)
(321)	(321)	e	(123)	(13)	(12)	(23)
(23)	(23)	(13)	(12)	e	(123)	(321)
(13)	(13)	(12)	(23)	(321)	e	(123)
(12)	(12)	(23)	(13)	(123)	(321)	e

Another group F of order 6 is formed by the functions

$$f_1(x) = x;\; f_2(x) = (1 - x)^{-1};\; f_3(x) = 1 - x^{-1};$$
$$f_4(x) = x^{-1};\; f_5(x) = 1 - x;\; f_6(x) = x(x - 1)^{-1}; \tag{2.12}$$

with the multiplication law of substitution;

$$f_k(x)f_l(x) = f_k(f_l(x)). \tag{2.13}$$

This means that f_1 is the identity element and that, say, $f_2^{-1} = f_3$. The following multiplication table is created:

F	f_1	f_2	f_3	f_4	f_5	f_6
f_1	f_1	f_2	f_3	f_4	f_5	f_6
f_2	f_2	f_3	f_1	f_6	f_4	f_5
f_3	f_3	f_1	f_2	f_5	f_6	f_4
f_4	f_4	f_5	f_6	f_1	f_2	f_3
f_5	f_5	f_6	f_4	f_3	f_1	f_2
f_6	f_6	f_4	f_5	f_2	f_3	f_1

It is interesting to note that with the following identifications or mappings:

$$E \leftrightarrow e \leftrightarrow f_1;\; C_3 \leftrightarrow (123) \leftrightarrow f_2;\; C_3^2 \leftrightarrow (321) \leftrightarrow f_3;$$
$$\sigma_1 \leftrightarrow (23) \leftrightarrow f_4;\; \sigma_2 \leftrightarrow (13) \leftrightarrow f_5;\; \sigma_3 \leftrightarrow (12) \leftrightarrow f_6; \tag{2.14}$$

the multiplication tables of the groups C_{3v}, S_3, and F are identical. These three groups are different realizations of one and the same non-Abelian group of order 6. There is actually only one such group, but, as is obvious from these examples, there can be many realizations pertaining to specific applications. Therefore, the usefulness and efficacy of the study of abstract groups can be substantiated without reference to specific applications.

2.1 SUBGROUPS

When a subset $H \subset G$ forms a group with the same multiplication law as the group G, then it is said to be a subgroup of G.

Example 2.4 The point group C_{3v} contains the subgroups

$$H_1\!: \{E, \sigma_1\};\; H_2 : \{E, \sigma_2\};\; H_3 : \{E, \sigma_3\};\; H_4 : \{E, C_3, C_3^2\}. \tag{2.15}$$

These are called *proper* subgroups. In addition the identity element by itself, $\{E\}$, and the entire group C_{3v} are subgroups of C_{3v}. They are called *improper* subgroups. It is noteworthy that the orders of the subgroups are integer divisors in the order of the group. This is an example of a general result.

The Order of an Element

For a finite group G we can form the powers of a single element $g, g^2, g^3, \ldots,$ and this procedure cannot keep producing new elements, but eventually the powers have to repeat themselves, say, $g^k = g^m$ for $m < k$. Multiplying by g^{-1} m times yields $g^{k-m} = e = g^p$. The smallest $p = k - m$ for which this is valid is said to be the order of the element g. An element g is then a *generator* of a cyclic subgroup $\{g, g^2, \ldots, g^p = e\}$. A cyclic group is a group generated by the powers of a single element. Such a group is obviously Abelian. Again considering the point group C_{3v}, one can conclude that the order of any of the reflections σ is 2, while the order of C_3 or C_3^2 is 3. It is also clear that the subgroup

$$\{E, C_3, C_3^2\} \tag{2.16}$$

is generated by C_3 or C_3^2.

It is useful to have some general criterion for finding subgroups that would apply to both finite and infinite groups. Theorem 2.1 provides such a criterion.

Theorem 2.1 A necessary and sufficient condition for $H \subset G$ to be a subgroup of G is that $h_1 h_2^{-1} \in H$ for all $h_1, h_2 \in H$ (sometimes called "closure under division").

Proof

- *Necessary condition*: If H is a subgroup, then obviously $h_1 h_2^{-1} \in H$ for all $h_1, h_2 \in H$.
- *Sufficient condition:*
  1. Put $h_1 = h_2$, then $h_1 h_1^{-1} = e \in H$ satisfying postulate III.[1]
  2. Put $h_1 = e$, then $e h_2^{-1} = h_2^{-1} \in H$ satisfying postulate IV.
  3. Associativity (postulate II) satisfied for all elements of G.
  4. Name $h_2^{-1} = h_3$, then $h_1 h_3 \in H$ satisfying postulate I.

QED

For a finite group, satisfying the closure postulate alone is sufficient for a subset to be a subgroup. The proof of this statement is a recommended exercise.

For $H \subset G$ a proper subgroup

$$H\colon \{h_1, h_2, \ldots, h_m\} \tag{2.17}$$

and a an element of G but $a \notin H$ the subset

$$Ha\colon \{h_1 a, h_2 a, \ldots, h_m a\} \subset G \tag{2.18}$$

[1] See list of postulates earlier in this chapter, preceding Eq. (2.3).

is called a *right coset* of H by a. Analogously, left cosets and even two-sided cosets can be defined. A coset has $m = |H|$ distinct elements and is not a subgroup, since it does not contain the identity element.

Theorem 2.2 Two cosets Ha and Hb have either no elements or all elements in common.

Proof Assume that the two cosets have one element in common, that is, $h_k a = h_l b$, then $b = h_l^{-1} h_k a \in Ha$ and for any element of Hb it holds that $h_i b = h_i h_l^{-1} h_k a \in Ha$. The conclusion is that two cosets cannot be partially overlapping.

QED

Thus, the group G can be partitioned into nonoverlapping subsets (including the subgroup H)

$$G: \left\{ |\underbrace{H}_{m}|\underbrace{Ha}_{m}|\underbrace{Hb}_{m}|\cdots| \right\}, \tag{2.19}$$

all containing $m = |H|$ distinct elements. This must exhaust the finite group G, and consequently one must conclude that the order of the subgroup is an integer divisor in the order of the group, $|G| = k|H|$, where the integer k is the *index* of H in G.

A corollary of this conclusion is that a group of prime order has no proper subgroups, can be generated by a single element, and thus is cyclic, and therefore Abelian.

Among all possible subgroups that exist, some with special properties have been given specific names. The following theorem serves as a means of defining some of these kinds of subgroups.

Theorem 2.3 Let $S \subset G$ be an arbitrary subset. Then define the set $C(S) = \{a \in G | as = sa, \text{ all } s \in S\}$, specifically, the subset of G with elements that commute with all the elements of the set S. Then $C(S)$ is a subgroup.

Proof Using the result of Theorem 2.1, it suffices to show that $C(S)$ is "closed under division," that is, $a_1 a_2^{-1} \in C(S)$ for all $a_1, a_2 \in C(S)$, which means

$$\begin{aligned} a_1 s &= s a_1, \\ a_2 s &= s a_2. \end{aligned} \tag{2.20}$$

From the second of these relations it follows that $s a_2^{-1} = a_2^{-1} s$. It is then obvious that

$$a_1 a_2^{-1} s = a_1 s a_2^{-1} = s a_1 a_2^{-1} \tag{2.21}$$

QED

In general, $C(S)$ is called the *centralizer* of S in G. Different choices of the set S lead to different names of the subgroup. For $S = s$, namely, a single element, $C(s) \equiv N(s)$ is said to be the *normalizer* of s in G. Should $S = G$, then $C(S)$ is the *center* of G. Obviously, these considerations apply to both finite and infinite groups and $C(S)$ can be either proper or improper.

Group Generators

Multiplication tables are less useful for groups of large order and not suitable for infinite groups. The entire group can be generated by a finite number of elements, and a choice of generating relations can then be exploited. We have already shown that a cyclic group is generated by a single element (generator) g and a generating relation of the type

$$g^p = e. \tag{2.22}$$

Another example is provided by the non-Abelian group of order 6, various realizations of which have been discussed. The generic multiplication table of this group is as follows:

G	e	a	b	c	d	f
e	e	a	b	c	d	f
a	a	b	e	f	c	d
b	b	e	a	d	f	c
c	c	d	f	e	a	b
d	d	f	c	b	e	a
f	f	c	d	a	b	e

It is readily seen from the generating relations

$$\begin{aligned} a^3 &= c^2 = e \\ b &= a^2 \\ d &= ca \\ f &= ac \end{aligned} \tag{2.23}$$

that a and c may be chosen as generators.

In the chapters on permutation groups it will be shown that the symmetric group S_n of order $n!$ is generated by the $n - 1$ interchanges (2-cycles) $(i, i + 1)$.

2.2 CONJUGATE CLASSES

A group G can be partitioned into nonoverlapping subsets such as a subgroup and its right cosets. Another possible partitioning of a group that reveals some of its structure can be done in terms of conjugate classes. The conjugate class of an element $g \in G$ is defined as the set

$$C_g = \{s^{-1}gs | g \text{ fixed, all } s \in G\}. \tag{2.24}$$

Two group elements h, g related by

$$h = s^{-1}gs \tag{2.25}$$

are said to be conjugate. This is an example of an equivalence relation, which is more general than an equality. One can write for an equivalence relation

$$h \sim g. \tag{2.26}$$

Such relations have the following properties:

- *Reflexive*, that is, $g \sim g$, which certainly is true for conjugation since for, say, $s = e$ g is conjugate with itself.
- *Symmetric*, that is $h \sim g$, which implies $g \sim h$. For conjugation this holds since $g = (s^{-1})^{-1}hs^{-1}$.
- *Transitive*, which means that $g \sim h$, and $h \sim k$ implies $g \sim k$. Since for $g = s^{-1}hs$ and $h = t^{-1}kt$, it follows that $g = s^{-1}t^{-1}kts = (ts)^{-1}k(ts)$, and transitivity holds for conjugation.

The transitivity property means that two classes cannot be partially overlapping. This can easily be seen by assuming a non-empty intersection $C_g \cap C_k$. Elements h in the intersection are conjugate to g and also conjugate to k, and one must conclude that $C_g = C_k$. This means that the group can be partitioned into nonoverlapping conjugate classes.

Theorem 2.4 The order of a class is a divisor in the order of the group.

Proof The number of elements s that yield the element g itself on conjugation $s^{-1}gs$ is m, that is, the elements of the normalizer of g, namely, the subgroup $N(g) = \{n_1, n_2, \ldots, n_m\}$. When s runs over the elements of $N(g)$, the element g is generated m times. The right coset $N(g)a$ has m distinct elements that, if used in the conjugation

$$(n_l a)^{-1}g(n_l a) = a^{-1}n_l^{-1}gn_l a = a^{-1}ga, \tag{2.27}$$

yield the element $a^{-1}ga \in C_g$ m times. Similarly, $b^{-1}gb \in C_g$ is generated m times by the distinct coset $N(g)b$, and so on.

QED

The multiplication table of the permutation group S_3 can be used to generate its conjugate classes:

$$C_e = \{e\}; C_{(123)} = \{(123), (321)\}; C_{(12)} = \{(12), (13), (23)\}. \tag{2.28}$$

The identity element is obviously always in a class by itself. Elements of the same cycle structure are in the same class. This is a general result and is discussed further in the chapters on permutation groups. The classes of C_{3v} are similarly found to be

$$C_E = \{E\}; C_{C_3} = \{C_3, C_3^2\}; C_{\sigma_1} = \{\sigma_1, \sigma_2, \sigma_3\}. \tag{2.29}$$

Here it is noteworthy that geometric operations of different kinds do not belong to the same class.

2.3 MAPPINGS OF GROUPS

The mapping f of a set A on a set B

$$A \xrightarrow[f]{} B \tag{2.30}$$

is in general many-to-one as illustrated in Figure 2.3.

This means that the inverse of f is not a mapping. Let $a \in A$ and $b \in B$, and let $b = f(a)$, which means that b is the image element of a in the mapping f. The set of all image elements of A in B is denoted $f(A)$. Mappings are classified as

- *Into* when $f(A) \subset B$
- *Onto* when $f(A) = B$

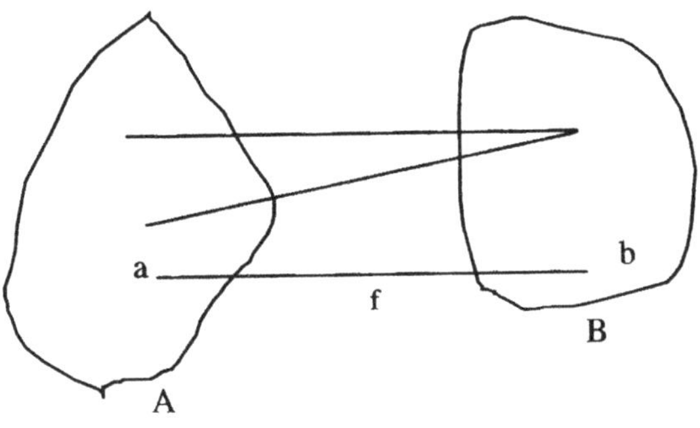

Figure 2.3 A mapping f of a set A on a set B.

When a mapping is one-to-one and onto, then

$$B \underset{f^{-1}}{\rightarrow} A \tag{2.31}$$

is also a mapping. The subset of elements of A with the same image element $b \in B$ is called the *fiber* of b.

For sets with algebraic structure such as groups, the mapping is said to preserve structure if the mapping and then multiplying yield the same result as first multiplying and then mapping. This can be illustrated by the Figure 2.4.

Thus, a structure preserving mapping can be expressed by the relation

$$f(a)f(b) = f(ab) \tag{2.32}$$

and is called a *homomorphism*, and a mapping that is one-to-one and onto is an *isomorphism*.

When A and B are groups f is said to be a group homomorphism or a group isomorphism, but when the context is clear just the words homomorphism or isomorphism are sufficient.

Some general properties of group homomorphisms can now be discussed.

Theorem 2.5 If $G \underset{f}{\rightarrow} G'$ is a homomorphism, then the image set $f(G)$ is a subgroup of G' with the identity element $e' = f(e)$ and with $f(g^{-1}) = (f(g))^{-1}$, that is, the identity is mapped on the identity and an inverse is mapped on the inverse of the image element.

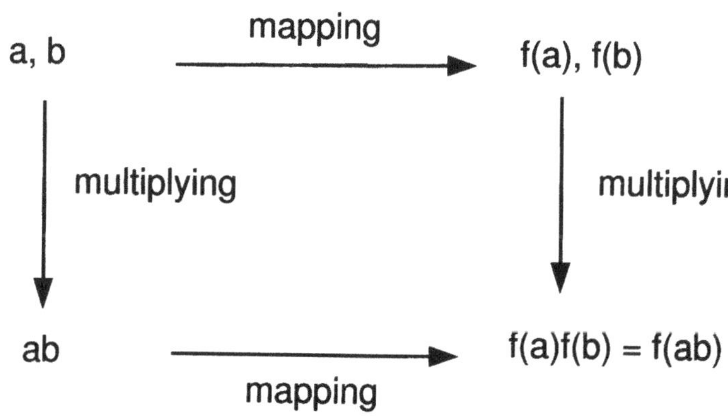

Figure 2.4 First mapping and then multiplying the image elements is the same as first multiplying the elements and then mapping the result in a structure preserving mapping or homomorphism.

Proof The homomorphism demands that the group structure be preserved, so $f(G) \subset G'$ is a subgroup (proper or improper). Furthermore, since for elements of G the relation $g = ge$ must hold and the homomorphism requires $f(g) = f(ge) = f(g)f(e)$, $e' = f(e)$. Obviously, the nature of the elements and their combination laws can be quite different in G and in G'. Since for elements of G $e = g^{-1}g$ the homomorphism requires that $f(e) = f(g^{-1}g) = f(g^{-1})f(g)$, which means that $f(g^{-1}) = (f(g))^{-1}$. From these results we could use the result of Theorem 2.1 and prove by showing "closure under division" that $f(G)$ is a subgroup:

$$f(g_1)(f(g_2))^{-1} = f(g_1)f(g_2^{-1}) = f(g_1g_2^{-1}) = f(g_3) \in f(G). \tag{2.33}$$

QED

The fiber of the identity element $e' \in G'$ is called the *kernel* K of the homomorphism.

Theorem 2.6 The kernel K of the homomorphism is a subgroup (proper or improper) of G.

Proof Let $k_1, k_2 \in K$, then by definition $f(k_1) = f(k_2) = e'$. Again we use the result of Theorem 2.1 and the "closure under division" property to show that K is a subgroup. Consider $k_1k_2^{-1} \rightarrow f(k_1k_2^{-1}) = f(k_1)f(k_2^{-1}) = e'$, that is, $k_1k_2^{-1} \in K$.

QED

When a right coset Ha of a subgroup $H \subset G$ equals the left coset aH, then the subgroup H is said to be *normal* or *invariant*. This is the same as saying that for such a subgroup $a^{-1}ha \in H$ for all elements $h \in H$ and $a \in G$. From this statement it follows that a normal or invariant subgroup H contains complete conjugate classes of G.

Example 2.5 The subgroup

$$\{E, C_3, C_3^2\} \tag{2.34}$$

of the point group C_{3v} is normal.

Theorem 2.7 The kernel K of a homomorphism

$$G \underset{f}{\rightarrow} G' \tag{2.35}$$

is an invariant or normal subgroup of G.

Proof Consider $k \in K$ and $a \in G$, then

$$f(a^{-1}ka) = f(a^{-1})f(k)f(a) = f(a^{-1})e'f(a) = e' \tag{2.36}$$

which means that $a^{-1}ka \in K$.

QED

It follows straightforwardly that all the elements of a right coset Ka of the kernel have the same image element, since

$$f(ka) = f(k)f(a) = f(a). \tag{2.37}$$

The kernel and all its distinct right cosets exhaust the group G and since each coset has the same number of elements $m = |K|$, it follows that the homomorphism has the multiplicity $m \to 1$, as shown in Figure 2.5.

Examples are

- G = all nonzero real numbers under multiplication.
  G' = all positive real numbers under multiplication.
  Multiplicity $2 \to 1$.
- G = arbitrary finite group.
  G' = the number 1 under multiplication.
  Multiplicity $|G| \to 1$.
- $G = C_{3v}$
  G' = the numbers 1 and -1.
  Multiplicity $3 \to 1$.
- $G = C_{3v}$
  $G' = S_3$
  Multiplicity $1 \to 1$ (isomorphism).

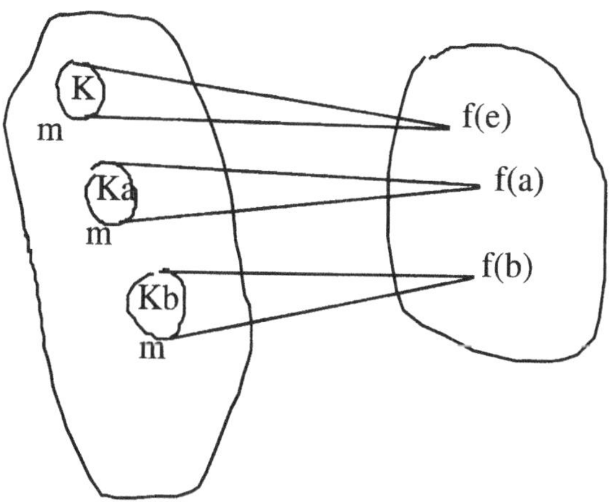

Figure 2.5 Illustration of the multiplicity of a homomorphism.

3

PRELIMINARIES ABOUT THE SYMMETRIC GROUP

3.1 CYCLE STRUCTURE

All the $n!$ permutations of n objects form a group called the *symmetric group* S_n where the multiplication law is successive application. As already mentioned, any permutation can be expressed as a product of nonoverlapping cycles such as

$$\begin{pmatrix} 1234567 \\ 2367514 \end{pmatrix} = (1236)(47)(5) \in S_7 . \tag{3.1}$$

Each cycle is a permutation in its own right. Another way of factoring a given permutation is in terms of 2-cycles or simple *interchanges*. It is sufficient to consider a single cycle (the r-cycle $(i_1 i_2 \cdots i_r)$):

$$\begin{aligned} (i_1 i_2 \cdots i_r) &= (i_1 i_2)(i_2 i_3) \cdots (i_{r-1} i_r) \\ &= (i_r i_{r-1})(i_r i_{r-2}) \cdots (i_r i_1) \\ &= (i_1 i_r)(i_1 i_{r-1}) \cdots (i_1 i_2) , \end{aligned} \tag{3.2}$$

which shows that there are a number of ways to express a given permutation as a product of simple (overlapping) interchanges.

Example 3.1 (123) = (12)(23) = (32)(31) = (13)(12).

The only distinguishing feature is that some permutations are expressed as a product of an odd number of interchanges and are called *odd permutations*, while others are expressed as an even number of simple interchanges and are called *even permutations*. All even permutations form a subgroup of S_n called the *alternating group* A_n: This is so since the product of two even permutations is an even permutation (closure, postulate I). The associative law holds for all elements of S_n (postulate II). The identity is an even permutation (postulate III). The inverse of an even permutation is an even permutation, for instance if $s = (ab)(cd)(ef)(gh)$ then $s^{-1} = (gh)(ef)(cd)(ab)$, since an interchange is its own inverse.

The permutations can be classified according to their *cycle structure*. The symmetric group S_3, for example, has one permutation (the identity) consisting of three 1-cycles, two permutations consisting of one 3-cycle, and three permutations consisting of one 1-cycle and one 2-cycle. We note that the permutations with the same cycle structure belong to the same class. This is a general result. The permutation

$$\begin{pmatrix} 12345678 \\ 21346785 \end{pmatrix} = (12)(3)(4)(5678) \in S_8 \tag{3.3}$$

has two 1-cycles, one 2-cycle, one 4-cycle, and zero 3-cycles, 5-cycles, and so on. Leaving out the zeros for the cycles not present one can label this permutation as belonging to the conjugate class $(1^2, 2, 4)$.

Theorem 3.1 The cycle structure of a permutation is invariant under conjugation; that is, all permutations with the same cycle structure are in the same conjugate class.

Proof Consider the permutation $s = C_1 C_2 \cdots C_r$ as a product of r nonoverlapping cycles C_i. Conjugation means that one can write $t^{-1}st = t^{-1}C_1 t t^{-1} C_2 t \cdots t^{-1} C_r t$, and conclude that each cycle undergoes the same conjugation. It is then sufficient to consider a single cycle, say, a p-cycle:

$$t^{-1}(a_1 a_2 \cdots a_p)t\,. \tag{3.4}$$

The element t is a general permutation of S_n, which can be expressed as

$$t = \begin{pmatrix} b_1 b_2 \cdots b_p b_{p+1} \cdots b_n \\ a_1 a_2 \cdots a_p a_{p+1} \cdots a_n \end{pmatrix}, \tag{3.5}$$

where the freedom of arranging the columns has been used to put the entries of the p-cycle in the first p columns (bottom row). In Eq. (3.4) t acts first, then the p-cycle, and last t^{-1}. Thus, the action of

- t is $b_i \to a_i$, $i = 1, 2, \ldots, n$.
- $(a_1 a_2 \cdots a_p)$ is $a_i \to a_{i+1}$, $i = 1, 2, \ldots, p$, modulus p, and $a_i \to a_i$, $i = (p+1), (p+2), \ldots, n$.
- t^{-1} is $a_i \to b_i$, $i = 1, 2, \ldots, n$.

The net result of all this is that

$$t^{-1}(a_1a_2\cdots a_p)t = (b_1b_2\cdots b_p), \tag{3.6}$$

and a p-cycle remains a p-cycle under conjugation.

QED

A general permutation of S_n has α_1 1-cycles, α_2 2-cycles, and so on up to α_n (of course, at most one) n-cycles, with the condition that

$$\alpha_1 + 2\alpha_2 + 3\alpha_3 + \cdots + n\alpha_n = n. \tag{3.7}$$

The order of the class of such a permutation can be readily calculated from its cycle structure. The permutation written as a product of nonoverlapping cycles can be depicted as

$$\underbrace{(\cdot)(\cdot)\cdots(\cdot)}_{\alpha_1}\underbrace{(\cdot\cdot)(\cdot\cdot)\cdots(\cdot\cdot)}_{\alpha_2}\cdots\underbrace{(\cdot\cdot\cdot\cdot\cdot)(\cdot\cdot\cdot\cdot\cdot)\cdots(\cdot\cdot\cdot\cdot\cdot)}_{\alpha_5}\cdots \tag{3.8}$$

The dots in each cycle indicate where to place the labels $1, 2, \ldots, n$ to obtain all the different elements in the class with this particular cycle structure. Disregarding redundancies, there are $n!$ elements. However, elements differing in only the arrangements of the 1-cycles among themselves, the 2-cycles among themselves, and so on, are identical, leaving only[1]

$$\frac{n!}{\alpha_1!\,\alpha_2!\,\cdots\alpha_n!} \tag{3.9}$$

elements. But even among these there are redundancies, since elements that differ only by cyclic rearrangements of the labels inside a cycle are identical. For an r-cycle there are r such cyclic rearrangements, and for the α_r r-cycles there are a total of r^{α_r} identical elements. Thus, the number of distinct elements of a given cycle structure or the order of the class $(1^{\alpha_1}, 2^{\alpha_2}, \ldots, n^{\alpha_n})$ is

$$\frac{n!}{\alpha_1!\,1^{\alpha_1}\alpha_2!\,2^{\alpha_2}\cdots\alpha_n!\,n^{\alpha_n}} \tag{3.10}$$

with the constraint

$$\alpha_1 + 2\alpha_2 + 3\alpha_3 + \cdots + n\alpha_n = n. \tag{3.11}$$

[1] The α_i that equal zero can obviously be included in this notation.

3.2 CAYLEY'S THEOREM

Permutation groups are perhaps the finite groups that has been most thoroughly studied. All finite groups can be related to permutation groups and a useful way to illustrate this is the following theorem, also called *Cayley's theorem.*

Theorem 3.2 Any finite group G is isomorphic to a subgroup of the symmetric group S_n with $n = |G|$.

Proof Consider G: $\{g_1, g_2, \ldots, g_n\}$ and a general element $g \in G$, and form

$$g\{g_1, g_2, \ldots, g_n\} = \{gg_1, gg_2, \ldots, gg_n\} = P_g\{g_1, g_2, \ldots, g_n\}, \tag{3.12}$$

where in the last step we have used the fact (seen, e.g., from a group multiplication table) that the multiplication of all the group elements by a given group element rearranges the group elements equivalent to the action of a permutation P_g. Such a permutation can be expressed as

$$P_g = \begin{pmatrix} g_1 \; g_2 \; \cdots \; g_n \\ gg_1 gg_2 \cdots gg_n \end{pmatrix} = \begin{pmatrix} g_i \\ gg_i \end{pmatrix}. \tag{3.13}$$

Since the order of the columns in this symbol for the permutation is immaterial, it follows that

$$P_g P_h = \begin{pmatrix} g_i \\ gg_i \end{pmatrix}\begin{pmatrix} g_i \\ hg_i \end{pmatrix} = \begin{pmatrix} hg_i \\ ghg_i \end{pmatrix}\begin{pmatrix} g_i \\ hg_i \end{pmatrix} = \begin{pmatrix} g_i \\ ghg_i \end{pmatrix} = P_{gh} \tag{3.14}$$

and since then $g \rightarrow P_g, h \rightarrow P_h, gh \rightarrow P_{gh}$ the mapping is a homomorphism. The permutations $\{P_g\}$ are all distinct, which makes the mapping one-to-one.

QED

The characteristics of the image group are further revealed by considering a group element g of a particular order r, so that $g^r = e$. The action of g on the r distinct group elements $g^p g_i$, $p = 0, 1, 2, \ldots, r - 1$ yields

$$g_i \rightarrow gg_i, \; gg_i \rightarrow g^2 g_i \; \cdots \; g^{r-1} g_i \rightarrow g^r g_i = g_i \tag{3.15}$$

which is an r-cycle. This means that the permutation P_g, which is the image element of g, is a product of nonoverlapping r-cycles. Such a permutation, which is a product of cycles all of the same length, is called a *regular permutation.*

Example 3.2 The point group C_{3v} with the elements originally arranged as

$$\{E, C_3, C_3^2, \sigma_1, \sigma_2, \sigma_3\} \tag{3.16}$$

is isomorphic to a subgroup of S_6, such that

$$\begin{aligned} E &\rightarrow (1)(2)(3)(4)(5)(6) \\ C_3 &\rightarrow (123)(465) \\ C_3^2 &\rightarrow (132)(456) \\ \sigma_1 &\rightarrow (14)(25)(36) \\ \sigma_2 &\rightarrow (15)(26)(34) \\ \sigma_3 &\rightarrow (16)(24)(35) \end{aligned} \tag{3.17}$$

This can readily be verified by referring to the group multiplication table of C_{3v}.

4

ALGEBRA STRUCTURE

4.1 GROUP ALGEBRA

The application of symmetry to problems in physics or chemistry is concerned not only with the structure and properties of groups but also with so-called group algebras. This topic falls within the realm of the theory of linear associative algebras. The full theory of such mathematical structures is far more elaborate than what is needed here, and we will make only casual contact with this general mathematical formalism. The material about group algebra and convolution algebra presented here follows closely a paper by P. O. Löwdin in *Reviews of Modern Physics* **39**, 259 (1967).

There is always a finite group G: $\{g_1, g_2, \ldots, g_n\}$ in the background, which most of the time is left unspecified.

4.1.1 Functions over the Group

The concept of a function over the group is a mapping of the group elements on the set of complex numbers. The group elements serve as the independent variable, and the functions can be defined by simply listing the function values that are the images of the group elements, such that a complex number α_i corresponds to the group element g_i. In this manner a function over the group is

$$\alpha(g) = \langle \alpha_1, \alpha_2, \ldots, \alpha_n \rangle . \tag{4.1}$$

The sum of two functions $\alpha(g) + \beta(g)$ is also a function over the group. The operation of addition of functions is defined via the addition of complex numbers, such that

$$\alpha(g) + \beta(g) = \langle \alpha_1 + \beta_1, \alpha_2 + \beta_2, \ldots, \alpha_n + \beta_n \rangle. \tag{4.2}$$

Similarly, the multiplication of a function by a complex number c also yields a function over the group

$$c\alpha(g) = \langle c\alpha_1, c\alpha_2, \ldots, c\alpha_n \rangle, \tag{4.3}$$

where again this operation is defined via the multiplication of complex numbers. It is obvious that the functions over the group form a linear space. There exists a zero function $0(g) = \langle 0, 0, \ldots, 0 \rangle$, with all the group elements mapped on the number 0.

A standard set of basis functions in this linear space can be defined as

$$\begin{aligned} f_1 &= \langle 1, 0, \ldots, 0 \rangle, \\ f_2 &= \langle 0, 1, \ldots, 0 \rangle, \\ &\vdots \\ f_n &= \langle 0, 0, \ldots, 1 \rangle, \end{aligned} \tag{4.4}$$

showing that the linear space is of dimension $n = |G|$. It follows straightforwardly that an arbitrary function over the group can be expanded in this standard basis:

$$\alpha(g) = \sum_{k=1}^{n} \alpha_k f_k. \tag{4.5}$$

4.1.2 Group Algebra and Convolution Algebra

The concept of a linear algebra has a number of formal defining properties. It might be useful to give these basic definitions or postulates before dealing with the specifics of interest to us.

A set of elements V is said to form a *linear associative algebra* if the operations of addition, scalar multiplication, and multiplication, can be defined for its elements to satisfy the following postulates:

- The elements form an Abelian group under addition:
  1. The sum of any two elements A and B of V is unique and also belongs to V, namely, $A + B \in V$.
  2. The sum of three elements is associative: $A + (B + C) = (A + B) + C$.
  3. There is an identity element O, such that $A + O = O + A = A$.
  4. For any element A there exists a -A in V, such that $A + (-A) = A - A = O$.
  5. Addition is commutative: $A + B = B + A$.

- Multiplication scalar is:
  1. The product of any element A with a complex number c is in V, specifically, $cA \in V$.
  2. The scalar multiplication is associative and distributive: $c(c'A) = (cc')A$; $c(A + B) = cA + cB$; $(c + c')A = cA + c'A$.
  3. Multiplication by unity leaves the element unchanged: $1A = A$.
- Multiplication is linear and associative:
  1. The product of any two elements is unique and also belongs to V, $A \times B = C \in V$.
  2. The product of three elements is associative: $A \times (B \times C) = (A \times B) \times C$.
  3. The product of any two elements can be expressed as a linear combination of elements, $A \times B = cC + c'D + \cdots$.
  4. There exists an identity element I, such that $I \times A = A \times I = A$.

A linear associative algebra with a group in the background is called a *group algebra*. The ordinary group multiplication under which the group is closed is always present. Following the idea of an algebra, a larger set is sought that is closed under addition of group elements

$$g_k + g_l = g_l + g_k \tag{4.6}$$

and under multiplication of a group element by a complex number

$$cg_k\,. \tag{4.7}$$

This means that a set with elements

$$A = \sum_{k=1}^{n} a_k g_k\,, \tag{4.8}$$

where the coefficients $\{a_k\}$ are complex numbers, is considered and is closed under addition

$$A + B = \sum_{k=1}^{n} (a_k + b_k) g_k \tag{4.9}$$

(i.e., again the addition of elements is defined via the addition of complex numbers) and under multiplication by a complex number

$$cA = \sum_{k=1}^{n} (ca_k) g_k\,, \tag{4.10}$$

satisfying the distributive laws

$$\begin{aligned} c(A + B) &= cA + cB\,, \\ (c_1 + c_2)A &= c_1A + c_2A\,. \end{aligned} \tag{4.11}$$

Therefore the elements $\{A\}$ form a linear space of dimension $n = |G|$. Obviously the elements form an Abelian group under addition. The identity element in that group is the element with all the coefficients in Eq. (4.8) equal to zero (the zero element), and the inverse of A is $(-1)A = -A$.

In order to obtain an algebra, the set must also be closed under multiplication of one element with another. Such an operation is defined as

$$\begin{aligned} A \times B &= \left(\sum_{k=1}^{n} a_k g_k\right) \times \left(\sum_{l=1}^{n} b_l g_l\right) = \sum_{k=1}^{n} \sum_{l=1}^{n} a_k b_l g_k g_l \\ &= \sum_{m=1}^{n} c_m g_m = C\,, \end{aligned} \tag{4.12}$$

that is, the distributive law of multiplication is applied and the operation $\times$ between group elements becomes the ordinary group multiplication. The group combination law yields $g_k g_l = g_m$, which is a single group element, yielding an algebra element C in the same set. Without specifying the group G, the detailed calculation of the coefficients of the element C from those of the elements A and B can, of course, not be done.

The complex coefficients of an algebra element define a function over the group, since there is one function value assigned to each group element. One can then establish a one-to-one mapping between the elements of the group algebra and the functions over the group, since the linear spaces of both are of the same dimension. In order to follow the convention used in most texts on the subject, the following particular mapping is chosen:

$$A = \frac{1}{|G|} \sum_{k=1}^{n} \alpha_k g_k^{-1} = \frac{1}{|G|} \sum_{g \in G} \alpha(g) g^{-1} = \mathrm{M}_g \alpha(g) g^{-1}. \tag{4.13}$$

The complex number that is the function value at group element g is multiplied with the inverse group element g^{-1}, and all the group elements are also multiplied with $1/|G|$. The special summation symbol M is called an *invariant mean*. This appellation is justified since we can write the average or mean of all function values as

$$\frac{\alpha_1 + \alpha_2 + \cdots + \alpha_n}{n} = \mathrm{M}_g \alpha(g)\,, \tag{4.14}$$

which satisfies

$$\mathrm{M}_g \alpha(g) = \mathrm{M}_g \alpha(g^{-1}) = \mathrm{M}_g \alpha(hg) = \mathrm{M}_g \alpha(gh)\,, \tag{4.15}$$

where h is a fixed arbitrary element of G. These relations are trivially true since taking the inverse of all the group elements or multiplying them with a fixed group element only rearranges their order and since the addition of complex numbers is commutative. The invariant mean notation is also justified in Eq. (4.13), since the addition of algebra elements is commutative and obviously the following relations hold:

$$\mathrm{M}_g \alpha(g) g^{-1} = \mathrm{M}_g \alpha(g^{-1}) g = \mathrm{M}_g \alpha(gh)(gh)^{-1}. \tag{4.16}$$

Since the one-to-one mapping

$$A \leftrightarrow \alpha \tag{4.17}$$

exists between the elements of the group algebra and the functions over the group there must also be a multiplication operation ($*$) among the functions over the group that corresponds to the group algebra multiplication ($\times$). This can be found by considering

$$\begin{aligned} A \times B &= (\mathrm{M}_g \alpha(g) g^{-1}) \times (\mathrm{M}_h \beta(h) h^{-1}) \\ &= \mathrm{M}_g \mathrm{M}_h \alpha(g) \beta(h) g^{-1} h^{-1} = \mathrm{M}_g \mathrm{M}_u \alpha(g) \beta(ug^{-1}) u^{-1} \\ &= \mathrm{M}_u \gamma(u) u^{-1} = C, \end{aligned} \tag{4.18}$$

where, in the last step, the *convolution product*

$$\alpha * \beta = \mathrm{M}_g \alpha(g) \beta(ug^{-1}) = \gamma \tag{4.19}$$

has been introduced. The convolution product of two functions over the group is obviously a function over the group. Again, the particular multiplication of complex numbers to form the new function cannot be obtained until a particular group G is specified. It should be clear from this discussion that there are two algebras: (1) the *group algebra*, with elements $\{A\}$ that are linear combinations of group elements; and (2) the *convolution algebra*, with elements $\{\alpha\}$ that are functions over the group. These two structures are entirely equivalent, and either can be used as the need arises. The relation

$$A = \mathrm{M}_g \alpha(g) g^{-1} \tag{4.20}$$

establishes the mapping between the two algebras, and the following correspondences are readily established: It is easy to show that the associative law of

Group Algebra	Convolution Algebra
A, B	α, β
$A + B$	$\alpha + \beta$
cA	$c\alpha$
$A \times B$	$\alpha * \beta$

multiplication hold in both algebras. In particular, for the convolution algebra, this means that

$$(\alpha * \beta) * \gamma = \alpha * (\beta * \gamma), \tag{4.21}$$

which can be shown to hold by writing the left-hand side (LHS) as

$$\mathrm{M}_g \mathrm{M}_h \alpha(g)\beta(hg^{-1})\gamma(sh^{-1}) = \delta(s), \tag{4.22}$$

and compare that to the right-hand side (RHS), which is

$$\mathrm{M}_g \mathrm{M}_u \alpha(g)\beta(u)\gamma(sg^{-1}u^{-1}) = \lambda(s) \tag{4.23}$$

by straightforward application of the convolution product. Renaming the dummy summation index $u = hg^{-1}$ and summing over h instead of u yields

$$\mathrm{M}_g \mathrm{M}_h \alpha(g)\beta(hg^{-1})\gamma(sh^{-1}) \tag{4.24}$$

and

$$\lambda = \delta, \tag{4.25}$$

as we set out to prove. It is also clear that, in general, the convolution product is not commutative, and $\alpha * \beta \neq \beta * \alpha$ as well as

$$A \times B \neq B \times A. \tag{4.26}$$

4.1.3 Stable Subspaces

The linear space V_G of the group algebra has both proper and improper subspaces (see Fig. 4.1).

The entire space $W = V_G$ and the zero element $W = \{O\}$ are *improper* subspaces, while any other subspace is *proper*.

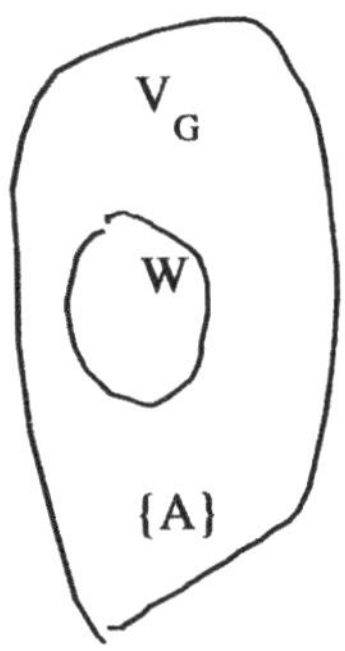

Figure 4.1 The linear space of the group algebra with elements $\{A\}$ and subspaces W.

A subspace is said to be *stable* under G when $g \times W \subset W$ for all $g \in G$. Let the $\{X_1, X_2, \ldots, X_f\}$ be a basis for the stable subspace W $(1 \leq f \leq n)$; thus, any element in the subspace can be expressed as a linear combination of these basis elements. In terms of the basis stability under $g \in G$, this means

$$g \times X_l = \sum_{k=1}^{f} X_k \Gamma_{kl}(g), \quad l = 1, 2, \ldots, f\,. \tag{4.27}$$

The complex numbers $\{\Gamma_{kl}(g)\}$ can be arranged in a square matrix array, $\mathbf{\Gamma}(g)$, and the basis elements in a row matrix array

$$\mathbf{X} = (X_1 \quad X_2 \quad \cdots \quad X_f) \tag{4.28}$$

so that

$$g \times \mathbf{X} = \mathbf{X\Gamma}(g)\,, \tag{4.29}$$

which maps each group element on a square matrix of complex numbers. It follows that for a group element expressed as a product of two group elements

$$(gh) \times \mathbf{X} = \mathbf{X\Gamma}(gh)\,, \tag{4.30}$$

while for successive operation of the same two group elements[1]

$$(g \times h) \times \mathbf{X} = g \times (h \times \mathbf{X}) = g \times \mathbf{X\Gamma}(h) = \mathbf{X\Gamma}(g)\mathbf{\Gamma}(h)\,, \tag{4.31}$$

with ordinary matrix multiplication combining the symbols on the right. Since by definition $g \times h = gh$ for the multiplication of individual group elements, it follows that

$$\mathbf{\Gamma}(g)\mathbf{\Gamma}(h) = \mathbf{\Gamma}(gh)\,, \tag{4.32}$$

and a group homomorphism has been established between the group G and a group of square f-dimensional matrices $\{\mathbf{\Gamma}(g)\}$ under matrix multiplication. This group of matrices is called a matrix representation of the group G or simply a *representation* of G.

The homomorphism

$$\mathbf{\Gamma}(g)\mathbf{\Gamma}(e) = \mathbf{\Gamma}(ge) = \mathbf{\Gamma}(g) \tag{4.33}$$

forces the image element of the identity to be the unit matrix $\mathbf{\Gamma}(e) = \mathbf{1}_f$ of dimension f. Similarly

$$\mathbf{\Gamma}(g)\mathbf{\Gamma}(g^{-1}) = \mathbf{\Gamma}(gg^{-1}) = \mathbf{\Gamma}(e) \tag{4.34}$$

[1] Individual group elements are, of course, elements of the group algebra.

forces the relation

$$\Gamma(g^{-1}) = \Gamma^{-1}(g), \qquad (4.35)$$

where the image matrix of g^{-1} is the inverse of the image matrix of g. The representation matrices must all have inverses. This is, of course, assumed by the introduction of a basis in the stable subspace W. A representation matrix without an inverse means that there are linear dependencies among the elements $\{X_l\}$ and they do then *not* form a basis.

No particular set of basis elements is preferred over any other, and one must ask what happens to the representation matrices when the basis suffers a transformation

$$\mathbf{Y} = \mathbf{X}\boldsymbol{\Omega}. \qquad (4.36)$$

For this to be a relation between basis sets, the inverse relation must also hold:

$$\mathbf{X} = \mathbf{Y}\boldsymbol{\Omega}^{-1}. \qquad (4.37)$$

The representation matrices in the two cases satisfy

$$\begin{aligned} g \times \mathbf{X} &= \mathbf{X}\boldsymbol{\Gamma}_X(g), \\ g \times \mathbf{Y} &= \mathbf{Y}\boldsymbol{\Gamma}_Y(g). \end{aligned} \qquad (4.38)$$

The second relation can be expressed in more detail as

$$g \times \mathbf{Y} = \mathbf{Y}\boldsymbol{\Gamma}_Y(g) = g \times \mathbf{X}\boldsymbol{\Omega} = \mathbf{X}\boldsymbol{\Gamma}_X(g)\boldsymbol{\Omega} = \mathbf{Y}\boldsymbol{\Omega}^{-1}\boldsymbol{\Gamma}_X(g)\boldsymbol{\Omega}, \qquad (4.39)$$

yielding

$$\boldsymbol{\Gamma}_Y(g) = \boldsymbol{\Omega}^{-1}\boldsymbol{\Gamma}_X(g)\boldsymbol{\Omega} \qquad (4.40)$$

for all $g \in G$. All the representation matrices undergo the same similarity transformation by the fixed matrix $\boldsymbol{\Omega}$. The representations $\{\boldsymbol{\Gamma}_X(g)\}$ and $\{\boldsymbol{\Gamma}_Y(g)\}$ are said to be equivalent.

The Character

The trace of a representation matrix

$$\mathrm{Tr}\{\boldsymbol{\Gamma}(g)\} = \sum_{k=1}^{f} \Gamma_{kk}(g) = \chi(g) \qquad (4.41)$$

is called the *character* at g. The traces of all the matrices $\{\boldsymbol{\Gamma}(g)\}$ of a representation defines the character as a function over the group. The trace of a square matrix **RS** expressed as the product of two square matrices **R** and **S** satisfies the relation

$$\mathrm{Tr}\{\mathbf{RS}\} = \sum_{k=1}^{f} \sum_{l=1}^{f} R_{kl} S_{lk} = \sum_{k=1}^{f} \sum_{l=1}^{f} S_{lk} R_{kl} = \mathrm{Tr}\{\mathbf{SR}\}. \tag{4.42}$$

Applied to the representation matrices in two different bases, this yields

$$\mathrm{Tr}\{\mathbf{\Gamma}_Y(g)\} = \mathrm{Tr}\{\mathbf{\Omega}^{-1}\mathbf{\Gamma}_X(g)\mathbf{\Omega}\} = \mathrm{Tr}\{\mathbf{\Gamma}_X(g)\mathbf{\Omega}\mathbf{\Omega}^{-1}\} = \mathrm{Tr}\{\mathbf{\Gamma}_X(g)\}, \tag{4.43}$$

which means that the character is *invariant* to basis transformations. Furthermore, by considering

$$\begin{aligned} \chi(t^{-1}gt) - \mathrm{Tr}\{\mathbf{\Gamma}(t^{-1}gt)\} &= \mathrm{Tr}\{\mathbf{\Gamma}(t^{-1}g)\mathbf{\Gamma}(t)\} = \mathrm{Tr}\{\mathbf{\Gamma}(t)\mathbf{\Gamma}(t^{-1}g)\} \\ &= \mathrm{Tr}\{\mathbf{\Gamma}(tt^{-1}g)\} = \mathrm{Tr}\{\mathbf{\Gamma}(g)\} = \chi(g), \end{aligned} \tag{4.44}$$

where we have used the property of the homomorphism, and the invariance of the trace as expressed in Eq. (4.42), it follows that the value of the character is the same at all elements of a conjugate class. The character as a function over the group is said to be a *class constant.*

The value of the character at the identity element of the group equals the dimension of the stable subspace of the linear space of the group algebra that *carries* the representation:

$$\chi(e) = \mathrm{Tr}\{\mathbf{\Gamma}(e)\} = \mathrm{Tr}\{\mathbf{1}_f\} = f. \tag{4.45}$$

4.1.4 The Regular Representation

The entire linear space of the group algebra V_G is a stable subspace (improper) and carries a representation called the *regular representation.* The group elements themselves can be chosen as the basis in this case, V_G: $\{g_1, g_2, \ldots, g_n\}$, and the condition of stability becomes

$$g \times g_k = \sum_{m=1}^{n} g_m \Gamma^R_{mk}(g) \tag{4.46}$$

for all $g \in G$. Since the algebra multiplication $\times$ between group elements simply becomes the group combination law, it follows that the result of $g \times g_k = gg_k = g_m$ must be a single group element. This determines the representation matrices, since

$$\Gamma_{mk}(g) = \begin{cases} 1 & \text{if} \quad g = g_m g_k^{-1} \\ 0 & \quad \text{otherwise.} \end{cases} \tag{4.47}$$

The representation matrices can then be obtained from a modified form of the group multiplication table.

Example 4.1 Consider the permutation group S_3 and write its multiplication table as

S_3	e	(321)	(123)	(23)	(13)	(12)
e	e	(321)	(123)	(23)	(13)	(12)
(123)	(123)	e	(321)	(12)	(23)	(13)
(321)	(321)	(123)	e	(13)	(12)	(23)
(23)	(23)	(12)	(13)	e	(123)	(321)
(13)	(13)	(23)	(12)	(321)	e	(123)
(12)	(12)	(13)	(23)	(123)	(321)	e

where the rows are labeled by g and the columns by g^{-1}. The representation matrix for a particular group element is then obtained by putting a 1 in the position that corresponds to its appearance in this modified multiplication table and putting 0 everywhere else. Thus

$$\mathbf{\Gamma}^R((12)) = \begin{bmatrix} 0 & 0 & 0 & 0 & 0 & 1 \\ 0 & 0 & 0 & 1 & 0 & 0 \\ 0 & 0 & 0 & 0 & 1 & 0 \\ 0 & 1 & 0 & 0 & 0 & 0 \\ 0 & 0 & 1 & 0 & 0 & 0 \\ 1 & 0 & 0 & 0 & 0 & 0 \end{bmatrix}, \tag{4.48}$$

and

$$\mathbf{\Gamma}^R((123)) = \begin{bmatrix} 0 & 0 & 1 & 0 & 0 & 0 \\ 1 & 0 & 0 & 0 & 0 & 0 \\ 0 & 1 & 0 & 0 & 0 & 0 \\ 0 & 0 & 0 & 0 & 1 & 0 \\ 0 & 0 & 0 & 0 & 0 & 1 \\ 0 & 0 & 0 & 1 & 0 & 0 \end{bmatrix}, \tag{4.49}$$

and so on. It can be seen that

$$\mathbf{\Gamma}^R(e) = \mathbf{1}_n \tag{4.50}$$

and that, say, $\mathbf{\Gamma}^R((321))$ is the inverse of $\mathbf{\Gamma}^R((123))$. Since only $\mathbf{\Gamma}^R(e) = \mathbf{1}_n$ has nonzero elements on the diagonal, the character of the regular representation is

$$\chi^R(g) = \begin{cases} n = |G| & \text{for} \quad g = e, \\ 0 & \text{otherwise.} \end{cases} \tag{4.51}$$

4.1.5 Reducible and Irreducible Subspaces

A stable subspace $W \subset V_G$ is said to be *reducible* if it contains a proper subspace, which is also stable under the group G. Otherwise it is said to be *irreducible*. These concepts are meaningful since, for instance, the one-dimensional subspace consisting of the single element $A = (g_1 + g_2 + \cdots + g_n)/n = \mathrm{M}_g g^{-1}$ is a stable proper subspace of V_G:

$$g \times A = A = A1\,. \tag{4.52}$$

This subspace generates the so-called identity representation or totally symmetric representation of 1×1 matrices all equal the number 1:

$$\boldsymbol{\Gamma}(g) = \mathbf{J}(g) = 1\,. \tag{4.53}$$

This representation must be irreducible, since a space of dimension 1 can contain no proper subspaces, and it is a representation that exists for all groups. Furthermore, since the entire linear space of the group algebra, V_G, contains this proper stable subspace, it must, in turn, be reducible.

Consider the subspaces

$$W' \subset W \subset V_G, \tag{4.54}$$

where W' is a proper subspace of the space W. We introduce a basis for the reducible space W

$$W\colon \{X_1, X_2, \ldots, X_f\} \tag{4.55}$$

and a basis for W'

$$W'\colon \left\{X_1', X_2', \ldots, X_{f'}'\right\} \tag{4.56}$$

where $f' < f$. The basis of W can be transformed by a nonsingular transformation ($\boldsymbol{\Omega}$) such that the first f' basis elements form a basis for W'. Using this transformed basis, we obtain

$$g \times \left(X_1' \quad X_2' \quad \cdots \quad X_{f'}'; \quad X_{f'+1}' \quad X_{f'+2}' \quad \cdots \quad X_f'\right)$$
$$= \left(X_1' \quad X_2' \quad \cdots \quad X_{f'}'; \quad X_{f'+1}' \quad X_{f'+2}' \quad \cdots \quad X_f'\right)\begin{bmatrix} \boldsymbol{\Gamma}_{11}(g) & \boldsymbol{\Gamma}_{12}(g) \\ \mathbf{0} & \boldsymbol{\Gamma}_{22}(g) \end{bmatrix}, \tag{4.57}$$

where the 11-block is a square matrix of dimension f', and the 22-block, a square matrix of dimension $f - f'$. All the representation matrices have a rectangular zero matrix of dimension $(f - f') \times f$ in the 21-block. The content of the 12-block is presently immaterial. The product of two such representation matrices can be expressed in block form as

$$\mathbf{\Gamma}(g)\mathbf{\Gamma}(h) = \begin{bmatrix} \mathbf{\Gamma}_{11}(g)\mathbf{\Gamma}_{11}(h) & \mathbf{R} \\ \mathbf{0} & \mathbf{\Gamma}_{22}(g)\mathbf{\Gamma}_{22}(h) \end{bmatrix}. \tag{4.58}$$

The homomorphism imposes the equality of this matrix and

$$\mathbf{\Gamma}(gh) = \begin{bmatrix} \mathbf{\Gamma}_{11}(gh) & \mathbf{R} \\ \mathbf{0} & \mathbf{\Gamma}_{22}(gh) \end{bmatrix}. \tag{4.59}$$

The net result is that two new representations have been formed from the *reducible representation* $\mathbf{\Gamma}$, since it follows that

$$\begin{aligned} \mathbf{\Gamma}_{11}(g)\mathbf{\Gamma}_{11}(h) &= \mathbf{\Gamma}_{11}(gh), \\ \mathbf{\Gamma}_{22}(g)\mathbf{\Gamma}_{22}(h) &= \mathbf{\Gamma}_{22}(gh). \end{aligned} \tag{4.60}$$

If these new representations, in turn, are reducible, the same procedure can be repeated for each of them, and so on, until all the matrices of the original reducible representation have become *completely reduced*, and of the form

$$\mathbf{\Gamma}(g) = \begin{bmatrix} \mathbf{\Gamma}^1 & / & / & / & \cdots & / \\ \mathbf{0} & \mathbf{\Gamma}^2 & / & / & \cdots & / \\ \mathbf{0} & \mathbf{0} & \mathbf{\Gamma}^2 & / & \cdots & / \\ \mathbf{0} & \mathbf{0} & \mathbf{0} & \mathbf{\Gamma}^3 & \cdots & / \\ \vdots & \vdots & \vdots & \vdots & \ddots & \vdots \\ \mathbf{0} & \mathbf{0} & \mathbf{0} & \mathbf{0} & \cdots & \mathbf{\Gamma}^{\alpha} \end{bmatrix} \tag{4.61}$$

with the *irreducible representations*, or *irreps* for short, that are contained in $\mathbf{\Gamma}$ appearing in the diagonal blocks. For finite groups the upper off-diagonal matrix blocks can also be brought to zero by a similarity transformation.

Unitary Representation Matrices

The representation matrices (for finite groups) are general nonsingular complex matrices. However, one can show that any representation can be brought to unitary form by a similarity transformation and thus is equivalent to a unitary representation, specifically, a representation where the representation matrices are unitary. A unitary matrix $\mathbf{U}$ is defined to satisfy

$$\mathbf{U}^{\dagger}\mathbf{U} = \mathbf{U}\mathbf{U}^{\dagger} = \mathbf{1}, \tag{4.62}$$

which is the same as saying that the adjoint or transpose complex conjugate matrix is the inverse:

$$\mathbf{U}^{-1} = \mathbf{U}^{\dagger}. \tag{4.63}$$

A general representation $\{\mathbf{\Gamma}(g)\}$ can be transformed to unitary form by a similarity transformation with a matrix

$$\mathbf{S} = \mathrm{M}_g \mathbf{\Gamma}(g)\mathbf{\Gamma}^\dagger(g), \tag{4.64}$$

which is simply a sum of matrices divided by the order of the group $|G|$. Should $\mathbf{\Gamma}$ be unitary already, then $\mathbf{S} = \mathbf{1}$.

The matrix $\mathbf{S}$ satisfies the relations

1. $\mathbf{S}^\dagger = \mathbf{S}$.
2. $\mathbf{S} > 0$.
3. $\mathbf{\Gamma}(h)\mathbf{S}\mathbf{\Gamma}^\dagger(h) = \mathbf{S}$, for all $h \in G$.

The first of these relations is obvious, since

$$\{\mathbf{\Gamma}\mathbf{\Gamma}^\dagger\}^\dagger = \mathbf{\Gamma}\mathbf{\Gamma}^\dagger. \tag{4.65}$$

The second relation means that for any nonzero vector

$$\mathbf{a} = \begin{bmatrix} a_1 \\ a_2 \\ \vdots \\ a_f \end{bmatrix} \tag{4.66}$$

it holds that $\mathbf{a}^\dagger\mathbf{S}\mathbf{a} > 0$, which is true since

$$\mathbf{a}^\dagger\mathbf{\Gamma}\mathbf{\Gamma}^\dagger\mathbf{a} = \sum_{l=1}^{f}\left(\sum_{k=1}^{f} a_k^*\Gamma_{kl}\right)\left(\sum_{m=1}^{f} a_m\Gamma_{ml}^*\right) = \sum_{l=1}^{f}|b_l|^2 > 0, \tag{4.67}$$

where in the last step $b_l = \sum_{k=1}^{f} a_k^*\Gamma_{kl}$. This means that, in particular, the eigenvalues of $\mathbf{S}$ are positive. Since $\mathbf{S}$ is self-adjoint, it can be diagonalized by a unitary transformation $\mathbf{U}$, such as

$$\mathbf{U}^\dagger\mathbf{S}\mathbf{U} = \mathbf{s}, \tag{4.68}$$

where $\mathbf{s}$ is diagonal. This permits the definition of $\mathbf{s}^{1/2}$ as the diagonal matrix with the positive square roots of the eigenvalues of $\mathbf{S}$ on the diagonal. Similarly its inverse $\mathbf{s}^{-1/2}$ is the diagonal matrix with the inverse positive square roots of the same eigenvalues on the diagonal. The inverse transforms are

$$\mathbf{S} = \mathbf{U}\mathbf{s}\mathbf{U}^\dagger, \qquad \mathbf{S}^{1/2} = \mathbf{U}\mathbf{s}^{1/2}\mathbf{U}^\dagger, \qquad \mathbf{S}^{-1/2} = \mathbf{U}\mathbf{s}^{-1/2}\mathbf{U}^\dagger, \tag{4.69}$$

which define the square root and its inverse of any positive definite matrix.

The third relation can be understood from

$$\boldsymbol{\Gamma}(h)\mathbf{S}\boldsymbol{\Gamma}^{\dagger}(h) = \mathbf{M}_g \boldsymbol{\Gamma}(hg)\boldsymbol{\Gamma}^{\dagger}(hg) = \mathbf{S}\,. \tag{4.70}$$

In order to transform the arbitrary representation to unitary form, we choose the transformation matrix $\boldsymbol{\Omega} = \mathbf{U}\mathbf{s}^{1/2}$ and perform the similarity transformation on all representation matrices, specifically, $\bar{\boldsymbol{\Gamma}}(g) = \boldsymbol{\Omega}^{-1}\boldsymbol{\Gamma}(g)\boldsymbol{\Omega}$ for all $g \in G$. It follows that

$$\bar{\boldsymbol{\Gamma}}(g)\bar{\boldsymbol{\Gamma}}^{\dagger}(g) = \mathbf{s}^{-1/2}\underbrace{\mathbf{U}^{\dagger}\underbrace{\boldsymbol{\Gamma}(g)\underbrace{\mathbf{U}\underbrace{\mathbf{s}^{1/2}\mathbf{s}^{1/2}}_{\mathbf{s}}\mathbf{U}^{\dagger}}_{\mathbf{S}}\boldsymbol{\Gamma}^{\dagger}(g)}_{\mathbf{S}}\mathbf{U}}_{\mathbf{s}}\mathbf{s}^{-1/2} = \mathbf{1}\,, \tag{4.71}$$

where the relations satisfied by the matrix $\mathbf{S}$ have been used. This proves that any representation of a finite group can be brought to unitary form.

When the representation matrices are unitary, it holds that

$$\Gamma_{kl}(g) = \Gamma^{*}_{lk}(g^{-1}) \tag{4.72}$$

since for such a representation matrix

$$\boldsymbol{\Gamma}^{\dagger}(g) = \boldsymbol{\Gamma}^{-1}(g) = \boldsymbol{\Gamma}(g^{-1})\,. \tag{4.73}$$

Putting $k = l$ in this relation and summing over l to obtain the trace or the value of the character at $g \in G$, one finds

$$\chi(g^{-1}) = \chi^{*}(g)\,. \tag{4.74}$$

This result, however, is general and not restricted to unitary representations, since the character is invariant to a basis transformation or similarity transformation. From this result we can conclude that when the character is a real function over the group, that is, when all $\chi(g)$ are real numbers, then g and g^{-1} belong to the same class. When this is the case, we call the conjugate class an *ambivalent class*. As we will see in Chapter 7, the classes that have real characters contain both g and g^{-1}, while for those that have complex characters, an element and its inverse are never in the same class.

4.2 SCHUR'S LEMMA

The cornerstone of representation theory is a theorem from the general theory of linear spaces called *Schur's lemma*. The consequences of Schur's lemma are far reaching and fundamental for the application of group theory and group algebra to physical problems.

The setting of Schur's lemma in the general theory of finite linear spaces contains two linear spaces V_1 and V_2, and a linear mapping L of V_1 on V_2. The space V_1 has elements $\{x\}$ and a set of p linear operators $\{A_i\}$

acting on these elements, while the space V_2, similarly, has elements $\{y\}$ and a set of p operators $\{B_i\}$ (see Fig. 4.2).

Theorem 4.1 (Schur's Lemma) If $LA_i = B_iL$, for $i = 1, 2, \ldots, p$, and V_1 and V_2 are both irreducible under the operations $\{A_i\}$ and $\{B_i\}$, respectively, then either (1) $L = 0$ or (2) L^{-1} exists.

Proof Consider the mapping $Lx = y$, and let $K \subset V_1$ be the subspace that is mapped on the zero element $0_2 \in V_2$, that is, $K = \{x | Lx = 0_2\} \subset V_1$. K is a linear space, since if x_1 and x_2 belongs to K, so does $c_1x_1 + c_2x_2$. Furthermore, K is stable under action of any of the operators $\{A_i\}$, since for any $x \in K$,

$$L(A_ix) = LA_ix = B_iLx = B_i0_2 = 0_2\,, \tag{4.75}$$

which means that $A_ix \in K$ for $i = 1, 2, \ldots, p$. But V_1 is irreducible under $\{A_i\}_1^p$, which means that $K \subset V_1$ must be an improper subspace. Either (1) $K = V_1$, which means that L maps the entire space V_1 on the zero element 0_2, that is, $L = 0$; or (2) $K = 0_1$, which means that the mapping is one-to-one and onto, so that L^{-1} exists.

This can be seen by assuming the converse to be true, namely, that for $x_1 \neq x_2$, the mapping is *not* one-to-one, but that two different elements of V_1 have the same image element, that is, $Lx_1 = y$ and $Lx_2 = y$. If this is so, then $Lx_1 - Lx_2 = L(x_1 - x_2) = y - y = 0_2$; thus, $x_1 - x_2 \in K = 0_1$. Thus, $x_1 = x_2$, which is a contradiction, and it follows that the mapping must be one-to-one.

Whether the mapping is onto can be decided by considering the image space $W = LV_1 \subset V_2$. W is obviously linear. It is also stable under the action of any of the operators $\{B_i\}$, since for any $y \in W$,

$$B_iy = B_iLx = LA_ix = L(A_ix) = Lx' \in W\,, \tag{4.76}$$

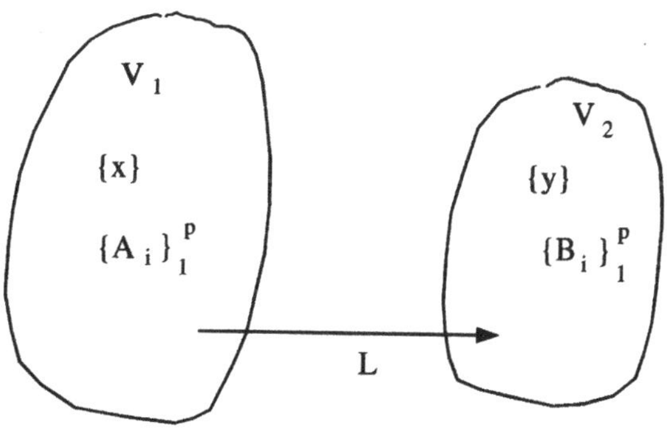

Figure 4.2 The setting of Schur's lemma for the mapping of one finite vector space on another.

and since V_2 is irreducible under $\{B_i\}_1^p$, W must be an improper subspace. Either (1) $W = 0_2$, which, as we have already seen, means that $L = 0$; or (2) $W = V_2$, and the mapping is onto.

QED

There is also need for a special case of Schur's lemma, namely, when $V_1 = V_2 = V$, and $A_i = B_i$ for $i = 1, 2, \ldots, p$.

Theorem 4.2 (Special Case of Schur's Lemma) If L is a linear mapping of the linear space V on itself, and V is irreducible under the linear operators $\{A_i\}$ for $i = 1, 2, \ldots, p$, and $LA_i = A_iL$ for all operators $\{A_i\}$, then L must be a multiple of the identity operator I.

Proof With λ a complex number the relation

$$(L - \lambda I)A_i = A_i(L - \lambda I), \tag{4.77}$$

for $i = 1, 2, \ldots, p$ must hold. Schur's lemma assures that either (1) $L - \lambda I = 0$ or (2) $(L - \lambda I)^{-1}$ exists. For a general L, consider the secular problem with the determinant $|L - cI| = 0$. This equation always has a solution, according to "the fundamental theorem of algebra," or according to the fact that the field of complex numbers is algebraically closed; in other words, one can always find at least one root to any polynomial equation with complex coefficients. For such a root obviously $(L - cI)^{-1}$ does not exist. Thus, L cannot be general, but must be of the form $L = \lambda I$.

4.3 CONVOLUTION RELATIONS

The results of Schur's lemma can now be applied to two irreducible subspaces W^α and W^β of the linear space of the group algebra. The bases of these two spaces are $\{X_l^\alpha\}$ and $\{X_l^\beta\}$, respectively. This means that a typical element of, say, W^α can be written as

$$A = \sum_{l=1}^{f^\alpha} X_l^\alpha a_l = [X_1^\alpha \quad X_2^\alpha \quad \cdots \quad X_{f^\alpha}^\alpha] \begin{bmatrix} a_1 \\ a_2 \\ \vdots \\ a_{f^\alpha} \end{bmatrix} = \mathbf{X}^\alpha \mathbf{a}. \tag{4.78}$$

The action of a group element $g \in G$ on such an element is

$$g \times A = \mathbf{X}^\alpha \boldsymbol{\Gamma}^\alpha(g)\mathbf{a}, \tag{4.79}$$

which illustrates that the action of the representation matrices $\{\boldsymbol{\Gamma}^\alpha(g)\}$ can be viewed in two complementary ways. They operate from either (1) the right on row arrays of basis elements (as we have already used) or (2) the left on column arrays of expansion coefficients of the particular group algebra element in question (see Fig. 4.3).

The irreducible representation matrices assume the role of the operators $\{A_i\}$ and $\{B_i\}$ of Schur's lemma, and for now we choose to let them operate to the right on column arrays of expansion coefficients. It is clear that $\mathbf{\Gamma}^\alpha(g)$ generates a new column array $\mathbf{\Gamma}^\alpha(g)\mathbf{a}$ and thus maps one group algebra element in W^α on another in the same subspace. The linear mapping $\mathbf{L}$ is then, in general, a rectangular $f^\alpha \times f^\beta$ matrix, which we choose to write as the matrix sum

$$\mathbf{L} = \mathrm{M}_s\mathbf{\Gamma}^\alpha(s)\mathbf{A}\mathbf{\Gamma}^\beta(s^{-1}), \tag{4.80}$$

where $\mathbf{A}$ is an arbitrary $f^\alpha \times f^\beta$ matrix. It follows that

$$\begin{aligned} \mathbf{L}\mathbf{\Gamma}^\beta(g) &= \mathrm{M}_s\mathbf{\Gamma}^\alpha(s)\mathbf{A}\mathbf{\Gamma}^\beta(s^{-1}g) \\ &= \mathrm{M}_u\mathbf{\Gamma}^\alpha(gu)\mathbf{A}\mathbf{\Gamma}^\beta(u^{-1}) = \mathbf{\Gamma}^\alpha(g)\mathbf{L}, \end{aligned} \tag{4.81}$$

which satisfies the condition of Schur's lemma. Then either (1) $\mathbf{L} = \mathbf{0}$ or (2) $\mathbf{L}^{-1}$ exists, which means that $W^\alpha \neq W^\beta$, $f^\alpha = f^\beta$, and $\mathbf{\Gamma}^\beta(g) = \mathbf{L}^{-1}\mathbf{\Gamma}^\alpha(g)\mathbf{L}$ for all $g \in G$ (equivalent representations). This case is *not* of interest in what follows in this section.

Introducing a Kronecker delta

$$\delta^{\alpha\beta} = \begin{cases} 1 & \text{if} \quad W^\alpha = W^\beta,\ \mathbf{\Gamma}^\alpha = \mathbf{\Gamma}^\beta,\ (\alpha = \beta), \\ 0 & \text{if} \quad W^\alpha \neq W^\beta,\ \mathbf{\Gamma}^\alpha \neq \mathbf{\Gamma}^\beta. \end{cases} \tag{4.82}$$

Schur's lemma and the special case of Schur's lemma can now be summarized by writing

$$\mathbf{L} = \mathrm{M}_s\mathbf{\Gamma}^\alpha(s)\mathbf{A}\mathbf{\Gamma}^\beta(s^{-1}) = \lambda\mathbf{1}\delta^{\alpha\beta}. \tag{4.83}$$

It is useful to generalize this expression by writing $\mathbf{A} = \mathbf{B}\mathbf{\Gamma}^\beta(g)$, where $\mathbf{B}$ is an arbitrary $f^\alpha \times f^\beta$ matrix, such that

$$\mathbf{L} = \mathrm{M}_s\mathbf{\Gamma}^\alpha(s)\mathbf{B}\mathbf{\Gamma}^\beta(gs^{-1}) = \lambda\mathbf{1}\delta^{\alpha\beta}. \tag{4.84}$$

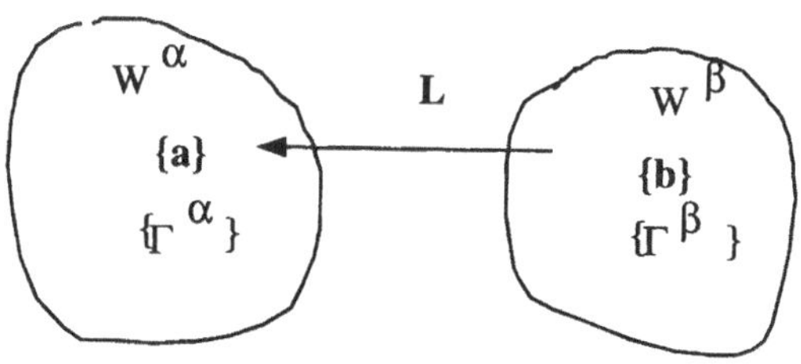

Figure 4.3 The setting of Schur's lemma for the mapping of one subspace of the linear space of the group algebra on another.

Putting $\alpha = \beta$ and taking the trace of both sides of this equation yields

$$\mathrm{M}_s \mathrm{Tr}\{\mathbf{\Gamma}^\alpha(s)\mathbf{B}\mathbf{\Gamma}^\alpha(gs^{-1})\} = \lambda \mathrm{Tr}\{\mathbf{1}\}, \tag{4.85}$$

which means that

$$\lambda = (f^\alpha)^{-1}\mathrm{Tr}\{\mathbf{B}\mathbf{\Gamma}^\alpha(g)\}, \tag{4.86}$$

and we can write

$$\mathrm{M}_s \mathbf{\Gamma}^\alpha(s)\mathbf{B}\mathbf{\Gamma}^\alpha(gs^{-1}) = (f^\alpha)^{-1}\mathrm{Tr}\{\mathbf{B}\mathbf{\Gamma}^\beta(g)\}\delta^{\alpha\beta}\mathbf{1}. \tag{4.87}$$

This equation can now be used to obtain a number of fundamental relationships among elements of irreducible representation matrices. For instance, the kl element of this matrix equation is

$$\mathrm{M}_s \sum_{m=1}^{f^\alpha}\sum_{n=1}^{f^\beta} \Gamma^\alpha_{km}(s)B_{mn}\Gamma^\beta_{nl}(gs^{-1}) = (f^\alpha)^{-1}\delta^{\alpha\beta}\delta_{kl}\sum_{m=1}^{f^\alpha}\sum_{n=1}^{f^\beta} B_{mn}\Gamma^\beta_{nm}(g), \tag{4.88}$$

where the Kronecker delta δ_{kl} is the kl element of the unit matrix. Since $\mathbf{B}$ is arbitrary, the coefficient of B_{mn} on the left and on the right must be the same:

$$\mathrm{M}_s \Gamma^\alpha_{km}(s)\Gamma^\beta_{nl}(gs^{-1}) = (f^\alpha)^{-1}\delta^{\alpha\beta}\delta_{kl}\Gamma^\beta_{nm}(g). \tag{4.89}$$

Multiplying this equation by $f^\alpha f^\beta$ and using the convolution product it yields

$$f^\alpha \Gamma^\alpha_{km} * f^\beta \Gamma^\beta_{nl} = \delta^{\alpha\beta}\delta_{kl} f^\beta \Gamma^\beta_{nm}. \tag{4.90}$$

The elements of the representation matrices are treated as functions over the group; thus, the function Γ^α_{km} is defined by the function values that are the complex numbers appearing in the km position, one in each of the $|G|$ matrices $\{\mathbf{\Gamma}^\alpha(g)\}$. Equation (4.90) called the *master convolution relation* (MCR), is fundamental for representation theory. Explicitly, it reads

$$\mathrm{M}_s f^\alpha \Gamma^\alpha_{km}(s) f^\beta \Gamma^\beta_{nl}(gs^{-1}) = \delta^{\alpha\beta}\delta_{kl} f^\beta \Gamma^\beta_{nm}(g). \tag{4.91}$$

All the special relations for functions over the group that are central to representation theory can be obtained from the MCR:

1. Put $\alpha = \beta$, $n = m$ and sum over m to obtain

$$\mathrm{M}_s f^\alpha \Gamma^\alpha_{kl}(sgs^{-1}) = \delta_{kl}\chi^\alpha(g), \tag{4.92}$$

and, in particular, for $k = l$ we obtain

$$\chi^{\alpha}(g) = f^{\alpha} \mathrm{M}_s \Gamma^{\alpha}_{kk}(sgs^{-1}) = \frac{f^{\alpha}}{h_g} \sum_{t \in C_g} \Gamma^{\alpha}_{kk}(t), \tag{4.93}$$

where we have used the notation $h_g = |C_g|$ for the order of the class of the element $g \in G$. This reveals a relationship between the trace of a particular irreducible representation matrix and the sum over the corresponding class of any diagonal matrix element.

2. In MCR put $m = k$, $n = l$, and sum over k and l to get

$$\mathrm{M}_s f^{\alpha} \chi^{\alpha}(s) f^{\beta} \chi^{\beta}(gs^{-1}) = \delta^{\alpha\beta} f^{\beta} \chi^{\beta}(g), \tag{4.94}$$

which can be expressed as

$$f^{\alpha} \chi^{\alpha} * f^{\beta} \chi^{\beta} = \delta^{\alpha\beta} f^{\beta} \chi^{\beta} \tag{4.95}$$

again invoking the convolution product. This is referred to as the master convolution relation for characters.

3. In MCR put $g = e$ (the identity element of the group) to get

$$\mathrm{M}_s f^{\alpha} \Gamma^{\alpha}_{km}(s) f^{\beta} \Gamma^{\beta}_{nl}(s^{-1}) = f^{\beta} \delta^{\alpha\beta} \delta_{kl} \delta_{mn}, \tag{4.96}$$

which is called the *master orthogonality relation* (MOR).

4. In MOR put $m = k$, $n = l$ and sum over k and l to get

$$\mathrm{M}_s \chi^{\alpha}(s) \chi^{\beta}(s^{-1}) = \delta^{\alpha\beta}, \tag{4.97}$$

which can be written as the convolution product of two characters of irreducible representations (irreps) evaluated at the identity:

$$(\chi^{\alpha} * \chi^{\beta})_e = \delta^{\alpha\beta}. \tag{4.98}$$

This relation is often referred to as the *orthogonality relation for characters of irreps.*

4.4 SPECTRAL ANALYSIS OF REPRESENTATIONS

The decomposition of an arbitrary representation Γ of G into its irreps was discussed in Section 4.1.5. All the representation matrices $\mathbf{\Gamma}(g)$ have the matrices of the irreps it contains in diagonal blocks. The calculation of the character can thus be expressed as

$$\chi(g) = \mathrm{Tr}\{\mathbf{\Gamma}(g)\} = \sum_{\alpha} \nu^{\alpha} \chi^{\alpha}(g), \tag{4.99}$$

where $\nu^\alpha = 0, 1, 2, \ldots$, the frequency number, is the number of times a particular irrep Γ^α is contained in Γ. The frequency number can be obtained using the character orthogonality relation such that

$$(\chi * \chi^\alpha)_e = \sum_\beta \nu^\beta (\chi^\beta * \chi^\alpha)_e = \nu^\alpha . \tag{4.100}$$

Similarly, one can show that

$$(\chi * \chi)_e = \sum_\alpha \sum_\beta \nu^\alpha \nu^\beta (\chi^\alpha * \chi^\beta)_e = \sum_\alpha (\nu^\alpha)^2, \tag{4.101}$$

a result that can be used to decide whether a given representation is irreducible.

Theorem 4.3 A necessary and sufficient condition for a representation Γ to be irreducible is that its character satisfies the relation $(\chi * \chi)_e = 1$.

Proof

1. The necessary condition is obvious, since if Γ is irreducible, it contains only one irrep, itself, once, and from Eq. (4.101), it follows that $(\chi * \chi)_e = 1$.
2. The sufficient condition also follows from $(\chi * \chi)_e = \sum_\alpha (\nu^\alpha)^2 = 1$, since a sum of integers ≥ 0 can be 1 only if all except one are zero and one $\nu^\alpha = 1$.

QED

The spectral content of the regular representation Γ^R is particularly interesting. The character

$$\chi^R(g) = \begin{cases} |G| & \text{for} \quad g = e \\ 0 & \text{otherwise} \end{cases} \tag{4.102}$$

for this reducible representation can be used to show that

$$(\chi^R * \chi^\alpha)_e = \chi^\alpha(e) = f^\alpha = \nu^\alpha . \tag{4.103}$$

This means that the frequency number of an irrep in the regular representation equals its dimension. Since the dimension of an irrep cannot be zero, this means that all irreps of the group are contained in Γ^R. The spectral resolution in this case is

$$\chi^R(g) = \sum_\alpha f^\alpha \chi^\alpha(g) . \tag{4.104}$$

For $g = e$, this equation becomes

$$|G| = \sum_\alpha (f^\alpha)^2, \tag{4.105}$$

which is referred to as the *completeness relation for irreps*.

4.5 THE CHARACTER TABLE

The character table of a group lists the values of the characters of the irreps of the group. Such character tables of point groups are particularly useful in the theory of molecular spectroscopy. For any (finite) group, the columns of the character table are labeled by the conjugate classes and the rows by the irreps. The character tables are square; the number of distinct conjugate classes N equals the number of irreps M. The character tables, particularly for point groups, often contains additional information, which is useful for molecular applications. The use of this information will become clear in the section on point groups. A generic character table then looks as follows:

G	C_1	C_2	$\cdots$	C_N
α_1	$\chi^{\alpha_1}(C_1)$	$\chi^{\alpha_1}(C_2)$	$\cdots$	$\chi^{\alpha_1}(C_N)$
α_2	$\chi^{\alpha_2}(C_1)$	$\chi^{\alpha_2}(C_2)$	$\cdots$	$\chi^{\alpha_2}(C_N)$
$\vdots$	$\vdots$	$\vdots$	$\vdots$	$\vdots$
α_M	$\chi^{\alpha_M}(C_1)$	$\chi^{\alpha_M}(C_2)$	$\cdots$	$\chi^{\alpha_M}(C_N)$

The proof that the number of irreps equals the number of conjugate classes can be done in several steps using the relations obtained from the MCR. The columns and rows of the character table are considered as vectors with complex components.

Considering the rows as M vectors in a vector space of dimension N and starting from the orthogonality relation for characters

$$\mathrm{M}_s \chi^\alpha(s)\chi^\beta(s^{-1}) = \mathrm{M}_s \chi^\alpha(s)(\chi^\beta(s))^* = \delta^{\alpha\beta}, \tag{4.106}$$

one obtains

$$\sum_{k=1}^{N} \left(\frac{h_k}{|G|}\right)^{1/2} \chi^\alpha(C_k) \left\{ \left(\frac{h_k}{|G|}\right)^{12} \chi^\beta(C_k) \right\}^* = \delta^{\alpha\beta}. \tag{4.107}$$

We have used the fact that the characters are class constants, that $\chi(s^{-1}) = \chi^*(s)$, and, as before, denoted the order of the class C_k by h_k. This relation shows that the rows of the character table are M orthogonal N-dimensional complex vectors, and thus $M \leq N$.

If we similarly could show that the columns of the character table are N orthogonal M-dimensional vectors, and thus $N \leq M$, then we could conclude that $M = N$ and that the character table is square. This can be proved in several steps starting from the MCR to obtain Eq. (4.92), which reads

$$\mathrm{M}_s f^\alpha \Gamma^\alpha_{kl}(sgs^{-1}) = \delta_{kl}\chi^\alpha(g). \tag{4.108}$$

Multiplying both sides of this equation by $\Gamma^\alpha_{lk}(t^{-1})$ and summing over k yields

$$\mathrm{M}_s f^\alpha \Gamma^\alpha_{ll}(t^{-1}sgs^{-1}) = \chi^\alpha(g)\Gamma^\alpha_{ll}(t^{-1}), \tag{4.109}$$

which can now be summed over l on both sides to obtain

$$\mathrm{M}_s f^{\alpha} \chi^{\alpha}(t^{-1}sgs^{-1}) = \chi^{\alpha}(g)\chi^{\alpha}(t^{-1}). \tag{4.110}$$

Summing both sides over irreps α and recalling the relation Eq. (4.104) for the character of the regular representation, we can now write $\sum_{\alpha} \chi^{\alpha}(g)\chi^{\alpha}(t^{-1}) = M_s \chi^R(t^{-1}sgs^{-1})$, and use the result of Eq. (4.102) for the character of the regular representation to conclude that

$$\mathrm{M}_s \chi^R(t^{-1}sgs^{-1}) = \delta_{gt} \frac{1}{|G|} |G| \frac{|G|}{h_g} = \delta_{gt} \frac{|G|}{h_g}, \tag{4.111}$$

where

$$\delta_{gt} = \begin{cases} 1 & \text{if} \quad C_g = C_t, \text{ or } t = sgs^{-1}, \\ 0 & \text{otherwise}. \end{cases} \tag{4.112}$$

This result means that we can write the expression

$$\sum_{\alpha=1}^{M} \left(\frac{h_g}{|G|} \right)^{1/2} \chi^{\alpha}(C_g) \left\{ \left(\frac{h_t}{|G|} \right)^{1/2} \chi^{\alpha}(C_t) \right\}^{*} = \delta_{gt}, \tag{4.113}$$

which shows the desired result, and we can conclude that $M = N$ and the character table is square.

4.6 SPLITTING OF THE GROUP ALGEBRA

The decomposition of the linear space V_G of the group algebra into irreducible subspaces can be accomplished by using the isomorphism to the convolution algebra. Recalling that the mapping of an element A of the group algebra on a function α over the group is done by the relation

$$A = \mathrm{M}_s \alpha(s) s^{-1}, \tag{4.114}$$

one introduces the elements

$$P^{\alpha}_{km} = \mathrm{M}_g f^{\alpha} \Gamma^{\alpha}_{km}(g) g^{-1} \tag{4.115}$$

of the group algebra. Since the master convolution relation (MCR) is

$$f^{\alpha} \Gamma^{\alpha}_{km} * f^{\beta} \Gamma^{\beta}_{nl} = \delta^{\alpha\beta} \delta_{kl} f^{\beta} \Gamma^{\beta}_{nm}, \tag{4.116}$$

it follows that the relation

$$P^{\alpha}_{km} \times P^{\beta}_{nl} = \delta^{\alpha\beta} \delta_{kl} P^{\beta}_{nm} \tag{4.117}$$

must hold among these elements of the group algebra.

Special cases are

$$P^{\alpha}_{kk} \times P^{\beta}_{ll} = \delta^{\alpha\beta}\delta_{kl}P^{\beta}_{ll} \tag{4.118}$$

and

$$P^{\alpha}_{kk} \times P^{\alpha}_{kk} = P^{\alpha}_{kk}\,. \tag{4.119}$$

These relations show that the (diagonal) elements $\{P^{\alpha}_{kk}\}$ are idempotent and mutually exclusive (orthogonal) and thus satisfy the requirements of projectors. In addition, they are a complete set of projectors and satisfy the resolution of the identity

$$\sum_{\alpha}\sum_{k=1}^{f^\alpha} P^{\alpha}_{kk} = e\,, \tag{4.120}$$

where e is the identity element of the group and of the group algebra (for the $\times$ multiplication). This completeness relation is readily proved by considering the definition of the projectors

$$\sum_{\alpha}\sum_{k=1}^{f^\alpha} P^{\alpha}_{kk} = \sum_{\alpha}\sum_{k=1}^{f^\alpha} \mathrm{M}_s f^\alpha \Gamma^{\alpha}_{kk}(s)s^{-1} = \mathrm{M}_s \sum_{\alpha} f^\alpha \chi^\alpha(s)s^{-1} = \mathrm{M}_s \chi^R(s)s^{-1}, \tag{4.121}$$

and again remembering the values of the character of the regular representation.

The result

$$e = \mathrm{M}_s \chi^R(s)s^{-1} \equiv \mathrm{M}_s E(s)s^{-1}, \tag{4.122}$$

clearly establishes the character of the regular representation $\chi^R(s) = E(s)$ as the identity element of the convolution algebra. It is straightforward to show that

$$E * \alpha = \alpha * E = \alpha \tag{4.123}$$

for any function α over the group.

There are $(f^\alpha)^2$ elements $\{P^{\alpha}_{km}\}$ for each irrep Γ^α, and since

$$\sum_{\alpha}(f^\alpha)^2 = |G|, \tag{4.124}$$

there are exactly the right number of elements $\{P^{\alpha}_{km}\}$ to form a basis of the linear space of the group algebra V_G. The question is whether these elements are linearly independent so they are suitable as a basis. That this is so can be shown by considering a particular element

$$A = \sum_{\beta}\sum_{l,n} P^{\beta}_{nl} a^{\beta}_{nl} \tag{4.125}$$

and forming the new element of the group algebra

$$P^{\alpha}_{km} \times A = \sum_{\beta}\sum_{n,l} \delta^{\alpha\beta}\delta_{kl}P^{\beta}_{nm}a^{\beta}_{nl} = f^{\alpha}\sum_{n} \mathrm{M}_s\Gamma^{\alpha}_{nm}(s)a^{\alpha}_{nk}s^{-1}. \tag{4.126}$$

In this new element the coefficient of the identity element is

$$\{P^{\alpha}_{km} \times A\}_e = f^{\alpha}a^{\alpha}_{mk}, \tag{4.127}$$

since $\Gamma^{\alpha}_{nm}(e) = \delta_{nm}$. Choosing the particular element A to be the zero element of the group algebra, that is, $A = O$ and

$$O = \sum_{\beta}\sum_{l,n} P^{\beta}_{nl}a^{\beta}_{nl}, \tag{4.128}$$

it follows from Eq. (4.127) that all $a^{\alpha}_{mk} = 0$. This is the same as saying that all the elements $\{P^{\alpha}_{km}\}$ of the group algebra are linearly independent and can serve as a basis of V_G.

We arrange these basis elements for each irrep in a square array

$$\begin{bmatrix} P^{\alpha}_{11} & P^{\alpha}_{12} & \cdots & P^{\alpha}_{1f^{\alpha}} \\ P^{\alpha}_{21} & P^{\alpha}_{22} & \cdots & P^{\alpha}_{2f^{\alpha}} \\ \vdots & \vdots & \vdots & \vdots \\ P^{\alpha}_{f^{\alpha}1} & P^{\alpha}_{f^{\alpha}2} & \cdots & P^{\alpha}_{f^{\alpha}f^{\alpha}} \end{bmatrix} \tag{4.129}$$

so that a particular element belongs not only to a given irrep but also to a row and a column of that irrep. A subspace W^{α}_k: $\{P^{\alpha}_{k1}, P^{\alpha}_{k2}, \ldots, P^{\alpha}_{kf^{\alpha}}\}$ spanned by the basis elements of a particular row of an irrep Γ^{α} is stable under G and generates the irrep Γ^{α} [cf. Eq. (4.27)]. This can be seen from

$$\begin{aligned} g \times P^{\alpha}_{km} &= f^{\alpha}\mathrm{M}_s\Gamma^{\alpha}_{km}(s)gs^{-1} = f^{\alpha}\mathrm{M}_u\Gamma^{\alpha}_{km}(ug)u^{-1} \\ &= \sum_{l=1}^{f^{\alpha}}\{f^{\alpha}\mathrm{M}_s\Gamma^{\alpha}_{kl}(u)u^{-1}\}\Gamma^{\alpha}_{lm}(g) = \sum_{l=1}^{f^{\alpha}} P^{\alpha}_{kl}\Gamma^{\alpha}_{lm}(g), \end{aligned} \tag{4.130}$$

where only basis elements of the subspace $W^{\alpha}_k \subset V_G$ appear (stable subspace), and where the entire representation matrices are generated for $m = 1, 2, \ldots, f^{\alpha}$.

As a consequence of this result the linear space of the group algebra can be decomposed into irreducible subspaces $W^{\alpha}_k \subset V_G$, and each of these f^{α} subspaces for $k = 1, 2, \ldots, f^{\alpha}$ span the irrep Γ^{α}. This is, of course, the origin of the fact that the frequency number of an irrep Γ^{α} in the regular representation is f^{α}. In this way a single row

$$\{\Gamma^{\alpha}_{k1}, \Gamma^{\alpha}_{k2}, \ldots, \Gamma^{\alpha}_{kf^{\alpha}}\} \tag{4.131}$$

of functions over the group determines the full irrep. The details of construction of the matrix elements from those of a single row follows from the convolution relation

$$\Gamma^{\alpha}_{nm}(g) = f^{\alpha}\Gamma^{\alpha}_{km} * \Gamma^{\alpha}_{nk} = \mathrm{M}_s f^{\alpha}\Gamma^{\alpha}_{km}(s)\Gamma^{\alpha}_{nk}(gs^{-1}) = \mathrm{M}_s f^{\alpha}\Gamma^{\alpha}_{km}(s)\Gamma^{\alpha *}_{kn}(sg^{-1}), \tag{4.132}$$

where in the last step the unitarity of the representation matrices has been used.

4.6.1 Character Projectors

The nature of the elements Q^{α} of the group algebra that corresponds to the characters is evident from the mapping

$$Q^{\alpha} = f^{\alpha}\mathrm{M}_s \chi^{\alpha}(s)s^{-1} = \sum_{k=1}^{f^{\alpha}} f^{\alpha}\mathrm{M}_s \Gamma^{\alpha}_{kk}(s)s^{-1} = \sum_{k=1}^{f^{\alpha}} P^{\alpha}_{kk}. \tag{4.133}$$

The isomorphism between the convolution algebra and the group algebra means that the convolution relation for characters in Eq. (4.95) yields the relation

$$Q^{\alpha} \times Q^{\beta} = \delta^{\alpha\beta} Q^{\beta}. \tag{4.134}$$

This shows that these elements of the group algebra are projectors, since they are idempotent

$$Q^{\alpha} \times Q^{\alpha} = Q^{\alpha} \tag{4.135}$$

and they satisfy the resolution of the identity

$$e = \sum_{\alpha} Q^{\alpha} = \sum_{\alpha} \sum_{k=1}^{f^{\alpha}} P^{\alpha}_{kk}. \tag{4.136}$$

It is noteworthy that

$$g \times Q^{\alpha} = f^{\alpha}\mathrm{M}_s \chi^{\alpha}(s) g s^{-1} = f^{\alpha}\mathrm{M}_t \chi^{\alpha}(g^{-1}tg)t^{-1}g = Q^{\alpha} \times g, \tag{4.137}$$

that is, all $\{Q^{\alpha}\}$ commute with the elements of the group and then also with all elements of the group algebra. In fact, the projector Q^{α} projects out a subspace

$$Q^{\alpha} V_G Q^{\alpha} = Q^{\alpha} V_G = V_G Q^{\alpha} = V^{\alpha}, \tag{4.138}$$

which is the union or direct sum of all the subspaces W^{α}_k for $k = 1, 2, \ldots, f^{\alpha}$. This means that the character projectors $\{Q^{\alpha}\}$ provide a less detailed subspace structure than does the $\{P^{\alpha}_{kk}\}$ projectors. Since the characters are invariant to basis set transformations, it would be satisfying if it was possible only use them in actual applications and not have to resort to the full representation matrices, which do not possess such invariance. This question is addressed in Chapter 9.

5

OUTER AND INNER DIRECT PRODUCTS

Let G' and G'' be two arbitrary finite groups. They could be of the same kind, for instance, point groups whose elements operate on the same kind of objects; or they could be different, say, a point group and a permutation group whose elements operate on different objects (such as coordinate axes and particle labels). If $g' \in G'$ and $g'' \in G''$, the "double entry" elements $g = (g', g'')$ form a group $G = G' \otimes G''$ called the *outer direct product* group of G' and G'' of order $|G| = |G'||G''|$. The combination law for this group is $g_1 g_2 = (g'_1, g''_1)(g'_2, g''_2) = (g'_1 g'_2, g''_1 g''_2) = (g'_3, g''_3) = g_3$; thus, in the first position the combination law of G' works and in the second, that of G''. This demonstrates the closure property of G (postulate I[1]). Obviously the associative law holds (postulate II). The identity element is $e = (e', e'')$ where e' and e'' are the identity elements of G' and G'', respectively, since

$$eg = ge = (e'g', e''g'') = (g', g'') = g \tag{5.1}$$

(postulate III). The inverse of $g = (g', g'')$ is $g^{-1} = (g'^{-1}, g''^{-1})$, since

$$g^{-1}g = gg^{-1} = (g', g'')(g'^{-1}, g''^{-1}) = (e', e'') = e \tag{5.2}$$

(postulate IV).

If Γ' is a representation of G' and Γ'' is a representation of G'', then $\Gamma = \Gamma' \otimes \Gamma''$ (the Krönecker product of the representation matrices) is a

[1] Chapter 2, preceding equation (2.3).

representation of the outer direct product group $G = G' \otimes G''$. This is true since

$$\begin{aligned}\boldsymbol{\Gamma}(g_1)\boldsymbol{\Gamma}(g_2) &= [\boldsymbol{\Gamma}'(g_1') \otimes \boldsymbol{\Gamma}''(g_1'')][\boldsymbol{\Gamma}'(g_2') \otimes \boldsymbol{\Gamma}''(g_2'')] \\ &= [\boldsymbol{\Gamma}'(g_1')\boldsymbol{\Gamma}'(g_2')] \otimes [\boldsymbol{\Gamma}''(g_1'')\boldsymbol{\Gamma}''(g_2'')] \\ &= \boldsymbol{\Gamma}'(g_1'g_2') \otimes \boldsymbol{\Gamma}''(g_1''g_2'') = \boldsymbol{\Gamma}(g_1g_2),\end{aligned} \tag{5.3}$$

where a well-known property of the matrix Kronecker product (direct product) has been used in going from the first line to the second line. The direct product matrix has the matrix elements $\Gamma_{ij;kl} = \Gamma'_{ik}\Gamma''_{jl}$ such that the row index is a double index made up of the row indices of the two factor matrices and similarly for the column index. This means that the character of the direct product representation is the product of the characters of its factors, which can be seen from

$$\chi(g) = \sum_i \sum_j \Gamma_{ij;ij}(g) = \sum_i \Gamma'_{ii}(g') \sum_j \Gamma''_{jj}(g'') = \chi'(g')\chi''(g''). \tag{5.4}$$

Theorem 5.1 If Γ' is an irrep of G' and Γ'' is an irrep of G'', then $\Gamma = \Gamma' \otimes \Gamma''$ is an irrep of the outer direct product group $G = G' \otimes G''$.

Proof From Theorem 4.3 it is known that the necessary and sufficient condition for a representation to be an irrep is that its character satisfies $(\chi * \chi)_e = 1$. For the direct product representation, it follows that

$$\begin{aligned}(\chi * \chi)_e &= \frac{1}{|G|} \sum_{g \in G} \chi(g)\chi(g^{-1}) \\ &= \frac{1}{|G'|} \sum_{g' \in G'} \chi'(g')\chi'(g'^{-1}) \frac{1}{|G''|} \sum_{g'' \in G''} \chi''(g')\chi''(g''^{-1}) \\ &= (\chi' * \chi')_e(\chi'' * \chi'')_e = 1 \cdot 1 = 1.\end{aligned} \tag{5.5}$$

QED

The outer direct product can also be formed by a group with itself $G \otimes G$. Such a direct product group contains elements (g, h) with both g and h belonging to G. The subset of elements of the "diagonal" form (h, h) form a subgroup

$$G[\times]G \subset G \otimes G \tag{5.6}$$

called the *inner direct product* of G with itself. The outer direct product can be generalized to an arbitrary number of factors. In particular, the outer direct product of a group with itself can be taken N times

$$\underbrace{G \otimes G \otimes G \otimes \cdots \otimes G}_{N} \tag{5.7}$$

and similarly for the corresponding inner direct product

$$\underbrace{G[\times]G[\times]G[\times]\cdots[\times]G}_{N}. \tag{5.8}$$

The inner direct product is always isomorphic to G itself.

Let Λ and Γ be representations of G. Then $\Lambda \otimes \Gamma$ is a representation of $G \otimes G$. Those matrices of $\Lambda \otimes \Gamma$ that correspond to elements of $G[\times]G \subset G \otimes G$ form the *inner product representation* $\Lambda[\times]\Gamma$ of $G[\times]G \simeq G$, where $\simeq$ means "isomorphic to."

6

SUBDUCED AND INDUCED REPRESENTATIONS

Let Γ be a representation of G and let $H \subset G$ be a subgroup, then for $h \in H$, those matrices $\{\Gamma(h)\}$ of Γ that are images of the elements of H form a representation of the subgroup H called the *representation of H subduced by* Γ, and denoted by $\Gamma \downarrow H$. In this way subducing produces a representation of a subgroup from a representation of the larger group. An example of a subduced representation already mentioned above is

$$\Lambda[\times]\Gamma = (\Lambda \otimes \Gamma) \downarrow (H[\times]H) \tag{6.1}$$

where Λ and Γ are two representations of H and

$$H \simeq H[\times]H \subset H \otimes H. \tag{6.2}$$

The procedure to create a representation of the larger group G from a representation Δ of one of its subgroups H is a bit more involved. Such a representation is called a *representation of G induced by* Δ and is denoted

$$\Delta \uparrow G. \tag{6.3}$$

An induced representation is constructed as follows:

- Decompose G in right cosets with respect to the subgroup H, such that G: $\{Hs_1, Hs_2, \ldots, Hs_k\}$, where the coset representatives s_i stay fixed once they are chosen, and where H itself is also included.

- It follows that for $g \in G$, $s_\sigma g \in Hs_\tau$ for some τ. When this is so, we write $s_\tau = s_{\sigma g}$.
- The matrices $\mathbf{A}(g)$ of the induced representation are blocked with $|G|/|H|$ blocks, each of the dimension of the matrices of the representation Δ.
- A block of $\mathbf{A}(g)$ can be expressed as

$$\mathbf{A}_{\sigma\tau}(g) = \mathbf{\Delta}(s_\sigma g s_\tau^{-1})\delta_{\tau(\sigma g)}, \tag{6.4}$$

  in other words, the block is $\mathbf{0}$ if $\tau \neq \sigma g$ and equal to $\mathbf{\Delta}(s_\sigma g s_{\sigma g}^{-1})$ for $\tau = \sigma g$, which is well defined since $s_\sigma g s_{\sigma g}^{-1} \in H$.

In order to prove that the matrices $\mathbf{A}(g)$ defined in this manner form a representation of G, we consider

$$\begin{aligned}
[\mathbf{A}(g_1)\mathbf{A}(g_2)]_{\sigma\mu} &= \sum_\tau \mathbf{A}_{\sigma\tau}(g_1)\mathbf{A}_{\tau\mu}(g_2) \\
&= \sum_\tau \mathbf{\Delta}(s_\sigma g_1 s_\tau^{-1})\delta_{\tau(\sigma g_1)}\mathbf{\Delta}(s_\tau g_2 s_\mu^{-1})\delta_{\mu(\tau g_2)} \\
&= \mathbf{\Delta}(s_\sigma g_1 s_{\sigma g_1}^{-1})\mathbf{\Delta}(s_{\sigma g_1} g_2 s_\mu^{-1})\delta_{\mu(\sigma g_1 g_2)} \\
&= \mathbf{\Delta}(s_\sigma g_1 g_2 s_\mu^{-1})\delta_{\mu(\sigma g_1 g_2)} = \mathbf{A}_{\sigma\mu}(g_1 g_2).
\end{aligned} \tag{6.5}$$

Another choice of coset representatives will lead to an equivalent representation. The character of the induced representation is

$$\chi(g) = \sum_\sigma \mathrm{Tr}\{\mathbf{A}_{\sigma\sigma}\} = \sum_\sigma \mathrm{Tr}\{\mathbf{\Delta}(s_\sigma g s_\sigma^{-1})\delta_{\sigma(\sigma g)} \tag{6.6}$$

or

$$\chi(g) = \sum_{\sigma'} \psi(s_{\sigma'} g s_{\sigma'}^{-1}) \quad \text{for} \quad s_{\sigma'} g s_{\sigma'}^{-1} \in \boldsymbol{H}\,, \tag{6.7}$$

where $\psi(u) = \mathrm{Tr}\{\mathbf{\Delta}(u)\}$. If H is a normal subgroup, which, as we know, contains complete classes of the larger group G, then $s_{\sigma'} g s_{\sigma'}^{-1}$, does not belong to H unless $g \in H$. This means that $\chi(g) = 0$ unless $g \in H$, and when this is the case, the sum over σ' runs over all coset representatives.

Example 6.1 $G = S_3$, and $H = A_3$, the so-called alternating group, which contains all even permutations A_3: $\{e, (123), (321)\}$, and which is normal.

We choose the representation Δ of the subgroup to be

$$\mathbf{\Delta}(e) = \begin{bmatrix} 1 & 0 \\ 0 & 1 \end{bmatrix}, \qquad \mathbf{\Delta}(123) = \begin{bmatrix} -\frac{1}{2} & -\frac{\sqrt{3}}{2} \\ \frac{\sqrt{3}}{2} & -\frac{1}{2} \end{bmatrix},$$
$$\mathbf{\Delta}(321) = \begin{bmatrix} -\frac{1}{2} & \frac{\sqrt{3}}{2} \\ -\frac{\sqrt{3}}{2} & -\frac{1}{2} \end{bmatrix}, \tag{6.8}$$

and we select the coset representatives $s_1 = e$, and $s_2 = (12)$. It then follows that $Hs_1 = H$ and $Hs_2 = \{(12), (13), (23)\}$. The construction of the induced representation for each element of G can then be done as follows:

- For $g = e$: $s_1 e \in Hs_1 = H$, $s_2 e = s_2 \in Hs_2$, and thus

$$\begin{aligned} \mathbf{A}_{12}(e) - \mathbf{A}_{21}(e) &= \mathbf{0}; \\ \mathbf{A}_{11}(e) = \mathbf{A}_{22}(e) &= \mathbf{\Delta}(e), \end{aligned} \tag{6.9}$$

that is

$$\mathbf{A}(e) = \begin{bmatrix} 1 & 0 & 0 & 0 \\ 0 & 1 & 0 & 0 \\ 0 & 0 & 1 & 0 \\ 0 & 0 & 0 & 1 \end{bmatrix}. \tag{6.10}$$

- For $g = (12)$: $s_1(12) \in Hs_2$, $s_2(12) = e \in Hs_1$, and thus

$$\begin{aligned} \mathbf{A}_{11}(12) &= \mathbf{A}_{22}(12) = \mathbf{0}; \\ \mathbf{A}_{12}(12) &= \mathbf{\Delta}(s_1(12)s_2^{-1}) = \mathbf{\Delta}(e) = \mathbf{A}_{21}(12), \end{aligned} \tag{6.11}$$

that is

$$\mathbf{A}(12) = \begin{bmatrix} 0 & 0 & 1 & 0 \\ 0 & 0 & 0 & 1 \\ 1 & 0 & 0 & 0 \\ 0 & 1 & 0 & 0 \end{bmatrix}. \tag{6.12}$$

For $g = (13)$:

- $s_1(13) \in Hs_2$, $s_2(13) = (321) \in Hs_1$ and

$$\begin{aligned} \mathbf{A}_{11}(13) &= \mathbf{A}_{22}(13) = \mathbf{0}; \\ \mathbf{A}_{12}(13) &= \mathbf{\Delta}(s_1(13)s_2^{-1}) = \mathbf{\Delta}(123), \\ \mathbf{A}_{21}(13) &= \mathbf{\Delta}(s_2(13)s_1^{-1}) = \mathbf{\Delta}(321), \end{aligned} \tag{6.13}$$

that is

$$\mathbf{A}(13) = \begin{bmatrix} 0 & 0 & -\frac{1}{2} & -\frac{\sqrt{3}}{2} \\ 0 & 0 & \frac{\sqrt{3}}{2} & -\frac{1}{2} \\ -\frac{1}{2} & \frac{\sqrt{3}}{2} & 0 & 0 \\ -\frac{\sqrt{3}}{2} & -\frac{1}{2} & 0 & 0 \end{bmatrix}, \tag{6.14}$$

and so on.

There is a theorem due to Frobenius called the *Frobenius reciprocity theorem*, which states that if G is a group with an irrep Γ, and $H \subset G$ is subgroup with an irrep Δ, then the frequency of Γ in $\Delta \uparrow G$ equals the frequency of Δ in $\Gamma \downarrow H$. The proof of this is beyond the scope of this text. However, as an example of a consequence of this theorem, we consider an arbitrary group G and its improper subgroup H: $\{e\}$. If we choose the irrep Δ of H to be the identity representation ($\Delta = J(H)$), then $\Delta \uparrow G$ is the regular representation of G. Each element $g \in G$ is in a right coset $Hg = g$ by itself, and the matrices have $|G|$ one-dimensional blocks, with

$$(\Delta \uparrow G)_{ij}(g) = \begin{cases} 1 & \text{if} \quad g = g_i^{-1} g_j \\ 0 & \text{otherwise.} \end{cases} \tag{6.15}$$

The reciprocity theorem then assures that the frequency of Γ in $\Delta \uparrow G$ equals the frequency of $J(H) = J(e)$ in $\Gamma \downarrow H = \Gamma \downarrow e = \Gamma(e)$, which is the dimension of Γ. So in this way the well-known result is obtained that the frequency of an irrep in the regular representation equals its dimension.

7

POINT GROUPS

The elements of point groups are geometric operations such as rotations about a symmetry axis or reflections in a symmetry plane, and the combinations are successive applications. As stated in the example of the point group C_{3v}, the operations can be carried out on the coordinate axis or on the symmetric object under consideration. For instance, the object can be a molecule in a symmetric conformation or a function of some coordinates. We use the notation of the elements most common in chemical applications.

7.1 NOTATIONS

The following different kinds of elements of point groups are used:

- E: the identity.
- C_n: rotation about an n-fold symmetry axis by $2\pi/n$ radians. In problems of molecular structure $n = 1, 2, 3, \ldots, 8$, and ∞. The latter leads to an infinite continuous group.
- σ: reflection in a symmetry plane.
- σ_h: reflection in a symmetry plane perpendicular to the principal symmetry axis (the one with the greatest n).
- σ_v: reflection in a symmetry plane containing the principal symmetry axis.
- C_2', (C_2''): rotations about twofold symmetry axis perpendicular to the principal symmetry axis.
- σ_d: reflection in a symmetry plane containing the principal symmetry axis and bisecting the angle between consecutive C_2' axes.

- S_n: improper rotation ($S_n = C_n\sigma_h = \sigma_h C_n$).
- i: inversion ($i = S_2$).

The point groups can be divided into those with one and those with more than one principal symmetry axis. The former can then be divided into *cyclic groups*, *dihedral groups*, and *improper cyclic* and *improper dihedral groups* (see Tables 7.1–7.3). The cyclic and improper cyclic groups are denoted by the symbol C with various subscripts and the dihedral and improper dihedral groups by the symbol D; similarly with various subscripts.

The irreps of point groups commonly have the following notations. Symbols A and B label one-dimensional irreps. The former label objects that transform into themselves under the operation C_n (rotation about the principal axis) and the latter label objects that transform into their negatives under the same operation. Symbols E, T, G, and H denote representations that are two-, three, four-, and five-dimensional, respectively. In addition, there are subscripts and

TABLE 7.1 Cyclic and Dihedral Groups

Order	Group	Elements
1	C_1	$\{E\}$
2	C_i	$\{E, i\}$
2	C_h	$\{E, \sigma_h\}$
n	C_n	$\{C_n^m, m = 1, 2, \ldots, n\}$
$2n$	D_n	$\{C_n^m, m = 1, 2, \ldots, n; nC_2'\}$

TABLE 7.2 Improper Cyclic and Dihedral Groups

Order	Group	Elements	n Even	n Odd
$2n$	C_{nh}		$C_n \otimes C_i$	$C_n \otimes C_h$
$2n$	S_{2n}	$\{S_{2n}^m, m = 1, 2, \ldots, 2n\}$		
$2n$	C_{nv}	$\{C_n^m, m = 1, 2, \ldots, n; k\sigma_v, l\sigma_d\}$	$k = l = n/2$	$k = n, l = 0$
$4n$	D_{nh}		$D_n \otimes C_i \simeq D_{2n}$	$D_n \otimes C_h$
$4n$	D_{nd}	$\{S_{2n}^m, m = 1, 2, \ldots, 2n; nC_2', n\sigma_d\}$		

TABLE 7.3 Groups with More than One Principal Symmetry Axis

Order	Group	Description
12	T	Symmetry group of a rigid regular tetrahedron
24	T_h	$T \otimes C_i$
24	O	Symmetry group of a rigid cube
48	O_h	$O \otimes C_i$
24	T_d	$\simeq O \simeq S_4$ (symmetric group of order 4!), T + improper rotations
60	I	Symmetry group of a rigid icosahedron
120	I_h	$I \otimes C_i$

superscripts that are generally assigned according to some rationale, which can be summarized as

Operation	σ_v	σ_h	i
Symmetric (+)	Subscript 1	Superscript ′	Subscript g
Antisymmetric (−)	Subscript 2	Superscript ″	Subscript u

In the next section the character tables for the most relevant point groups are displayed. It is well worth studying these tables in some detail. The identification of all the elements in each class of a particular point group is a worthwhile exercise. For instance, the group D_{6h} has a class called $2S_3$. Since the characters are real, this class must contain the improper rotation $S_3 = C_3\sigma_h$ and its inverse, which is $S_3^5 = C_3^2\sigma_h$. The tetrahedral group T_d has a class called $8C_3$. These eight elements can be visualized if the tetrahedron is inscribed in a cube such that a cube diagonal is a symmetry axis perpendicular to and going through the center of one of the regular triangles that form the sides of the tetrahedron. Rotations by $2\pi/6$ and $4\pi/6$ radians about such symmetry axes are a C_3, and a C_3^2 operation. There are four cube diagonals, thus eight operations.

Also note the functions of Cartesian coordinates listed in the right-hand columns of the character tables indicating how they transform. There are also symbols such as R_x, R_y, and R_z, which indicate rotations about the x, y, and z axes, respectively. This means that a small curved arrow indicating such a rotation will transform into itself or into its negative or a pair of such arrows will transform among themselves under the operations of the group. All this information is useful for the application of group theory to molecular structure and spectra.

7.1.1 Generating Relations for Point Groups

As has already been discussed, all groups can be generated by a few of its elements, called *generators*, used in suitable generating relations. Such generating relations can also be used to obtain explicit forms of irreps. Examples are given in Table 7.4.

Since the group C_n is Abelian, each element is a separate class, and the number of irreps must equal the order of the group. This means that there can

TABLE 7.4 Point Group Generation

Group	Generators	Generating Relations
C_n	$a = C_n$	$a^n = 1$
D_n	$a = C_n, b = C_2'$	$a^n = b^2 = 1, a^m b = ba^{-m}$
$O \simeq T_d$	$a = C_4, S_4, b = C_2, c = C_3$	$a^4 = b^2 = c^3 = 1, bca = 1$

be only one-dimensional irreps. The generating relation yields the solutions

$$a = \exp \frac{2\pi i m}{n}, \qquad m = 1, 2, \ldots, n, \tag{7.1}$$

in other words, the roots of unity generate n different (nonequivalent) irreps labeled by m.

The one-dimensional irreps of the group D_n can be obtained as follows. The generating relation $a^m b = ba^{-m}$ leads to $a^{2m} = 1$ (for one-dimensional irreps) for arbitrary integers m. This means that $a^2 = 1$, or $a = \pm 1$. Since, in addition, $a^n = 1$, only $a = 1$ can be true for n odd, while $b = \pm 1$ holds in general, giving two possible one-dimensional irreps for this case. For n even, both $a = 1$ and $a = -1$ are possible, and together with $b = \pm 1$, yield four possible one-dimensional irreps.

For $O \simeq T_d$, the values $a = \pm 1, b = \pm 1, c = 1$ follow from the generating relations, but only the combinations $a = b = c = 1$ and $a = b = -1, c = 1$ are solutions, since $bca = 1$, resulting in two one-dimensional irreps. Since for one-dimensional irreps the representation matrices are the characters, these results can be checked in the character tables.

It is perfectly possible in an analogous way to consider the two-dimensonal irreps and so on. However, as will become evident in Chapter 11, often the characters are sufficient for our purposes, and only rarely does one need the full representation matrices.

8

CHARACTER TABLES OF SOME POINT GROUPS

The Nonaxial Groups

C_1	E
A	1

C_s	E	σ_h		
A'	1	1	x, y, R_z	x^2, y^2, z^2, xy
A''	1	−1	z, R_x, R_y	xz, yz

C_i	E	i		
A_g	1	1	R_x, R_y, R_z	$x^2, y^2, z^2, xy, xz, yz$
A_u	1	−1	x, y, z	

The C_n Groups

C_2	E	C_2		
A	1	1	z, R_z	x^2, y^2, z^2, xy
B	1	−1	x, y, R_x, R_y	xz, yz

C_3	E	C_3	C_3^2		$a = \exp(2\pi i/3)$
A	1	1	1	z, R_z	$x^2 + y^2, z^2$
$E\{$	1	a	a^*		
	1	a^*	a	$(x, y); (R_x, R_y)$	$(x^2 - y^2, xy); (xz, yz)$

C_4	E	C_4	C_4^2	C_4^3		
A	1	1	1	1	z, R_z	$x^2 + y^2; z^2$
B	1	-1	1	-1		$x^2 - y^2; xy$
$E\{$	1	i	-1	$-i$		
	1	$-i$	-1	i	$(x, y); (R_x, R_y)$	(yz, xz)

C_5	E	C_5	C_5^2	C_5^3	C_5^4		$a = \exp(2\pi i/5)$
A	1	1	1	1	1	$z; R_z$	$x^2 + y^2; z^2$
$E_1\{$	1	a	a^2	a^3	a^*		
	1	a^*	a^{*2}	a^{*3}	a	$(x, y); (R_x, R_y)$	(yz, xz)
$E_2\{$	1	a^2	a^*	a	a^{*2}		
	1	a^{*2}	a	a^*	a^2		$(x^2 - y^2, xy)$

C_6	E	C_6	C_6^2	C_6^3	C_6^4	C_6^5		$a = \exp(2\pi i/6)$
A	1	1	1	1	1	1	$z; R_z$	$x^2 + y^2; z^2$
B	1	-1	1	-1	1	-1		
$E_1\{$	1	a	$-a^*$	-1	$-a$	a^*		
	1	a^*	$-a$	-1	$-a^*$	a	$(x, y); (R_x, R_y)$	(xz, yz)
$E_2\{$	1	$-a^*$	$-a$	1	$-a^*$	$-a$		
	1	$-a$	$-a^*$	1	$-a$	$-a^*$		$(x^2 - y^2, xy)$

The D_n Groups

D_2	E	$C_2(z)$	$C_2(y)$	$C_2(x)$		
A	1	1	1	1		x^2, y^2, z^2
B_1	1	1	-1	-1	z, R_z	xy
B_2	1	-1	1	-1	y, R_y	xz
B_3	1	-1	-1	1	x, R_x	yz

D_3	E	$2C_3$	$3C_2$		
A_1	1	1	1		$x^2 + y^2, z^2$
A_2	1	1	-1	(z, R_z)	
E	2	-1	0	$(x, y); (R_x, R_y)$	$(x^2 - y^2, xy);(xz, yz)$

D_4	E	$2C_4$	C_4^2	$2C_2'$	$2C_2''$		
A_1	1	1	1	1	1		$x^2 + y^2, z^2$
A_2	1	1	1	-1	-1	z, R_z	
B_1	1	-1	1	1	-1		$x^2 - y^2$
B_2	1	-1	1	-1	1		xy
E	2	0	-2	0	0	$(x, y); (R_x, R_y)$	(xy, yz)

D_5	E	$2C_5$	$2C_5^2$	$5C_2'$		
A_1	1	1	1	1		$x^2 + y^2, z^2$
A_2	1	1	1	-1	z, R_z	
E_1	2	$2\cos(2\pi/5)$	$2\cos(4\pi/5)$	0	$(x, y); (R_x, R_y)$	(xz, yz)
E_2	2	$2\cos(4\pi/5)$	$2\cos(2\pi/5)$	0		$(x^2 - y^2, xy)$

D_6	E	$2C_6$	$2C_6^2$	C_6^3	$3C_2'$	$3C_2''$		
A_1	1	1	1	1	1	1		$x^2 + y^2, z^2$
A_2	1	1	1	1	-1	-1	z, R_z	
B_1	1	-1	1	-1	1	-1		
B_2	1	-1	1	-1	-1	1		
E_1	2	1	-1	-2	0	0	$(x, y); (R_x, R_y)$	(xz, yz)
E_2	2	-1	-1	2	0	0		$(x^2 - y^2, xy)$

The C_{nv} Groups

C_{2v}	E	C_2	$\sigma_v(xz)$	$\sigma_v(yz)$		
A_1	1	1	1	1	z	x^2, y^2, z^2
A_2	1	1	−1	−1	R_z	xy
B_1	1	−1	1	−1	x, R_y	xz
B_2	1	−1	−1	1	y, R_x	yz

C_{3v}	E	$2C_3$	$3\sigma_v$		
A_1	1	1	1	z	$x^2 + y^2, z^2$
A_2	1	1	−1	R_z	
E	2	−1	0	$(x, y); (R_x, R_y)$	$(x^2 - y^2, xy); (xz, yz)$

C_{4v}	E	$2C_4$	C_4^2	$2\sigma_v$	$2\sigma_d$		
A_1	1	1	1	1	1	z	$x^2 + y^2, z^2$
A_2	1	1	1	−1	−1	R_z	
B_1	1	−1	1	1	−1		$x^2 - y^2$
B_2	1	−1	1	−1	1		xy
E	2	0	−2	0	0	$(x, y); (R_x, R_y)$	(xz, yz)

C_{5v}	E	$2C_5$	$2C_5^2$	$5\sigma_v$		
A_1	1	1	1	1	z	$x^2 + y^2, z^2$
A_2	1	1	1	−1	R_z	
E_1	2	$2\cos(2\pi/5)$	$2\cos(4\pi/5)$	0	$(x, y); (R_x, R_y)$	(xz, yz)
E_2	2	$2\cos(4\pi/5)$	$2\cos(2\pi/5)$	0		$(x^2 - y^2, xy)$

C_{6v}	E	$2C_6$	$2C_6^2$	C_6^3	$3\sigma_v$	$3\sigma_d$		
A_1	1	1	1	1	1	1	z	$x^2 + y^2, z^2$
A_2	1	1	1	1	−1	−1	R_z	
B_1	1	−1	1	−1	1	−1		
B_2	1	−1	1	−1	−1	1		
E_1	2	1	−1	−2	0	0	$(x, y); (R_x, R_y)$	(xz, yz)
E_2	2	−1	−1	2	0	0		$(x^2 - y^2, xy)$

The C_{nh} Groups

C_{2h}	E	C_2	i	σ_h		
A_g	1	1	1	1	R_z	x^2, y^2, z^2, xy
B_g	1	−1	1	−1	R_x, R_y	xz, yz
A_u	1	1	−1	−1	z	
B_u	1	−1	−1	1	x, y	

C_{3h}	E	C_3	C_3^2	σ_h	S_3	S_3^5	$a = \exp(2\pi i/3)$	
A'	1	1	1	1	1	1	R_z	$x^2 + y^2, z^2$
E' {	1	a	a^*	1	a	a^*		
	1	a^*	a	1	a^*	a	(x, y)	$(x^2 - y^2, xy)$
A''	1	1	1	−1	−1	−1	z	
E'' {	1	a	a^*	−1	$-a$	$-a^*$		
	1	a^*	a	−1	$-a^*$	$-a$	(R_x, R_y)	(xz, yz)

C_{4h}	E	C_4	C_4^2	C_4^3	i	S_4^3	σ_h	S_4		
A_g	1	1	1	1	1	1	1	1	R_z	$x^2 + y^2; z^2$
B_g	1	−1	1	−1	1	−1	1	−1		$x^2 - y^2; xy$
E_g {	1	i	−1	$-i$	1	i	−1	$-i$		
	1	$-i$	−1	i	1	$-i$	−1	i	(R_x, R_y)	(xz, yz)
A_u	1	1	1	1	−1	−1	−1	−1	z	
B_u	1	−1	1	−1	−1	1	−1	1		
E_u {	1	i	−1	$-i$	−1	$-i$	1	i		
	1	$-i$	−1	i	−1	i	1	$-i$	(x, y)	

C_{5h}	E	C_5	C_5^2	C_5^3	C_5^4	σ_h	S_5	S_5^7	S_5^3	S_5^9	$a = \exp(2\pi i/5)$	
A'	1	1	1	1	1	1	1	1	1	1	R_z	$q; z^2$
E_1'	1	a	a^2	a^{*2}	a^*	1	a	a^2	a^{*2}	a^*		
	1	a^*	a^{*2}	a^2	a	1	a^*	a^{*2}	a^2	a	(x, y)	
E_2'	1	a^2	a^*	a	a^{*2}	1	a^2	a^*	a	a^{*2}		
	1	a^{*2}	a	a^*	a^2	1	a^{*2}	a	a^*	a^2		(p, xy)
A''	1	1	1	1	1	-1	-1	-1	-1	-1	z	
E_1''	1	a	a^2	a^{*2}	a^*	-1	$-a$	$-a^2$	$-a^{*2}$	$-a^*$		
	1	a^*	a^{*2}	a^2	a	-1	$-a^*$	$-a^{*2}$	$-a^2$	$-a$	$(R_x, R_y)(xz, yz)$	
E_2''	1	a^2	a^*	a	a^{*2}	-1	$-a^2$	$-a^*$	$-a$	$-a^{*2}$		
	1	a^{*2}	a	a^*	a^2	-1	$-a^{*2}$	$-a$	$-a^*$	$-a^2$		

$q = x^2 + y^2; p = x^2 - y^2$

C_{6h}	E	C_6	C_6^2	C_6^3	C_6^4	C_6^5	i	S_3^5	S_6^5	σ_h	S_6	S_3	$a = \exp(2\pi i/6)$
A_g	1	1	1	1	1	1	1	1	1	1	1	1	$R_z, x^2 + y^2, z^2$
B_g	1	-1	1	-1	1	-1	1	-1	1	-1	1	-1	
E_{1g}	1	a	$-a^*$	-1	$-a$	a^*	1	a	$-a^*$	-1	$-a$	a^*	
	1	a^*	$-a$	-1	$-a^*$	a	1	a^*	$-a$	-1	$-a^*$	a	$(R_x, R_y);(xz, yz)$
E_{2g}	1	$-a^*$	$-a$	1	$-a^*$	$-a$	1	$-a^*$	$-a$	1	$-a^*$	$-a$	
	1	$-a$	$-a^*$	1	$-a$	$-a^*$	1	$-a$	$-a^*$	1	$-a$	$-a^*$	$(x^2 - y^2, xy)$
A_u	1	1	1	1	1	1	-1	-1	-1	-1	-1	-1	z
B_u	1	-1	1	-1	1	-1	-1	1	-1	1	-1	1	
E_{1u}	1	a	$-a^*$	-1	$-a$	a^*	-1	$-a$	a^*	1	a	$-a^*$	
	1	a^*	$-a$	-1	$-a^*$	a	-1	$-a^*$	a	1	a^*	$-a$	(x, y)
E_{2u}	1	$-a^*$	$-a$	1	$-a^*$	$-a$	-1	a^*	a	-1	a^*	a	
	1	$-a$	$-a^*$	1	$-a$	$-a^*$	-1	a	a^*	-1	a	a^*	

The D_{nh} Groups

D_{2h}	E	$C_2(z)$	$C_2(y)$	$C_2(x)$	i	$\sigma(xy)$	$\sigma(xz)$	$\sigma(yz)$		
A_g	1	1	1	1	1	1	1	1		$x^2; y^2; z^2$
B_{1g}	1	1	−1	−1	1	1	−1	−1	R_z	xy
B_{2g}	1	−1	1	−1	1	−1	1	−1	R_y	xz
B_{3g}	1	−1	−1	1	1	−1	−1	1	R_x	yz
A_u	1	1	1	1	−1	−1	−1	−1		
B_{1u}	1	1	−1	−1	−1	−1	1	1	z	
B_{2u}	1	−1	1	−1	−1	1	−1	1	y	
B_{3u}	1	−1	−1	1	−1	1	1	−1	x	

D_{3h}	E	$2C_3$	$3C_2$	σ_h	$2S_3$	$3\sigma_v$		
A_1'	1	1	1	1	1	1		$x^2 + y^2; z^2$
A_2'	1	1	−1	1	1	−1	R_z	
E'	2	−1	0	2	−1	0	(x, y)	$(x^2 - y^2, xy)$
A_1''	1	1	1	−1	−1	−1		
A_2''	1	1	−1	−1	−1	1	z	
E''	2	−1	0	−2	1	0	(R_x, R_y)	(xy, xz)

D_{4h}	E	$2C_4$	C_4^2	$2C_2'$	$2C_2''$	i	$2S_4$	σ_h	$2\sigma_v$	$2\sigma_d$		
A_{1g}	1	1	1	1	1	1	1	1	1	1		$q; z^2$
A_{2g}	1	1	1	−1	−1	1	1	1	−1	−1	R_z	
B_{1g}	1	−1	1	1	−1	1	−1	1	1	−1		p
B_{2g}	1	−1	1	−1	1	1	−1	1	−1	1		xy
E_g	2	0	−2	0	0	2	0	−2	0	0	(R_x, R_y)	(xz, yz)
A_{1u}	1	1	1	1	1	−1	−1	−1	−1	−1		
A_{2u}	1	1	1	−1	−1	−1	−1	−1	1	1	z	
B_{1u}	1	−1	1	1	−1	−1	1	−1	−1	1		
B_{2u}	1	−1	1	−1	1	−1	1	−1	1	−1		
E_u	2	0	−2	0	0	−2	0	2	0	0	(x, y)	

$p = x^2 - y^2; q = x^2 + y^2$

D_{5h}	E	$2C_5$	$2C_5^2$	$5C_2'$	σ_h	$2S_5$	$2S_5^3$	$5\sigma_v$		
A_1'	1	1	1	1	1	1	1	1		q
A_2'	1	1	1	−1	1	1	1	−1	R_z	
E_1'	2	2 cos 72°	2 cos 144°	0	2	2 cos 72°	2 cos 144°	0	t	
E_2'	2	2 cos 144°	2 cos 72°	0	2	2 cos 144°	2 cos 144°	0		s
A_1''	1	1	1	1	−1	−1	−1	−1		
A_2''	1	1	1	−1	−1	−1	−1	1	z	
E_1''	2	2 cos 72°	2 cos 144°	0	−2	−2 cos 72°	−2 cos 144°	0	r	u
E_2''	2	2 cos 144°	2 cos 72°	0	−2	−2 cos 144°	−2 cos 72°	0		

$q = x^2 + y^2; z^2$; $r = (R_x, R_y)$; $s = (x^2 - y^2, xy)$; $t = (x, y)$; $u = (xz, yz)$

D_{6h}	E	$2C_6$	$2C_6^2$	C_6^3	$3C_2'$	$3C_2''$	i	$2S_3$	$2S_6$	σ_h	$3\sigma_d$	$3\sigma_v$		
A_{1g}	1	1	1	1	1	1	1	1	1	1	1	1		q
A_{2g}	1	1	1	1	−1	−1	1	1	1	1	−1	−1	R_z	
B_{1g}	1	−1	1	−1	1	−1	1	−1	1	−1	1	−1		
B_{2g}	1	−1	1	−1	−1	1	1	−1	1	−1	−1	1		
E_{1g}	2	1	−1	−2	0	0	2	1	−1	−2	0	0	r	s
E_{2g}	2	−1	−1	2	0	0	2	−1	−1	2	0	0		t
A_{1u}	1	1	1	1	1	1	−1	−1	−1	−1	−1	−1		
A_{2u}	1	1	1	1	−1	−1	−1	−1	−1	−1	1	1	z	
B_{1u}	1	−1	1	−1	1	−1	−1	1	−1	1	−1	1		
B_{2u}	1	−1	1	−1	−1	1	−1	1	−1	1	1	−1		
E_{1u}	2	1	−1	−2	0	0	−2	−1	1	2	0	0	u	
E_{2u}	2	−1	−1	2	0	0	−2	1	1	−2	0	0		

$q = x^2 + y^2; z^2$; $r = (R_x, R_y)$; $s = (xz, yz)$; $t = (x^2 - y^2, xy)$; $u = (x, y)$

The D_{nd} Groups

D_{2d}	E	$2S_4$	C_2	$2C_2'$	$2\sigma_d$		
A_1	1	1	1	1	1		$x^2 + y^2; z^2$
A_2	1	1	1	−1	−1	R_z	
B_1	1	−1	1	1	−1		$x^2 - y^2$
B_2	1	−1	1	−1	1	z	xy
E	2	0	−2	0	0	$(x, y); (R_x, R_y)$	(xz, yz)

D_{3d}	E	$2C_3$	$3C_2'$	i	$2S_6$	$3\sigma_d$		
A_{1g}	1	1	1	1	1	1		$x^2 + y^2; z^2$
A_{2g}	1	1	−1	1	1	−1	R_z	
E_g	2	−1	0	2	−1	0	(R_x, R_y)	$(x^2 - y^2, xy); (xz, yz)$
A_{1u}	1	1	1	−1	−1	−1		
A_{2u}	1	1	−1	−1	−1	1	z	
E_u	2	−1	0	−2	1	0	(x, y)	

D_{4d}	E	$2S_8$	$2C_4$	$2S_8^3$	C_4^2	$4C_2'$	$4\sigma_d$		
A_1	1	1	1	1	1	1	1		$x^2 + y^2; z^2$
A_2	1	1	1	1	1	−1	−1	R_z	
B_1	1	−1	1	−1	1	1	−1		
B_2	1	−1	1	−1	1	−1	1	z	
E_1	2	$\sqrt{2}$	0	$-\sqrt{2}$	−2	0	0	(x, y)	
E_2	2	0	−2	0	2	0	0		$(x^2 - y^2, xy)$
E_3	2	$-\sqrt{2}$	0	$\sqrt{2}$	−2	0	0	(R_x, R_y)	(xz, yz)

D_{5d}	E	$2C_5$	$2C_5^2$	$5C_2'$	i	$2S_{10}^3$	$2S_{10}$	$5\sigma_d$		
A_{1g}	1	1	1	1	1	1	1	1		q
A_{2g}	1	1	1	−1	1	1	1	−1	R_z	
E_{1g}	2	2 cos 72°	2 cos 144°	0	2	2 cos 72°	2 cos 144°	0	r	s
E_{2g}	2	2 cos 144°	2 cos 72°	0	2	2 cos 144°	2 cos 72°	0		t
A_{1u}	1	1	1	1	−1	−1	−1	−1		
A_{2u}	1	1	1	−1	−1	−1	−1	1	z	
E_{1u}	2	2 cos 72°	2 cos 144°	0	−2	−2 cos 72°	−2 cos 144°	0	u	
D_{5d}	2	2 cos 144°	2 cos 72°	0	−2	−2 cos 144°	−2 cos 72°	0		

$q = x^2 + y^2; z^2$; $r = (R_x, R_y)$; $s = (xz, yz)$; $t = (x^2 - y^2, xy)$; $u = (x, y)$

D_{6d}	E	$2S_{12}$	$2C_6$	$2S_4$	$2C_6^2$	$2S_{12}^5$	C_6^3	$6C_2'$	$6\sigma_d$		
A_1	1	1	1	1	1	1	1	1	1		q
A_2	1	1	1	1	1	1	1	−1	−1	R_z	
B_1	1	−1	1	−1	1	−1	1	1	−1		
B_2	1	−1	1	−1	1	−1	1	−1	1	z	
E_1	2	$\sqrt{3}$	1	0	−1	$-\sqrt{3}$	−2	0	0	(x, y)	
E_2	2	1	−1	−2	−1	1	2	0	0		r
E_3	2	0	−2	0	2	0	−2	0	0		
E_4	2	−1	−1	2	−1	−1	2	0	0		
E_5	2	$-\sqrt{3}$	1	0	−1	$\sqrt{3}$	−2	0	0	s	t

$q = x^2 + y^2; z^2$; $r = (x^2 - y^2, xy)$; $s = (R_x, R_y)$; $t = (xz, yz)$

The S_n Groups

S_4	E	S_4	C_2	S_4^3		
A	1	1	1	1	R_z	$x^2 + y^2; z^2$
B	1	−1	1	−1	z	$x^2 - y^2; xy$
$E\{$	1	i	−1	$-i$		
	1	$-i$	−1	i	$(x, y); (R_x, R_y)$	(xz, yz)

S_6	E	C_3	C_3^2	i	S_6^5	S_6	$a = \exp(2\pi i/3)$	
A_g	1	1	1	1	1	1	R_z	$x^2 + y^2; z^2$
E_g	1	a	a^*	1	a	a^*		
	1	a^*	a	1	a^*	a	(R_x, R_y)	$(x^2 - y^2, xy); (xz, yz)$
A_u	1	1	1	−1	−1	−1	z	
E_u	1	a	a^*	−1	$-a$	$-a^*$		
	1	a^*	a	−1	$-a^*$	$-a$	(x, y)	

S_8	E	S_8	C_4	S_8^3	C_2	S_8^5	C_4^3	S_8^7	$a = \exp(2\pi i/8)$	
A	1	1	1	1	1	1	1	1	R_z	$x^2 + y^2; z^2$
B	1	−1	1	−1	1	−1	1	−1	z	
E_1	1	a	i	$-a^*$	−1	$-a$	$-i$	a^*		
	1	a^*	$-i$	$-a$	−1	$-a^*$	i	a	$(x, y); (R_x R_y)$	
E_2	1	i	−1	$-i$	1	i	−1	$-i$		
	1	$-i$	−1	i	1	$-i$	−1	i		$(x^2 - y^2, xy)$
E_3	1	$-a^*$	$-i$	a	−1	a^*	i	$-a$		
	1	$-a$	i	a^*	−1	a	$-i$	$-a^*$		(xz, yz)

The Cubic Groups

T	E	$4C_3$	$4C_3^2$	$3C_2$	$a = \exp(2\pi i/3)$	
A	1	1	1	1		$x^2 + y^2 + z^2$
E	1	a	a^*	1		
	1	a^*	a	1		$(2z^2 - x^2 - y^2, x^2 - y^2)$
T	3	0	0	−1	$(R_x, R_y, R_z); (x, y, z)$	(xy, xz, yz)

T_h	E	$4C_3$	$4C_3^2$	$3C_2$	i	$4S_6$	$4S_6^5$	$3\sigma_h$	$a = \exp(2\pi i/3)$	
A_g	1	1	1	1	1	1	1	1		$x^2 + y^2 + z^2$
A_u	1	1	1	1	−1	−1	−1	−1		
E_g	1	a	a^*	1	1	a	a^*	1		
	1	a^*	a	1	1	a^*	a	1		(r, s)
E_u	1	a	a^*	1	−1	$-a$	$-a^*$	−1		
	1	a^*	a	1	−1	$-a^*$	$-a$	−1		
T_g	3	0	0	−1	3	0	0	−1	(R_x, R_y, R_z)	(xy, xz, yz)
T_u	3	0	0	−1	−3	0	0	1	(x, y, z)	

$r = 2z^2 - x^2 - y^2$, $s = x^2 - y^2$

T_d	E	$8C_3$	$3C_2$	$6S_4$	$6\sigma_d$		
A_1	1	1	1	1	1		$x^2 + y^2 + z^2$
A_2	1	1	1	−1	−1		
E	2	−1	2	0	0		$(2z^2 - x^2 - y^2, x^2 - y^2)$
T_1	3	0	−1	1	−1	(R_x, R_y, R_z)	
T_2	3	0	−1	−1	1	(x, y, z)	(xy, xz, yz)

O	E	$6C_4$	$3C_4^2$	$8C_3$	$6C_2$		
A_1	1	1	1	1	1		$x^2 + y^2 + z^2$
A_2	1	−1	1	1	−1		
E	2	0	2	−1	0		$(2z^2 - x^2 - y^2, x^2 - y^2)$
T_1	3	1	−1	0	−1	$(R_x, R_y, R_z); (x, y, z)$	
T_2	3	−1	−1	0	1		(xy, xz, yz)

O_h	E	$8C_3$	$6C_2$	$6C_4$	$3C_4^2$	i	$6S_4$	$8S_6$	$3\sigma_h$	$6\sigma_d$		
A_{1g}	1	1	1	1	1	1	1	1	1	1		q
A_{2g}	1	1	−1	−1	1	1	−1	1	1	−1		
E_g	2	−1	0	0	2	2	0	−1	2	0		r
T_{1g}	3	0	−1	1	−1	3	1	0	−1	−1	s	
T_{2g}	3	0	1	−1	−1	3	−1	0	−1	1		t
A_{1u}	1	1	1	1	1	−1	−1	−1	−1	−1		
A_{2u}	1	1	−1	−1	1	−1	1	−1	−1	1		
E_u	2	−1	0	0	2	−2	0	1	−2	0		
T_{1u}	3	0	−1	1	−1	−3	−1	0	1	1	u	
T_{2u}	3	0	1	−1	−1	−3	1	0	1	−1		

$q = x^2 + y^2 + z^2$; $r = (2z^2 - x^2 - y^2, x^2 - y^2)$; $s = (R_x, R_y, R_z)$;
$t = (xy, xz, yz)$; $u = (x, y, z)$

The Icosahedral Groups[a]

I_h	E	$12C_5$	$12C_5^2$	$20C_3$	$15C_2$	i	$12S_{10}$	$12S_{10}^3$	$20S_6$	15σ		
A_g	1	1	1	1	1	1	1	1	1	1		q
T_{1g}	3	a	b	0	−1	3	b	a	0	−1	r	
T_{2g}	3	b	a	0	−1	3	a	b	0	−1		
G_g	4	−1	−1	1	0	4	−1	−1	1	0		
H_g	5	0	0	−1	1	5	0	0	−1	1		s
A_u	1	1	1	1	1	−1	−1	−1	−1	−1		
T_{1u}	3	a	b	0	−1	−3	$-b$	$-a$	0	1	t	
T_{2u}	3	b	a	0	−1	−3	$-a$	$-b$	0	1		
G_u	4	−1	−1	1	0	−4	1	1	−1	0		
H_u	5	0	0	−1	1	−5	0	0	1	−1		

$a = \frac{1}{2}(1 + \sqrt{5})$; $b = \frac{1}{2}(1 - \sqrt{5})$; $q = x^2 + y^2 + z^2$; $r = (R_x, R_y, R_z)$;
$s = (2z^2 - x^2 - y^2, x^2 - y^2, xy, xz, yz)$; $t = (x, y, z)$

[a] For the pure rotation group I, the actual character table is indicated in the upper left corner. In that case the subscripts g of the irrep labels should be dropped and the (x,y,z) assigned to the T_1 irrep.

The groups of linear molecules

$C_{\infty v}$	E	$2C(\alpha)$		$\infty\sigma_v$
Σ^+	1	1	$\cdots$	1
Σ^-	1	1	$\cdots$	-1
Π	2	$2\cos\alpha$	$\cdots$	0
Δ	2	$2\cos 2\alpha$	$\cdots$	0
Φ	2	$2\cos 3\alpha$	$\cdots$	0
$\vdots$	$\vdots$	$\vdots$	$\vdots$	$\vdots$

$D_{\infty h}$	E	$2C(\alpha)$	$\cdots$	$\infty\sigma_v$	i	$2S(\alpha)$	$\cdots$	∞C_2
Σ_g^+	1	1	$\cdots$	1	1	1	$\cdots$	1
Σ_g^-	1	1	$\cdots$	-1	1	1	$\cdots$	-1
Π_g	2	$2\cos\alpha$	$\cdots$	0	2	$-2\cos\alpha$	$\cdots$	0
Δ_g	2	$2\cos 2\alpha$	$\cdots$	0	2	$2\cos 2\alpha$	$\cdots$	0
$\vdots$	$\vdots$	$\vdots$	$\vdots$	$\vdots$	$\vdots$	$\vdots$	$\vdots$	$\vdots$
Σ_u^+	1	1	$\cdots$	1	-1	-1	$\cdots$	-1
Σ_u^-	1	1	$\cdots$	-1	-1	-1	$\cdots$	1
Π_u	2	$2\cos\alpha$	$\cdots$	0	-2	$2\cos\alpha$	$\cdots$	0
Δ_u	2	$2\cos 2\alpha$	$\cdots$	0	-2	$-2\cos 2\alpha$	$\cdots$	0
$\vdots$	$\vdots$	$\vdots$	$\vdots$	$\vdots$	$\vdots$	$\vdots$	$\vdots$	$\vdots$

9

CONNECTION TO QUANTUM CHEMISTRY

9.1 THE HYDROGEN ATOM

The dynamics of the simple atomic system of an electron with mass m_e and a proton with mass m_p can be treated as two separate problems: (1) a freely moving particle with the total mass of the system (the translational motion of the atom as a whole) and (2) an electron with a modified, so-called reduced, mass $\mu = m_e m_p/(m_e + m_p)$ moving in an attractive Coulomb potential due to the fixed proton. All hydrogenlike ions can be treated simultaneously by considering the potential energy

$$V(r) = -\frac{Ze^2}{r} \tag{9.1}$$

where Z is the atomic number, $-e$ is the electron charge, and $r = (x^2 + y^2 + z^2)^{1/2}$ is the distance between the proton and the electron.

In theoretical treatments of dynamical systems such as an atom,[1] one often looks for simplifications coming from exact (or approximate) separation of

[1] The hydrogen atom can be given a completely group theoretical treatment in terms of continuous groups. Angular momentum is directly related to the continuous group $SO(3)$ and diatomic molecules to the continuous point groups $C_{\infty v}$ and $D_{\infty h}$. These matters are discussed in later chapters. Here we want first to establish some simple quantum-chemical ideas without invoking the full machinery of groups.

coordinates. The Schrödinger equation[2]

$$H\psi(\vec{r}) = \left[-\frac{\hbar^2}{2\mu} \vec{\nabla}^2 - \frac{Ze^2}{r} \right] \psi(\vec{r}) = E\psi(\vec{r}) \tag{9.2}$$

for the electron in the hydrogen atom cannot be separated in terms of Cartesian coordinates because of the form of the potential energy that strongly couples x, y, and z. Therefore one introduces spherical polar coordinates (r, θ, φ) defined as

$$x = r \sin\theta \cos; \qquad y = r \sin\theta \sin\varphi; \qquad z = r\cos\theta. \tag{9.3}$$

When the Cartesian coordinates of the electron vary from $-\infty$ to $+\infty$ to cover the entire space, the spherical polar coordinates satisfy

$$0 \le r \le \infty; \qquad 0 \le \theta \le \pi; \qquad 0 \le \varphi < 2\pi. \tag{9.4}$$

The volume element $dx\,dy\,dz$ becomes $r^2 \sin\theta\, dr\, d\theta\, d\varphi$, and the kinetic-energy operator becomes

$$-\frac{\hbar^2}{2\mu}\vec{\nabla}^2 = -\frac{\hbar^2}{2\mu}\left[\frac{1}{r^2}\frac{\partial}{\partial r}\left(r^2 \frac{\partial}{\partial r}\right) + \frac{1}{r^2 \sin\theta}\frac{\partial}{\partial\theta}\left(\sin\theta \frac{\partial}{\partial\theta}\right) + \frac{1}{r^2\sin^2\theta}\frac{\partial^2}{\partial\varphi^2}\right]. \tag{9.5}$$

A derivation of this expression is presented in the next section.

The Schrödinger equation in spherical coordinates becomes

$$-\frac{1}{2}\left[\frac{1}{r^2}\frac{\partial}{\partial r}\left(r^2 \frac{\partial}{\partial r}\right) + \frac{1}{r^2 \sin\theta}\frac{\partial}{\partial\theta}\left(\sin\theta \frac{\partial}{\partial\theta}\right) + \frac{1}{r^2\sin^2\theta}\frac{\partial^2}{\partial\varphi^2} - \frac{Z}{r}\right]\psi(\vec{r}) = E\psi(\vec{r}), \tag{9.6}$$

where we have used atomic units $\hbar = 1$, $\mu = 1$, $e = 1$, yielding energy eigenvalues in hartrees. Here we are interested only in describing stable atoms, that is, situations in which the electron is bound to the nucleus. This means that the Schrödinger equation is solved with the boundary condition that $\psi(\vec{r}) \to 0$ when $r \to \infty$, leading to the so-called bound-state eigenfunctions. The probability intepretation that

$$|\psi(\vec{r})|^2 r^2 \sin\theta\, dr\, d\theta\, d\varphi \tag{9.7}$$

is the probability of finding the electron in volume element $r^2 \sin\theta\, dr\, d\theta\, d\varphi$, which means that the eigenfunctions have to be normalized so that

---

[2] $\hbar = h/2\pi$ is Planck's constant divided by 2π, and $\vec{\nabla}^2 = (\partial^2/\partial x^2 + \partial^2/\partial y^2 + \partial^2/\partial z^2)$. The quantum-mechanical operator for the linear momentum of the electron is $\hat{\vec{p}} = -i\hbar\vec{\nabla} = -i\hbar(\partial/\partial x \quad \partial/\partial y \quad \partial/\partial z)$, and the kinetic-energy operator then becomes $\hat{\vec{p}}^2/2\mu = -\hbar^2\vec{\nabla}^2/2\mu$.

$\int_0^\infty r^2 dr \int_0^\pi \sin\theta \, d\theta \int_0^{2\pi} d\varphi |\psi(\vec{r})|^2 = 1$, so that the probability of finding the electron somewhere in the entire space is unity.

The eigenfunctions of the Hamiltonian of the electron in the hydrogen atom can then be written as products of a radial function and an angular function such that

$$\psi_{nlm}(\vec{r}) = R_{nl}(r) Y_{lm}(\theta, \varphi), \tag{9.8}$$

with

$$R_{nl}(r) = \left\{ \frac{4Z^3(n-l-1)!}{n^4((n+l)!)^3} \right\}^{1/2} \left(\frac{2Zr}{n} \right)^l \exp\left(-\frac{Zr}{n} \right) L_{n+l}^{2l+1}\left(\frac{2Zr}{n} \right) \tag{9.9}$$

and the so-called spherical harmonics

$$Y_{lm}(\theta, \varphi) = \left\{ \frac{(2l+1)(l-|m|)!}{4\pi(l+|m|)!} \right\} P_l^{|m|}(\cos\theta) \exp(im\varphi). \tag{9.10}$$

The symbol $L_{n+l}^{2l+1}(2Zr/n)$ stands for a Laguerre polynomial (of degree $n - l - 1$, and $P_l^{|m|}(\cos\theta)$ is an associate Legendre function (a factor $\sin^{|m|/2}\theta$ times a polynomial in $\cos\theta$ of degree $l - |m|$). The subscripts of the eigenfunctions are the celebrated quantum numbers: the principal quantum number $n = 1, 2, 3, \ldots$, the angular momentum quantum number $l < n$, $l = 0, 1, 2, \ldots$, and the magnetic quantum number $m = -l, -l + 1, \ldots, +l$. The corresponding energy eigenvalues depend only on the principal quantum number

$$E_n = -\frac{Z^2}{2n^2} \text{ hartrees} \tag{9.11}$$

(1 hartree = 27.209 eV or 4.360×10^{-18}J).

For historical reasons, the atomic eigenfunctions with $l = 0$ are called *s orbitals* (an *orbital* is a wavefunction of a single electron), those with $l = 1$ *p* orbitals, and $l = 2$ *d* orbitals, then the notations continue with *f* orbitals for $l = 3$, then *g* and *h* orbitals.

The lowest energy of the hydrogen atom (atomic number $Z = 1$)

$$E_1 = -\frac{1}{2} \text{ hartrees} = -13.6 \text{ eV} \tag{9.12}$$

corresponds to the *ground-state* eigenfunction

$$1s = \psi_{100}(\vec{r}) = \exp(-r), \tag{9.13}$$

where we have left out the normalization constant, which can readily be calculted should we need it for, say, the probability interpretation. This would be done as follows:

$$\begin{aligned}(\psi_{100}|\psi_{100}) &= \int_{\text{all space}} \psi^*_{100}(\vec{r})\psi_{100}(\vec{r})d\tau \\ &= \int_0^\infty e^{-2r}r^2\,dr \int_0^\pi \sin\theta\,d\theta \int_0^{2\pi} d\varphi = \frac{2!}{2^3}4\pi = \pi,\end{aligned} \tag{9.14}$$

leading to the normalized eigenfunction

$$\frac{1}{\sqrt{\pi}}\exp(-r). \tag{9.15}$$

It is noteworthy that the zero of energy of the electron corresponds to the situation that it is "removed to $r = \infty$ and put at rest." A better way to state this is that the minimum energy required to remove the electron from the atom, that is, ionize the atom, is the ionization potential (IP) equal to $\text{IP} = -E_1 = 13.6$ eV.

The eigenfunctions corresponding to excited bound states of the electron in the hydrogen atom are all those with eigenvalues $-(1/2n^2)$ hartrees for $n = 2, 3, \ldots, < \infty$. Beyond $n = \infty$ the energy of the electron is positive, and this situation corresponds to a free electron. We do not discuss those eigenfunctions that obviously correspond to a continuum of possible energies given by the kinetic energy of the electron.

Corresponding to the eigenvalue $E_2 = -\frac{1}{8}$ hartrees are the four (unnormalized) eigenfunctions

$$\begin{aligned}2s &= \psi_{200}(\vec{r}) = (2 - r)e^{-r/2}, \\ 2p_x &= \frac{\psi_{211} + \psi_{21-1}}{2} = re^{-r/2}\sin\theta\cos\varphi, \\ 2p_y &= \frac{\psi_{211} - \psi_{21-1}}{2i} = re^{-r/2}\sin\theta\sin\varphi, \\ 2p_z &= \psi_{210} = re^{-r/2}\cos\theta,\end{aligned} \tag{9.16}$$

where the combinations of degenerate orbitals have been taken to get real wavefunctions. The fact that orbitals with different angular momentum quantum numbers such as $2s$ and $2p$ are degenerate is true only for hydrogenlike systems, not for many-electron atoms.

Corresponding to $E_3 = -\frac{1}{18}$ Hartrees are the $3s$, the $3p$ and the $3d$ orbitals. The $3s$ orbital is spherically symmetric (depending only on the distance r) as do all s orbitals, and the $3p$ orbitals have the same angular shape as do the

2p. The orbitals get more radial nodes and extend to greater r values as the principal quantum number and thus the energy of the electron grows. The 3d orbitals can also be combined to (unnormalized) real forms such as

$$\begin{aligned}
3d_{x^2-y^2} &= \frac{\psi_{322} + \psi_{32-2}}{2} = r^2 e^{-r/3} \sin^2\theta(\cos^2\varphi - \sin^2\varphi), \\
3d_{xy} &= \frac{\psi_{322} - \psi_{32-2}}{2i} = r^2 e^{-r/3} \sin^2\theta \cos\varphi \sin\varphi, \\
3d_{xz} &= \frac{\psi_{321} + \psi_{32-1}}{2} = r^2 e^{-r/3} \sin\theta \cos\theta \cos\varphi, \qquad (9.17) \\
3d_{yz} &= \frac{\psi_{321} - \psi_{32-1}}{2i} = r^2 e^{-r/3} \sin\theta \cos\theta \sin\varphi, \\
3d_{z^2} &= \psi_{320} = r^2 e^{-r/3}(3\cos^2\theta - 1).
\end{aligned}$$

Because the s orbitals are spherically symmetric, they are often depicted as a sphere for some fixed value of r. The p orbitals are depicted as dumbbells with a positive end in the positive x direction for a p_x orbital and so on. This comes from the fact that a polar graph of, say, a p_z orbital for a fixed value of r in a plane containing the z axis resembles the plot shown in Figure 9.1.

Since $p_z \propto \cos\theta$, the function value is plotted as a distance from the origin as indicated in Figure 9.1. Since the graph looks the same in all planes containing the z axis, the three-dimensional dumbbell is obtained by rotating this figure about the z axis. The p_x and p_y orbitals have the same dumbbell shape in the other coordinate directions. It is a useful exercise to draw the shapes of the d orbitals.

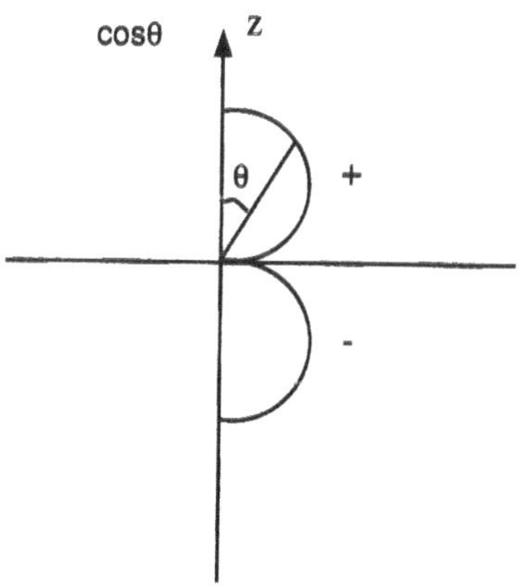

Figure 9.1 The polar representation of a p_z orbital.

9.1.1 Average Value

Any property that can be measured corresponds to a quantum-mechanical operator. When the system (atom or molecule) is in a particular state given by the wavefunction ψ, the value of the property for the system in that state can be calculated as the expectation value or *average value* of the operator over that wavefunction. The average value of an operator $\hat{A}$ is

$$\langle \hat{A} \rangle = \frac{(\psi | \hat{A} \psi)}{(\psi | \psi)}, \tag{9.18}$$

where the normalization integral in the denominator assures that the result is correct even for an unnormalized wavefunction.

Example 9.1 The average value of the distance of the electron from the proton in the hydrogen atom ground state is ($\hat{A} = r$)

$$\langle r \rangle = \frac{(1s|r1s)}{(1s|1s)} = \frac{\int_0^\infty r^3 e^{-2r} dr \int_0^\pi \sin\theta \, d\theta \int_0^{2\pi} d\varphi}{\int_0^\infty r^3 e^{-2r} dr \int_0^\pi \sin\theta \, d\theta \int_0^{2\pi} d\varphi} \tag{9.19}$$

$$= \frac{3!/2^4}{2!/2^3} = \frac{3}{2} \text{ au}$$

[1 au (atomic unit) = 1 bohr = 0.529177 Å = 5.29177×10^{-11} m].

9.2 KINETIC ENERGY IN SPHERICAL POLAR COORDINATES

Orbital angular momentum of a particle is the vector product of the position vector with the particle momentum

$$\vec{l} = \vec{r} \times \vec{p} = (yp_z - zp_y \quad zp_x - xp_z \quad xp_y - yp_x) \\ = (l_x \quad l_y \quad l_z). \tag{9.20}$$

Since the momentum operator is

$$\hat{\vec{p}} = (\hat{p}_x \quad \hat{p}_y \quad \hat{p}_z) = -i\hbar \left(\frac{\partial}{\partial x} \quad \frac{\partial}{\partial y} \quad \frac{\partial}{\partial z} \right), \tag{9.21}$$

the corresponding quantum-mechanical angular momentum operators are

$$\begin{aligned} \hat{l}_x &= -i\hbar \left(y \frac{\partial}{\partial z} - z \frac{\partial}{\partial y} \right) \\ \hat{l}_y &= -i\hbar \left(z \frac{\partial}{\partial x} - x \frac{\partial}{\partial z} \right) \\ \hat{l}_z &= -i\hbar \left(x \frac{\partial}{\partial y} - y \frac{\partial}{\partial x} \right). \end{aligned} \tag{9.22}$$

The commutator

$$[\hat{x},\ \hat{p}_x]_- \equiv \hat{x}\hat{p}_x - \hat{p}_x\hat{x} = -i\hbar\left(x\frac{\partial}{\partial x} - \frac{\partial}{\partial x}x\right) \tag{9.23}$$

is an operator that when acting on a function of the coordinate x satisfies

$$\begin{aligned}[\hat{x},\ \hat{p}_x]_- f(x) &= -i\hbar x\frac{\partial f(x)}{\partial x} + i\hbar f(x) + i\hbar x\frac{\partial f(x)}{\partial x} \\ &= i\hbar f(x),\end{aligned} \tag{9.24}$$

from which the operator identity

$$[\hat{x},\ \hat{p}_x]_- = i\hbar \tag{9.25}$$

follows. Similarly, $[\hat{y},\ \hat{p}_y]_- = i\hbar$ and $[\hat{z},\ \hat{p}_z]_- = i\hbar$, while all the "mixed" commutators $[\hat{x},\ \hat{p}_y]_-$, $[\hat{z},\ \hat{p}_y]_-$, and so on vanish. From this result one may show that

$$\begin{aligned}[\hat{l}_x,\ \hat{l}_y]_- &= [\hat{y}\hat{p}_z - \hat{z}\hat{p}_y,\ \hat{z}\hat{p}_x - \hat{x}\hat{p}_z]_- \\ &= \hat{y}\hat{p}_x[\hat{p}_z, \hat{z}]_- + \hat{x}\hat{p}_y[\hat{z},\ \hat{p}_z]_- = (\hat{x}\hat{p}_y - \hat{y}\hat{p}_x)[\hat{z},\ \hat{p}_z]_- \\ &= i\hbar(\hat{x}\hat{p}_y - \hat{y}\hat{p}_x) = i\hbar\hat{l}_z\end{aligned} \tag{9.26}$$

and one can similarly show that

$$\begin{aligned}[\hat{l}_y,\ \hat{l}_z]_- &= i\hbar\hat{l}_x\,, \\ [\hat{l}_z,\ \hat{l}_x]_- &= i\hbar\hat{l}_y\,.\end{aligned} \tag{9.27}$$

The adjoint of an operator $\hat{A}$ is an operator $\hat{A}^\dagger$ that satisfies the relation

$$(u|\hat{A}v) = (\hat{A}^\dagger u|v) \tag{9.28}$$

for two functions u and v that satisfy proper boundary conditions. This property as well as the bracketed notation for the integrals can be illustrated by the following example.

Example 9.2 Choose $\hat{A} = \partial/\partial x$ and let the functions $u(x)$ and $v(x)$ be defined on an interval (a, b) and with the boundary conditions that $u(a) = u(b) = v(a) = v(b) = 0$; then

$$\left(u \;\middle|\; \frac{\partial}{\partial x} v\right) = \int_a^b u^*(x) \frac{\partial v(x)}{\partial x}\, dx = u^*(x)v(x)\Big|_a^b - \int_a^b \left(\frac{\partial v(x)}{\partial x}\right)^* v(x)\, dx$$
$$= \left(-\frac{\partial}{\partial x} u \middle| v\right), \tag{9.29}$$

which shows that

$$\left(\frac{\partial}{\partial x}\right)^{\dagger} = -\frac{\partial}{\partial x} \tag{9.30}$$

or that $\partial/\partial x$ and $(-\partial/\partial x)$ are each other's adjoint operators. When an operator equals its adjoint, one calls it a *self-adjoint operator*. For example, $-i\hbar(\partial/\partial x)$ is self-adjoint, as are all quantum-mechanical operators that correspond to an observable. Self-adjoint operators have real eigenvalues. In particular, the components of angular momentum are self-adjoint. Since this is so, it follows that $\hat{l}_+^{\dagger} = \hat{l}_-$ and consequently $\hat{l}_-^{\dagger} = \hat{l}_+$, for $\hat{l}_+ = \hat{l}_x + i\hat{l}_y$ and $\hat{l}_- = \hat{l}_x - i\hat{l}_y$.

The square of the orbital angular momentum is

$$(\hat{\vec{l}})^2 = \hat{l}_x^2 + \hat{l}_y^2 + \hat{l}_z^2, \tag{9.31}$$

or, more specifically

$$(\hat{\vec{l}})^2 = \left\{\left[-i\hbar\left(y\frac{\partial}{\partial z} - z\frac{\partial}{\partial y}\right)\right]^2 + \left[-i\hbar\left(z\frac{\partial}{\partial x} - x\frac{\partial}{\partial z}\right)\right]^2 + \left[-i\hbar\left(x\frac{\partial}{\partial y} - y\frac{\partial}{\partial x}\right)\right]^2\right\}$$
$$= -\hbar^2\left[y^2\frac{\partial^2}{\partial z^2} + z^2\frac{\partial^2}{\partial y^2} - 2yz\frac{\partial^2}{\partial y\,\partial z} - y\frac{\partial}{\partial y} - z\frac{\partial}{\partial z}\right. \tag{9.32}$$
$$+ z^2\frac{\partial^2}{\partial x^2} + x^2\frac{\partial^2}{\partial z^2} - 2xz\frac{\partial^2}{\partial x\,\partial z} - x\frac{\partial}{\partial x} - z\frac{\partial}{\partial z}$$
$$\left. + x^2\frac{\partial^2}{\partial y^2} + y^2\frac{\partial^2}{\partial x^2} - 2yx\frac{\partial^2}{\partial y\,\partial x} - y\frac{\partial}{\partial y} - x\frac{\partial}{\partial x}\right]$$

$$
\begin{aligned}
&= -\hbar^2\Big[(x^2 + y^2 + z^2)\left(\frac{\partial^2}{\partial x^2} + \frac{\partial^2}{\partial y^2} + \frac{\partial^2}{\partial z^2}\right) \\
&\quad - x^2 \frac{\partial^2}{\partial x^2} - y^2 \frac{\partial^2}{\partial y^2} - z^2 \frac{\partial^2}{\partial z^2} - 2xy \frac{\partial^2}{\partial x\, \partial y} \\
&\quad - 2xz \frac{\partial^2}{\partial x\, \partial z} - 2yz \frac{\partial^2}{\partial y\, \partial z} \\
&\quad - 2\left(x\frac{\partial}{\partial x} + y\frac{\partial}{\partial y} + z\frac{\partial}{\partial z}\right)\Big] \\
&= -\hbar^2 r^2\Big[\frac{\partial^2}{\partial x^2} + \frac{\partial^2}{\partial y^2} + \frac{\partial^2}{\partial z^2} \\
&\quad - \frac{x^2}{r^2}\frac{\partial^2}{\partial x^2} - \frac{y^2}{r^2}\frac{\partial^2}{\partial y^2} - \frac{z^2}{r^2}\frac{\partial^2}{\partial z^2} - \frac{2xy}{r^2}\frac{\partial^2}{\partial x\, \partial y} \\
&\quad - \frac{2xz}{r^2}\frac{\partial^2}{\partial x\, \partial z} - \frac{2yz}{r^2}\frac{\partial^2}{\partial y\, \partial z} \\
&\quad - \frac{2}{r}\left(\frac{x}{r}\frac{\partial}{\partial x} + \frac{y}{r}\frac{\partial}{\partial y} + \frac{z}{r}\frac{\partial}{\partial z}\right)\Big],
\end{aligned}
\tag{9.33}
$$

where we have used the notation $r = (x^2 + y^2 + z^2)^{1/2}$. Since the Cartesian coordinates (x, y, z) in terms of spherical polar coordinates (r, θ, φ) are $(0 \le r < \infty,\ 0 \le \theta \le \pi,\ 0 \le \varphi < 2\pi)$

$$
\begin{aligned}
x &= r \sin\theta \cos\varphi \\
y &= r \sin\theta \sin\varphi \\
z &= r \cos\theta \\
r &= (x^2 + y^2 + z^2)^{1/2},
\end{aligned}
\tag{9.34}
$$

the chain rule of differentiation yields

$$
\frac{\partial}{\partial r} = \frac{\partial x}{\partial r}\frac{\partial}{\partial x} + \frac{\partial y}{\partial r}\frac{\partial}{\partial y} + \frac{\partial z}{\partial r}\frac{\partial}{\partial z} = \frac{x}{r}\frac{\partial}{\partial x} + \frac{y}{r}\frac{\partial}{\partial y} + \frac{z}{r}\frac{\partial}{\partial z}. \tag{9.35}
$$

Also

$$
\begin{aligned}
\frac{\partial^2}{\partial r^2} &= \left(\frac{x}{r}\frac{\partial}{\partial x} + \frac{y}{r}\frac{\partial}{\partial y} + \frac{z}{r}\frac{\partial}{\partial z}\right)^2 = \frac{x^2}{r^2}\frac{\partial^2}{\partial x^2} + \frac{y^2}{r^2}\frac{\partial^2}{\partial y^2} + \frac{z^2}{r^2}\frac{\partial^2}{\partial z^2} \\
&\quad + 2\left(\frac{xy}{r^2}\frac{\partial^2}{\partial x\, \partial y} + \frac{xz}{r^2}\frac{\partial^2}{\partial x\, \partial z} + \frac{yz}{r^2}\frac{\partial^2}{\partial y\, \partial z}\right),
\end{aligned}
\tag{9.36}
$$

which should be derived as an exercise [note that $(\partial/\partial x)(1/r) = -(x/r^3)$].
These results and Eq. (9.32) mean that

$$(\hat{\vec{l}})^2 = -\hbar^2 r^2\left[\frac{\partial}{\partial x^2} + \frac{\partial^2}{\partial y^2} + \frac{\partial^2}{\partial z^2} - \frac{\partial^2}{\partial r^2} - \frac{2}{r}\frac{\partial}{\partial r}\right]. \tag{9.37}$$

Then the Hamiltonian for the electron in the hydrogen atom can be expressed as {note that $(\partial^2/\partial r^2) + (2/r)(\partial/\partial r) = (1/r^2)(\partial/\partial r)[r^2(\partial/\partial r)]$}

$$\begin{aligned} H &= -\frac{\hbar^2}{2\mu}\vec{\nabla}^2 - \frac{Ze^2}{r} \\ &= -\frac{\hbar^2}{2\mu}\left[\frac{\partial^2}{\partial x^2} + \frac{\partial^2}{\partial y^2} + \frac{\partial^2}{\partial z^2}\right] - \frac{Ze^2}{r} \\ &= -\frac{\hbar^2}{2\mu}\frac{1}{r^2}\frac{\partial}{\partial r}\left(r^2\frac{\partial}{\partial r}\right) + \frac{(\hat{\vec{l}})^2}{2\mu r^2} - \frac{Ze^2}{r}. \end{aligned} \tag{9.38}$$

The unit vectors in the polar coordinate directions are (see Fig. 9.2)

$$\begin{aligned} \vec{e}_r &= \vec{e}_x \sin\theta\cos\varphi + \vec{e}_y \sin\theta\sin\varphi + \vec{e}_z\cos\theta \\ \vec{e}_\theta &= \vec{e}_x \cos\theta\cos\varphi + \vec{e}_y \cos\theta\sin\varphi - \vec{e}_z\sin\theta \\ \vec{e}_\varphi &= -\vec{e}_x\sin\varphi + \vec{e}_y\cos\varphi \end{aligned} \tag{9.39}$$

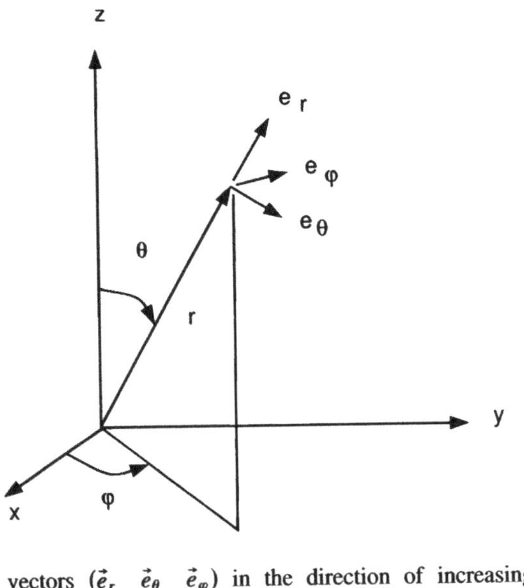

Figure 9.2 Unit vectors ($\vec{e}_r$ $\vec{e}_\theta$ $\vec{e}_\varphi$) in the direction of increasing spherical polar coordinates and their relationship to the Cartesian unit vectors ($\vec{e}_x$ $\vec{e}_y$ $\vec{e}_z$).

in terms of the unit vectors in the Cartesian coordinate directions. From Eq. (9.34) it follows that

$$\begin{aligned} dx &= dr \sin\theta \cos\varphi + d\theta \, r \cos\theta \cos\varphi - d\varphi \, r \sin\theta \sin\varphi \\ dy &= dr \sin\theta \sin\varphi + d\theta \, r \cos\theta \sin\varphi + d\varphi \, r \sin\theta \cos\varphi \\ dz &= dr \cos\theta - d\theta \, r \sin\theta \, , \end{aligned} \tag{9.40}$$

and thus a length element square is

$$\begin{aligned} (ds)^2 &= (dx)^2 + (dy)^2 + (dz)^2 \\ &= (dr)^2 + (r\, d\theta)^2 + (r \sin\theta \, d\varphi)^2, \end{aligned} \tag{9.41}$$

from which follows that the del (nabla) or gradient operator is

$$\begin{aligned} \vec{\nabla} &= \left(\frac{\partial}{\partial x} \; \frac{\partial}{\partial y} \; \frac{\partial}{\partial z} \right) = \vec{e}_x \frac{\partial}{\partial x} + \vec{e}_y \frac{\partial}{\partial y} + \vec{e}_z \frac{\partial}{\partial z} \\ &= \vec{e}_r \frac{\partial}{\partial r} + \vec{e}_\theta \frac{1}{r} \frac{\partial}{\partial \theta} + \vec{e}_\varphi \frac{1}{r \sin\theta} \frac{\partial}{\partial \varphi} \, . \end{aligned} \tag{9.42}$$

The orbital angular momentum operator is in spherical polar coordinates

$$\begin{aligned} \hat{l} &= \vec{r} \times (-i\hbar\vec{\nabla}) = r\vec{e}_r \times (-i\hbar\vec{\nabla}) \\ &= -i\hbar\left(\vec{e}_\varphi \frac{\partial}{\partial \theta} - \vec{e}_\theta \frac{1}{\sin\theta} \frac{\partial}{\partial \varphi} \right) \\ &= -i\hbar\left(-\sin\varphi \frac{\partial}{\partial \theta} - \cot\,\theta \cos\varphi \frac{\partial}{\partial \varphi} \right) \vec{e}_x \\ &\quad - i\hbar\left(\cos\varphi \frac{\partial}{\partial \theta} - \cot\,\theta \sin\varphi \frac{\partial}{\partial \varphi} \right) \vec{e}_y \\ &\quad - i\hbar\left(\frac{\partial}{\partial \varphi} \right) \vec{e}_z = \hat{l}_x \vec{e}_x + \hat{l}_y \vec{e}_y + \hat{l}_z \vec{e}_z \, , \end{aligned} \tag{9.43}$$

where Eq. (9.39) has been used.

Define

$$\begin{aligned} \hat{l}_+ &= \hat{l}_x + i\hat{l}_y = \hbar e^{i\varphi}\left(\frac{\partial}{\partial \theta} + i \cot\,\theta \frac{\partial}{\partial \varphi} \right), \\ \hat{l}_- &= \hat{l}_x - i\hat{l}_y = \hbar e^{-i\varphi}\left(-\frac{\partial}{\partial \theta} + i \cot\,\theta \frac{\partial}{\partial \varphi} \right), \end{aligned} \tag{9.44}$$

and form

$$\hat{l}_+\hat{l}_- = \hat{l}_x^2 + \hat{l}_y^2 - i[\hat{l}_x, \hat{l}_y]_- = \hat{l}_x^2 + \hat{l}_y^2 + \hbar\hat{l}_z, \tag{9.45}$$

from which it follows that

$$\begin{aligned}
(\hat{\vec{l}})^2 &= \hat{l}_+\hat{l}_- + \hat{l}_z^2 - \hbar\hat{l}_z \\
&= \hbar^2\left[e^{i\varphi}\left(\frac{\partial}{\partial\theta} + i\cot\theta\frac{\partial}{\partial\varphi}\right)e^{-i\varphi}\left(-\frac{\partial}{\partial\theta} + i\cot\theta\frac{\partial}{\partial\varphi}\right)\right. \\
&\qquad \left. - \frac{\partial^2}{\partial\varphi^2} + i\frac{\partial}{\partial\varphi}\right] \\
&= \hbar^2\left[-\frac{\partial^2}{\partial\theta^2} - i\frac{1}{\sin^2\theta}\frac{\partial}{\partial\varphi} + i\cot\theta\frac{\partial^2}{\partial\theta\,\partial\varphi}\right. \\
&\qquad + \cot\theta\left(-\frac{\partial}{\partial\theta} + i\cot\theta\frac{\partial}{\partial\varphi}\right) \\
&\qquad \left. - i\cot\theta\frac{\partial^2}{\partial\theta\,\partial\varphi} - \cot^2\theta\frac{\partial^2}{\partial\varphi^2} - \frac{\partial^2}{\partial\varphi^2} + i\frac{\partial}{\partial\varphi}\right]
\end{aligned} \tag{9.46}$$

$$\begin{aligned}
&= \hbar^2\left[-\frac{\partial^2}{\partial\theta^2} - \cot\theta\frac{\partial}{\partial\theta} + i\left(\cot^2\theta - \frac{1}{\sin^2\theta} + 1\right)\frac{\partial}{\partial\varphi}\right. \\
&\qquad \left. - (\cot^2\theta + 1)\frac{\partial^2}{\partial\varphi^2}\right] \\
&= -\hbar^2\left[\frac{1}{\sin\theta}\frac{\partial}{\partial\theta}\left(\sin\theta\frac{\partial}{\partial\theta}\right) + \frac{1}{\sin^2\theta}\frac{\partial^2}{\partial\varphi^2}\right].
\end{aligned} \tag{9.47}$$

From Eq. (9.38) it follows that

$$\begin{aligned}
-\frac{\hbar^2}{2\mu}\vec{\nabla}^2 = -\frac{\hbar^2}{2\mu}&\left[\frac{1}{r^2}\frac{\partial}{\partial r}\left(r^2\frac{\partial}{\partial r}\right) + \frac{1}{r^2\sin\theta}\frac{\partial}{\partial\theta}\left(\sin\theta\frac{\partial}{\partial\theta}\right)\right. \\
&\left. + \frac{1}{r^2\sin\theta}\frac{\partial^2}{\partial\varphi^2}\right].
\end{aligned} \tag{9.48}$$

The volume element in Cartesian coordinates can be transformed to the corresponding volume element in spherical polar coordinates with the appropriate

Jacobian as

$$dx\,dy\,dz \rightarrow \begin{vmatrix} \dfrac{\partial x}{\partial r} & \dfrac{\partial x}{\partial \theta} & \dfrac{\partial x}{\partial \varphi} \\ \dfrac{\partial y}{\partial r} & \dfrac{\partial y}{\partial \theta} & \dfrac{\partial y}{\partial \varphi} \\ \dfrac{\partial z}{\partial r} & \dfrac{\partial z}{\partial \theta} & \dfrac{\partial z}{\partial \varphi} \end{vmatrix} dr\,d\theta\,d\varphi = r^2 \sin\theta\,dr\,d\theta\,d\varphi \qquad (9.49)$$

This is also a useful exercise.

It can be shown that quantum-mechanical operators that commute have common eigenfunctions. The square of angular momentum $\vec{l}^2$ and anyone of the Cartesian components of angular momentum $\vec{l} = (l_x \quad l_y \quad l_z)$ commute. One normally chooses $\vec{l}^2$ and l_z as a pair of commuting operators. As we have already indicated, the eigenfunctions of these two operators are known as *spherical harmonics*:

$$Y_{lm}(\theta, \varphi) \propto P_l^{|m|}(\cos\theta)e^{im\varphi}. \qquad (9.50)$$

The eigenfunction relations are

$$\begin{aligned} \vec{l}^2 Y_{lm} &= l(l+1)Y_{lm}, \\ l_z Y_{lm} &= mY_{lm}, \end{aligned} \qquad (9.51)$$

where we have put $\hbar = 1$, which means that we are measuring angular momenta[3] in units of $\hbar$. We have also removed the ^ symbol from the quantum-mechanical operators. The associated Legendre functions $P_l^{|m|}(\omega)$ can be generated by the relation

$$P_l^{|m|}(\omega) \propto (1-\omega^2)^{|m|/2} \frac{\partial^{l+|m|}}{\partial \omega^{l+|m|}} (\omega^2 - 1)^l, \qquad (9.52)$$

for all the possible values of the magnetic quantum number $m = -l, -l+1, \ldots, l$ for any given angular momentum quantum number l.

Since the commutation relations between the Cartesian components of angular momentum can be used to show that $l_z l_+ = l_+(l_z + 1)$, one can readily find that $l_z(l_+ Y_{lm}) = l_+(l_z + 1)Y_{lm} = (m+1)(l_+ Y_{lm})$, which from the eigenvalue relations must mean that $(l_+ Y_{lm}) \propto Y_{lm+1}$. Similarly, one can show that $(l_- Y_{lm}) \propto Y_{lm-1}$.

[3] The term $\hbar$ has the dimension of energy × time = mass × (length)^2 × (time)$^{-1}$ = angular momentum.

The properties of the ladder operators can now be used to show that

$$
\begin{aligned}
0 < (l_+ Y_{lm} | l_+ Y_{lm}) &= (Y_{lm} | l_- l_+ Y_{lm}) \\
&= (Y_{lm} | (\vec{l}^2 - l_z^2 - l_z) Y_{lm}) = [l(l+1) - m^2 - m] \\
&= (l - m)(l + m + 1) .
\end{aligned}
\tag{9.53}
$$

Similarly

$$
(l_- Y_{lm} | l_- Y_{lm}) = (l + m)(l - m + 1) . \tag{9.54}
$$

This means that since the spherical harmonics are assumed to be normalized to unity, we can state that

$$
l_\pm Y_{lm} = [(l \mp m)(l \pm m + 1)]^{1/2} Y_{lm\pm 1} . \tag{9.55}
$$

9.3 ELECTRON SPIN

It is well known from spectroscopy that the electron has an intrinsic magnetic moment $-e\vec{s}/m$, which aligns with or is opposed to an applied magnetic field. Such a magnetic moment is associated with an angular momentum $\vec{s} = (s_x \quad s_y \quad s_z)$, for which one can construct the corresponding quantum-mechanical operators $\vec{s}^2$, and s_z as was done for orbital angular momentum. Note that we are still using the units with $\hbar = 1$. The common eigenfunctions of these two commuting operators for a single electron are very simple. Analogously to the case of orbital angular momentum, there is a spin angular momentum quantum number s and a spin magnetic quantum number m_s such that the eigenvalues of $\vec{s}^2$ and s_z are $s(s+1)$, and $m_s = -s, -s + 1, \ldots, +s$, respectively. For an electron the only possible values of the quantum numbers are

$$
s = \tfrac{1}{2} \quad \text{and} \quad m_s = -\tfrac{1}{2}, \tfrac{1}{2} , \tag{9.56}
$$

which means that there are two degenerate eigenfunctions of $\vec{s}^2$. They are commonly defined as

$$
\alpha(m_s) = \begin{cases} 1 & \text{for} \quad m_s = \dfrac{1}{2} \\ 0 & \text{for} \quad m_s = -\dfrac{1}{2} \end{cases} \qquad \beta(m_s) = \begin{cases} 0 & \text{for} \quad m_s = \dfrac{1}{2} \\ 1 & \text{for} \quad m_s = -\dfrac{1}{2} \end{cases} \tag{9.57}
$$

and they satisfy the eigenvalue relations

$$
\vec{s}^2 \alpha = \frac{3}{4} \alpha; \qquad \vec{s}^2 \beta = \frac{3}{4} \beta; \qquad s_z \alpha = \frac{1}{2} \alpha; \qquad s_z \beta = -\frac{1}{2} \beta. \tag{9.58}
$$

The quantum number m_s is used as a variable and since it can only assume the two discrete values it is simple to define "integrals" over these *spin functions* by using summations over m_s, such that

$$
\begin{aligned}
(\alpha|\alpha) &= \sum_{m_s} \alpha(m_s)\alpha(m_s) = 1, \\
(\beta|\beta) &= \sum_{m_s} \beta(m_s)\beta(m_s) = 1, \\
(\alpha|\beta) &= (\beta|\alpha) = \sum_{m_s} \alpha(m_s)\beta(m_s) = 0.
\end{aligned}
\tag{9.59}
$$

An atomic one-electron wavefunction, called an *atomic orbital*, which we have written as

$$\psi_{nlm}(\vec{r}) = R_{nl}(r)Y_{lm}(\theta, \varphi), \tag{9.60}$$

now has to be augmented with a spin factor in order to completely specify the state of the electron. One commonly introduces the combined space–spin variable $x = (\vec{r}, m_s)$ and writes the two possible *spin orbitals* for each choice of the quantum numbers n, l, and m:

$$\psi_{nlm}(x) = R_{nl}(r)Y_{lm}(\theta, \varphi)\begin{cases} \alpha(m_s) \\ \beta(m_s) \end{cases}. \tag{9.61}$$

Integration over all space and all spin variables then justifies the definition of integrals such as

$$(\phi|\psi) = \int \phi^*(x)\psi(x)dx = \int d\vec{r} \sum_{m_s} \phi^*(\vec{r}, m_s)\psi(\vec{r}, m_s). \tag{9.62}$$

9.4 MANY-ELECTRON WAVEFUNCTIONS AND ANTISYMMETRY

Consider the two electrons of a helium atom. The Hamiltonian must have the kinetic-energy operators for both electrons, the potential energy due to the attraction of each electron to the nucleus and the potential energy due to the Coulomb repulsion between the two electrons. Thus

$$H = -\frac{1}{2}\vec{\nabla}_1^{\,2} - \frac{2}{r_1} - \frac{1}{2}\vec{\nabla}_2^{\,2} - \frac{2}{r_2} + \frac{1}{r_{12}}, \tag{9.63}$$

where again we are using atomic units and r_k is the distance of electron k from the helium nucleus (of atomic number $Z = 2$), and where $r_{12} = |\vec{r}_1 - \vec{r}_2|$ is the distance between electron 1 and 2. The Schrödinger equation is

$$H\Psi(x_1, x_2) = E\Psi(x_1, x_2), \tag{9.64}$$

where the two-electron wavefunction depends on the position and spin coordinates $(\vec{r}_k, m_{s_k}) = x_k$ of the two electrons.

This eigenvalue problem is not exactly solvable, but one can find very accurate approximate solutions. We are interested in only very approximate solutions, so we proceed by first observing that if the electron–electron repulsion term of the potential energy is neglected, which is indeed a very bad approximation, then the problem separates into two one-electron problems, each of which can be written as

$$\left[-\frac{1}{2}\vec{\nabla}^2 - \frac{2}{r}\right]u(\vec{r}) = \epsilon u(\vec{r}). \tag{9.65}$$

This is the hydrogen atom problem all over again, and we can solve for the eigenfunctions and eigenvalues. The total wavefunction is simply the product of two such solutions in this approximation:

$$\Psi(x_1, x_2) = u_1(x_1)u_2(x_2), \tag{9.66}$$

belonging to the eigenvalue $\epsilon_1 + \epsilon_2$. However, any property that we might want to calculate with a two-electron wavefunction must be independent of the labeling of the two electrons. In particular, the interpretation of

$$|\Psi(x_1, x_2)| dx_1\, dx_2 \tag{9.67}$$

as the probability of finding one electron in $(\vec{r}_1, \vec{r}_1 + d\vec{r}_1)$ with spin m_{s_1} and the other electron in $(\vec{r}_2, \vec{r}_2 + d\vec{r}_2)$ with spin m_{s_2} means that, since the electrons are identical twins, $|\Psi(x_1,x_2)|^2 = |\Psi(x_2,x_1)|^2$. When applying the permutation (12) to the electron labels, we get $(12)\Psi(x_1,x_2) = c\Psi(x_2,x_1)$, where c must be a number of modulus unity. Applying the permutation again yields

$$(12)(12)\Psi(x_1,x_2) = c(12)\Psi(x_2,x_1) = c^2\Psi(x_1,x_2) = \Psi(x_1,x_2), \tag{9.68}$$

which means that $c = \pm 1$ for identical particles. Physically acceptable results can be obtained only if we choose the minus sign for many-electron wavefunctions.[4]

Returning to the approximate wavefunction in Eq. (9.66), we can conclude that it is not antisymmetric as it should. However, we can also conclude that we could just as well have chosen to write the solution as

$$u_1(x_2)u_2(x_1), \tag{9.69}$$

[4] In general, the minus signs leading to wavefunctions of space–spin varaibles that are antisymmetric under the permutation of two particle labels hold for identical *fermions*, which are particles with half-integral spin quantum numbers. The plus signs, and thus symmetric wavefunctions under the interchange of two space–spin variables, apply to identical particles, called *bosons*, with integer spin quantum numbers.

which has the same eigenvalue. Linear combinations of degenerate eigenfunctions are also eigenfunctions belonging to the same eigenvalue, so we can form the antisymmetric product

$$\Psi(x_1, x_2) = u_1(x_1)u_2(x_2) - u_1(x_2)u_2(x_1), \tag{9.70}$$

which, even if an approximate two-electron wavefunction, at least has the correct antisymmery property when the two electron labels are interchanged.

This kind of wavefunction is a determinant

$$\begin{vmatrix} u_1(x_1) & u_1(x_2) \\ u_2(x_1) & u_2(x_2) \end{vmatrix} \tag{9.71}$$

with rows labeled by the function index and the columns by the electron label. Interchange of two rows or two columns of a determinant changes its sign, which is the desired antisymmetry property.

This approximate wavefunction can now be used to calculate the average value of the helium atom Hamiltonian in Eq. (9.63). An important question to answer is which orbitals are the best ones to put into the determinantal wavefunction in order to obtain the best possible average value of the Hamiltonian for the ground state of the helium atom, or for any many-electron atom. This question has to be answered in several steps.

9.4.1 Basis

The familiar Fourier expansion

$$f(x) = \frac{1}{2} a_0 + \sum_{n=1}^{\infty} (a_n \cos nx + b_n \sin nx) \tag{9.72}$$

of a function $f(x)$ in one dimension on the interval $(-\pi, \pi)$ is an example of the use of a basis set. The known functions $\{\cos nx, \sin nx\}$ constitute the basis set, and the unknown coefficients $\{a_n, b_n\}$ are determined in some way to best represent the function. In practice, one can never use an infinite basis set but must settle for a finite number of basis functions. It is then important that the basis is such that a systematic increase of basis size yields a better representation of the function one seeks. A basis set that can, in principle, represent any function in the given space is said to be complete.

All the eigenfunctions of a Hamiltonian can be used as a basis to represent any function in the same space. This is not what one would normally do, since if one has all the eigenfunctions, the Schrödinger equation is already completely solved. However, if we use such a basis of Hamiltonian eigenfunctions, we can make an important statement about the average value of the Hamiltonian with respect to an arbitrary function f. Assume that

$$Hu_n = E_n u_n \tag{9.73}$$

and that the energy eigenvalues are $E_1 < E_2 < E_3 < \cdots$. Then the function can be expressed in the known basis as

$$f = \sum_n u_n c_n , \tag{9.74}$$

where the "best" unknown coefficients $\{c_n\}$ are to be determined. A basis set can consist of functions that are mutually orthogonal and normalized to unity. The eigenfunctions of a Hamiltonian can always be assumed to be such. This is so since from Eq. (9.73) it follows by complex conjugation that

$$Hu_m^* = E_m u_m^*. \tag{9.75}$$

Multiplying Eq. (9.73) from the left by u_m^* and Eq. (9.75) from the right by u_n and integrating overall space yields

$$\begin{aligned} (u_m|Hu_n) &= E_n(u_m|u_n), \\ (Hu_m|u_n) &= E_m(u_m|u_n). \end{aligned} \tag{9.76}$$

Since the Hamiltonian is self-adjoint, the two left-hand sides are equal and subtraction of the second equation from the first yields

$$0 = (E_n - E_m)(u_n|u_m), \tag{9.77}$$

which means that if $E_n \neq E_m$, the integral $(u_m|u_n) = 0$. We say that the two eigenfunctions are orthogonal. For eigenfunctions belonging to the same eigenvalue, it is always possible to take linear combinations to ensure orthogonality, so we can assume that energy eigenfunctions are orthogonal and they can, of course, always be normalized to unity. When our basis possesses these qualities, then we find

$$\langle H \rangle = \frac{(f|Hf)}{(f|f)} = \frac{\sum_k \sum_l c_k^*(u_k|Hu_l)c_l}{\sum_k \sum_l c_k^*(u_k|u_l)c_l} = \frac{\sum_k E_k|c_k|^2}{\sum_l |c_l|^2} > E_1. \tag{9.78}$$

This means that the average value in any basis is always greater than the exact ground-state energy. Since the average value depends on the unknown expansion coefficients, these coefficients can be varied in such a manner that the smallest possible average value is obtained, thus yielding the best approximate ground state possible to represent in the chosen basis.

The best possible basis functions to use in a determinantal wavefunction for a many-electron atom can be shown to be the eigenfunctions of an effective Hamiltonian that in addition to the kinetic-energy operator, and the potential-energy due to the presence of the nucleus, also has a potential-energy term that is the average potential energy $V(r)$ of the electron due to the presence of all the other electrons. Such an effective Hamiltonian for an electron in a many-electron atom is called a *Fock operator* and can be expressed as

$$F = -\frac{1}{2}\vec{\nabla}^2 - \frac{Z}{r} + V(r). \tag{9.79}$$

The eigenfunctions of this operator cannot be determined exactly but need to be expanded in a basis.

Since the kinetic-energy operator is the same as in the case of the hydrogen atom, one can conclude that the eigenfunctions will have the form

$$R_{nl}(r)Y_{lm}(\theta, \varphi) \tag{9.80}$$

in which the angular factor is the well-known spherical harmonics that are the eigenfunctions of orbital angular momentum. The eigenfunctions can again be labeled by the principal quantum number, the angular momentum and magnetic quantum numbers, but the radial wavefunctions are not the ones we found for the hydrogen atom but now have to be expressed in a basis. The most common basis functions used for that purpose are

1. Slater-type orbitals (STOs), which can be expressed as $R_{nl}(r) = r^{n-1}e^{-\zeta_{nl}r}$.
2. Gauss-type orbitals (GTOs), which can be expressed as[5] $R_{nl}(r) = r^l e^{-\zeta_{nl}r^2}$.

In both cases the orbital exponents ζ_{nl} are parameters, the values of which have often has been determined by extensive variational procedures and supplied as fixed numbers for a variety of standard basis sets.

9.4.2 Secular Equations

Forming the average value of the Fock operator (the effective Hamiltonian of an electron in a many-electron atom) with an approximate wavefunction that depends on a set of adjustable parameters is the starting point of finding the "best" approximate wavefunction of the particular form chosen. Since the average value is always an upper bound to the exact lowest energy eigenvalue, one seeks the minimum of the average value by varying the parameters.

When the wavefunction is expressed in terms of a number of known basis functions $\{u_k\}$ such that

$$\psi(\vec{r}) = \sum_k u_k(\vec{r})c_k \tag{9.81}$$

[5] The radial GTOs are often combined with the spherical harmonics in Cartesian form so that the total basis function simply is $e^{-\zeta_{nl}r^2}$ multiplied with various powers of the Cartesian coordinates.

in which the expansion coefficients are the adjustable parameters, then the problem of finding the minimum average value becomes a set of homogeneous linear equations to determine the optimal expansion coefficients. This set of equations is called the *secular equations*.

Substituting the expansion in the expression for the average value yields

$$\begin{aligned}\epsilon = \frac{(\psi|F\psi)}{(\psi|\psi)} &= \frac{\sum_{k=1}^{M}\sum_{l=1}^{M} c_k^*(u_k|Fu_l)c_l}{\sum_{k=1}^{M}\sum_{l=1}^{M} c_k^*(u_k|u_l)c_l} \\ &= \frac{\sum_{k=1}^{M}\sum_{l=1}^{M} c_k^* F_{kl} c_l}{\sum_{k=1}^{M}\sum_{l=1}^{M} c_k^* \Delta_{kl} c_l},\end{aligned} \tag{9.82}$$

where the integrals $F_{kl} = (u_k|Fu_l)$ and $\Delta_{kl} = (u_k|u_l)$ over the known basis functions are numbers that can be calculated once and for all for each given basis set. Rewriting this as

$$\epsilon \sum_{k=1}^{M}\sum_{l=1}^{M} c_k^* \Delta_{kl} c_l = \sum_{k=1}^{M}\sum_{l=1}^{M} c_k^* F_{kl} c_l \tag{9.83}$$

and taking the derivative on both sides with respect to, say, c_m^* results in

$$\sum_{k=1}^{M} (F_{ml} - \Delta_{ml}\epsilon)c_l = 0 \tag{9.84}$$

when the condition $\partial\epsilon/\partial c_m^* = 0$ for an extremum is met.

Differentiation with respect to c_m yieds an equation that is the complex conjugate to Eq. (9.84) and thus gives nothing new. Since there are M different coefficients c_m^* for $m = 1, 2, \ldots, M$, we can express the secular equations in matrix form as

$$\begin{bmatrix} F_{11} - \Delta_{11}\epsilon & F_{12} - \Delta_{12}\epsilon & \cdots & F_{1M} - \Delta_{1M}\epsilon \\ F_{21} - \Delta_{21}\epsilon & F_{22} - \Delta_{22}\epsilon & \cdots & F_{2M} - \Delta_{2M}\epsilon \\ \vdots & \vdots & \cdots & \vdots \\ F_{M1} - \Delta_{M1}\epsilon & F_{2M} - \Delta_{2M}\epsilon & \cdots & F_{MM} - \Delta_{MM}\epsilon \end{bmatrix} \begin{bmatrix} c_1 \\ c_2 \\ \vdots \\ c_M \end{bmatrix} = \begin{bmatrix} 0 \\ 0 \\ \vdots \\ 0 \end{bmatrix}. \tag{9.85}$$

A set of homogeneous linear equations such as this can have nontrivial solutions only when the determinant of the matrix $\mathbf{F} - \boldsymbol{\Delta}\epsilon$ vanishes:

$$\det\{\mathbf{F} - \boldsymbol{\Delta}\epsilon\} = 0. \tag{9.86}$$

This is a polynomial equation in ϵ of degree M, which has in general M roots

$$\epsilon_1 \leq \epsilon_2 \leq \epsilon_3 \leq \cdots \leq \epsilon_M. \tag{9.87}$$

Selecting the lowest root yields a set of homogeneous linear equations with a known coefficient matrix $\mathbf{F} - \boldsymbol{\Delta}\epsilon_1$, and the unknown expansion coefficients can now be determined to give the lowest atomic orbital, which will be of $1s$ type. In analogous fashion, the higher atomic orbitals can be obtained that correspond to the other energy roots.

When this is done, one would ideally find that the orbitals are ordered in energy according to the value of $n + l$ the sum of the principal and the angular momentum quantum number. When $n + l = n' + l'$, the orbital with the smaller principal quantum number has the lower energy. In this way the atomic orbitals are arranged in order $\epsilon_{1s} < \epsilon_{2s} < \epsilon_{2p} < \epsilon_{3s} < \epsilon_{3p} < \epsilon_{4s} < \epsilon_{3d} < \epsilon_{4p} < \epsilon_{5s} < \cdots$ and so on. The calculations are usually done in such a way that the typical $(2l + 1)$ degeneracies appear for the atomic orbitals.

9.5 MANY-ELECTRON ATOMS

9.5.1 Addition of Angular Momenta

The symmetry labels of the stationary states of one-electron atomic systems are, as we have seen, given by the angular momentum quantum numbers. For many-electron atoms this is also the case; however, the angular momentum operators, both orbital and spin, are now sums of angular momentum operators for the individual electrons. For a two-electron atom the orbital angular momentum is

$$\vec{L} = (L_x \quad L_y \quad L_z) = \vec{l}_1 + \vec{l}_2 \tag{9.88}$$

with obvious generalizations for an N-electron atom. It is not difficult to prove that such an operator commutes with the Hamiltonian in Eq. (9.63). Obviously, the only term that requires special consideration is the electron–electron interaction

$$\frac{1}{r_{12}} = \frac{1}{[(x_1 - x_2)^2 + (y_1 - y_2)^2 + (z_1 - z_2)^2]^{1/2}}. \tag{9.89}$$

We consider explicitly the component $L_x = l_{1x} + l_{2x}$, with

$$l_{kx} = -i\left(y_k \frac{\partial}{\partial z_k} - z_k \frac{\partial}{\partial y_k}\right), \quad (\hbar = 1), \tag{9.90}$$

and show by differentiation that

$$L_x \frac{1}{r_{12}} = \frac{1}{r_{12}} L_x . \tag{9.91}$$

It is similarly straightforward to show that

$$\begin{aligned} [L_x, L_y] &= iL_z, \\ [L_y, L_z] &= iL_x, \\ [L_z, L_x] &= iL_y, \quad (\hbar = 1), \end{aligned} \tag{9.92}$$

which are the defining relations of an angular momentum.

The orbitals describing the individual electrons are eigenfunctions of angular momentum. We write $Y_{lm} = |l, m\rangle$, and observe that

$$\vec{l}_k^2 |l_k, m_k\rangle = l_k(l_k + 1)|l_k, m_k\rangle,$$
$$l_{kz}|l_k, m_k\rangle = m_k|l_k, m_k\rangle, \ (\hbar = 1). \tag{9.93}$$

For a two-electron atom we would have

$$\vec{L}^2|L, M\rangle = L(L + 1)|L, M\rangle,$$
$$L_z|L, M\rangle = M|L, M\rangle, \tag{9.94}$$

where the eigenstates $|L, M\rangle$ are formed, in general, as linear combinations of products

$$|l_1, m_1\rangle|l_2, m_2\rangle, \tag{9.95}$$

where because $L_z = l_{1z} + l_{2z}$, it must necessarily follow that $M = m_1 + m_2$. Since

$$m_k = l_k, l_k - 1, \ldots, -l_k \tag{9.96}$$

it also follows that the maximum value of M is $l_1 + l_2$ and can occur in only one way, namely, for $m_1 = l_1$ and $m_2 = l_2$. The angular momentum quantum numbers must satisfy

$$M = L, L - 1, \ldots, -L, \tag{9.97}$$

which means that the maximum value of L is $l_1 + l_2$. The "next" value of M is $l_1 + l_2 - 1$, which can occur in one of two ways: (1) $m_1 = l_1$ and $m_2 = l_2 - 1$ or (2) vice versa. One of these possibilities must belong to $L = l_1 + l_2$ and the other to $L = l_1 + l_2 - 1$. Obviously, $M = l_1 + l_2 - 2$ can occur in three different ways, and we conclude that one belongs to $L = l_1 + l_2$, the second to $L = l_1 + l_2 - 1$, and the third to $L = l_1 + l_2 - 2$. If, say, $l_1 \geq l_2$, the smallest value of $m_2 = -l_2$ will decide when this set of steps ends. We reach a smallest possible value $M = l_1 + l_2 - m$, which can occur in a number of ways

$$\begin{aligned} &m_1 = l_1 - m, \ m_2 = l_2, \\ &m_1 = l_1 - m + 1, \ m_2 = l_2 - 1, \\ &\vdots \\ &m_1 = l_1, \ m_2 = l_2 - m = -l_2, \end{aligned} \tag{9.98}$$

which means that $m = 2l_2$ and the smallest value of L is $l_1 - l_2$. This means that the possible values of the angular momentum quantum number L are

$$L = l_1 + l_2, l_1 + l_2 - 1, \ldots, |l_1 - l_2|. \tag{9.99}$$

9.5.2 The Helium Atom

The helium atom ground-state wavefunction can then be approximately represented as the determinant

$$\begin{vmatrix} 1s(1)\alpha(1) & 1s(2)\alpha(2) \\ 1s(1)\beta(1) & 1s(2)\beta(2) \end{vmatrix} = 1s(1)1s(2)[\alpha(1)\beta(2) - \beta(1)\alpha(2)], \qquad (9.100)$$

where in comparison to the two-electron wavefunction in Eq. (9.71) $u_1 = 1s\alpha$ and $u_2 = 1s\beta$. Obviously the wavefunction with, say, $u_1 = u_2 = 1s\alpha$ will be zero, since a determinant with two rows identical vanishes. The spin function $\alpha\beta - \beta\alpha$ corresponds to the spin quantum number $S = 0$. Using the result for addition of angular momenta, we have in general that

$$S = s_1 + s_2, s_1 + s_2 - 1, \ldots, |s_1 - s_2| \qquad (9.101)$$

and $M_S = m_{s_1} + m_{s_2} = S, S - 1, \ldots, -S$. So for two electron with $s_1 = s_2 = \frac{1}{2}$, the possible values of the total spin quantum number is $S = 0, 1$. Angular momenta in general involve the ladder operators, and for each electron we have the spin ladder operators $s_\pm = s_x \pm is_y$. From the analogy of Eq. (9.55), we find that $s_\pm$ operating on a one-electron spin function (s, m_s) is a factor $[(s \mp m_s)(s \pm m_s + 1)]^{1/2}$ times a function with $s, m_s \pm 1$. This means that for $s = \frac{1}{2}$ and $m_s = \pm\frac{1}{2}$,

$$\begin{aligned} s_+\alpha &= 0; & s_+\beta &= \alpha; \\ s_-\alpha &= \beta; & s_-\beta &= 0; \end{aligned} \qquad (9.102)$$

For the total spin, similar operators exist such that

$$S_\pm = \sum_j s_\pm(j), \qquad (9.103)$$

that is, a sum over the spin operators of all electrons. Comparing with Eq. (9.32), we can write ($\hbar = 1$):

$$\vec{S}^2 = S_+S_- + S_z^2 - S_z. \qquad (9.104)$$

Since

$$S_z = \sum_j s_z(j) \qquad (9.105)$$

and thus for two electrons

$$S_\pm = s_\pm(1) + s_\pm(2) \qquad (9.106)$$

and

$$S_z = s_z(1) + s_z(2), \qquad (9.107)$$

we find that

$$\begin{aligned}\vec{S}^2[\alpha(1)\beta(2) - \beta(1)\alpha(2)] &= (S_+S_- + S_z^2 - S_z) \\ &\quad \times [\alpha(1)\beta(2) - \beta(1)\alpha(2)] \\ &= S(S+1)[\alpha(1)\beta(2) - \beta(1)\alpha(2)] = 0\end{aligned} \tag{9.108}$$

showing that this spin function has the total spin quantum number $S = 0$.

In general, we say that the helium ground-state *electron configuration* has two electrons in a $1s$ orbital, which is written $(1s)^2$. This corresponds to a state with total orbital angular momentum $L = 0$, since $l_1 = l_2 = 0$ for the two $1s$ orbitals.

The atomic term symbol that labels the corresponding state of the helium atom, or any atom for that matter, is written as

$$^{2S+1}L \tag{9.109}$$

where capital letters are used such that

$$\begin{aligned} L &= 0 \to S, \\ L &= 1 \to P, \\ L &= 2 \to D, \\ L &= 3 \to F \end{aligned} \tag{9.110}$$

and so on. This means that the ground-state term of the helium atom is ^1S, read as singlet S.

An excited state of the helium atom can be obtained by promoting one electron to, say, the $2s$ orbital. There are several corresponding determinantal wavefunctions. For instance

$$\begin{vmatrix} 1s(1)\alpha(1) & 1s(2)\alpha(2) \\ 2s(1)\alpha(1) & 2s(2)\alpha(2) \end{vmatrix} = \{1s(1)2s(2) - 2s(1)1s(2)\}\alpha(1)\alpha(2), \tag{9.111}$$

which corresponds to a total spin quantum numbers $S = 1, M_S = 1$. This is so, since

$$\begin{aligned}\vec{S}^2[\alpha(1)\alpha(2)] &= (S_+S_- + S_z^2 - S_z)[\alpha(1)\alpha(2)] \\ &= 2[\alpha(1)\alpha(2)] = 1(1+1)[\alpha(1)\alpha(2)],\end{aligned} \tag{9.112}$$

$$S_z[\alpha(1)\alpha(2)] = \{\tfrac{1}{2} + \tfrac{1}{2}\}[\alpha(1)\alpha(2)] = 1[\alpha(1)\alpha(2)].$$

The partner spin functions can then be obtained by using the ladder operator S_- once

$$S_-[\alpha(1)\alpha(2)] = [\alpha(1)\beta(2) + \beta(1)\alpha(2)] \tag{9.113}$$

and twice

$$S_-^2[\alpha(1)\alpha(2)] = S_-[\alpha(1)\beta(2) + \beta(1)\alpha(2)] \propto \beta(1)\beta(2). \tag{9.114}$$

These are the three spin triplet wavefunctions corresponding to the quantum numbers $S = 1, M_S = 1, 0, -1$, respectively, and with the spin multiplicity $2S + 1 = 3$. The corresponding atomic term symbol is ^3S. It is also possible to form the spin singlet wavefunction

$$\begin{aligned} &\begin{vmatrix} 1s(1)\alpha(1) & 1s(2)\alpha(2) \\ 2s(1)\beta(1) & 2s(2)\beta(2) \end{vmatrix} - \begin{vmatrix} 1s(1)\beta(1) & 1s(2)\beta(2) \\ 2s(1)\alpha(1) & 2s(2)\alpha(2) \end{vmatrix} \\ &= \{1s(1)2s(2) + 2s(1)1s(2)\}\{\alpha(1)\beta(2) - \beta(1)\alpha(2)\}, \end{aligned} \tag{9.115}$$

again showing the typical singlet spin function as in Eq. (9.108). This means that the electron configuration $(1s)(2s)$ corresponds to the two terms ^3S and ^1S.

9.5.3 Atomic Term Symbols

The lithium atom ground state has the electron configuration $(1s)^2(2s)$. This corresponds to the two different but degenerate wavefunctions

$$\begin{aligned} &\det\{1s\alpha 1s\beta 2s\alpha\}, \\ &\det\{1s\alpha 1s\beta 2s\beta\}, \end{aligned} \tag{9.116}$$

where we express the determinant by the term "det" and the "first diagonal." The *closed-shell* part of the wavefunction always is a spin singlet so the open-shell part determines the total spin quantum numbers that are possible. In this case there is a single electron in a 2*s* orbital so the total spin quantum number is $S = \frac{1}{2}$ and the term is a "doublet *S*" (^2S).

The beryllium atom again is a closed-shell ground state $(1s)^2(2s)^2$ with a ^1S term. The boron atom ground-state electron configuration is $(1s)^2(2s)^2(2p)$. The number of possible determinantal wavefunctions corresponding to this configuration is six, since there are three *p* orbitals and each can be associated with an α or a β spin function. Again the spin multiplicity is determined by the single electron in the open-shell part, as is the total orbital angular momentum, so the atomic term is ^2P.

The carbon atom ground-state electron configuration is $(1s)2(2s)^2(2p)^2$, and now the open-shell part consists of two electrons distributed in various ways among six 2*p* spin orbitals. The number of possible determinants is given by the binomial coefficient $\binom{6}{2} = 15$. Which atomic terms they form can be analyzed as follows. Since only the p orbitals matter we denote the determinants with the shorthand notation, such that

$$(1| - 1) = \det\{1s\alpha 1s\beta 2s\alpha 2s\beta 2p_+\alpha 2p_-\beta\}, \tag{9.117}$$

with α spin orbitals to the left of the vertical bar and β spin orbital to the right. All possible determinants are placed in a table such that

M_L/M_s	1	0	−1
2		$(1\|1)$	
1	$(10\|$	$(1\|0)$, $(0\|1)$	$\|10)$
0	$(1-1\|$	$(1\|-1)$, $(0\|0)$, $(-1\|1)$	$\|1-1)$
−1	$(-10\|$	$(-1\|0)$, $(0\|-1)$	$\|-10)$
−2		$(-1\|1)$	

It is obvious that the determinant $(1|1)$ has $M_L = 2$, and since there is no higher value of M_L, this determinant must also have $L = 2$ (remember that $M_L = -L, -L+1, \ldots, L$). This determinant thus also has a partner with M_L =1 (and $L = 2$). This must be a combination of the determinants $(1|0)$, $(0|1)$ and so on. Since this is basically only a counting scheme, we construct a table with only the number of determinants in each block.

M/M_s	1	0	−1
2		1	
1	1	2	1
0	1	3	1
−1	1	2	1
−2		1	

We note that there are 15 determinants and that we can decompose this table as follows:

$$\begin{matrix} 1 & & & & \\ 1 & & 1 & 1 & 1 & \\ 1 & + & 1 & 1 & 1 & + 1 \\ 1 & & 1 & 1 & 1 & \\ 1 & & & & \end{matrix} \tag{9.118}$$

where we can clearly discern the $^1D, ^3P, ^1S$ terms.

Among these atomic states for the carbon atom the

$$^3P \tag{9.119}$$

term is the lowest in energy[6]

Obviously, the $^3P, ^1D$ terms can both be represented by a single determinant, say, $(10|$, and $(1|1)$, respectively. On the other hand, the ^1S term has to be represented by a fixed linear combination of determinants as was the case for the singlet term of the $(1s)(2s)$ configuration of helium. The angular momentum ladder operators can again be used to decide which combination belongs to the ^1S. From the table above we know that it must be a combination of the determinants $(1|-1), (0|0), (-1|1)$. Using the notation

$$|LM_LSM_S\rangle \tag{9.120}$$

for an eigenstate of $\vec{L}^2, L_z, \vec{S}^2, S_z$ and using the angular momentum properties of ladder operators, we can write

$$L_\pm|LM_LSM_S\rangle = [(L \pm M_L)(L \mp M_L + 1)]^{1/2}|LM_L \pm 1SM_S\rangle \tag{9.121}$$

and an analogous expression for the spin operators. We also need the relations $L_\pm = \sum_j l_\pm(j)$ and the results $l_\pm Y_{lm} = [(l \pm m)(l \mp m + 1)]^{1/2} Y_{lm\pm1}$ for the orbitals.

Since

$$|2200\rangle = (1|1), \tag{9.122}$$

we have

$$\begin{aligned} L_-|2200\rangle &= [(2+2)(2-2+1)]^{1/2}|2100\rangle \\ &= [(1+1)(1-1+1)]^{1/2}(0|1) + [(1+1)(1-1+1)]^{1/2}(1|0), \end{aligned} \tag{9.123}$$

and

$$|2100\rangle = \frac{1}{\sqrt{2}}\{(1|0) + (0|1)\}. \tag{9.124}$$

Furthermore

$$L_-|2100\rangle = \sqrt{6}|2000\rangle = [(-1|1) + 2(0|0) + (1|-1)], \tag{9.125}$$

yielding one combination of these determinants that belongs to the ^1D term.

[6] The so-called Hund rule of spectroscopy states that the term with the lowest energy among those coming from the same electron configuration is the one with maximum spin multiplicity. If two terms have the same spin multiplicity, the one among them with the highest orbital multiplicity $(2L + 1)$ is the ground state.

The combination that belongs to the ^3P term can be derived using the spin ladder operator such that

$$S_-|1011\rangle = S_-(1\ -1|, \tag{9.126}$$

which yields

$$[(1+1)(1-1+1)]^{1/2}|1010\rangle \tag{9.127}$$

on the left, and

$$\left[\left(\tfrac{1}{2}+\tfrac{1}{2}\right)\left(\tfrac{1}{2}-\tfrac{1}{2}+1\right)\right]^{1/2}\{(1|-1)-(-1|1)\} \tag{9.128}$$

on the right, where the change of sign in the second determinant comes from the fact that the two rows of the determinant have been interchanged, that is, in going from $\det\{\cdots 2p_{+1}\alpha 2p_{-1}\alpha\}$ to $\det\{\cdots 2p_{+1}\alpha 2p_{-1}\beta\}$, there is no change of phase, but in going to $\det\{\cdots 2p_{-1}\alpha 2p_{+}\beta\}$, two rows have been interchanged before turning α into β.

In summary, the two states made up from these determinants belonging to the ^3P and ^1D terms are

$$|1010\rangle = \frac{1}{\sqrt{2}}\{(1|-1)-(-1|1)\} \tag{9.129}$$

and

$$|2000\rangle = \frac{1}{\sqrt{6}}[(-1|1)+2(0|0)+(1|-1)], \tag{9.130}$$

respectively. The combination of these determinants that is orthogonal to both of these must be the state belonging to the ^1S term. This is

$$|0000\rangle = \frac{1}{\sqrt{6}}\{(1|-1)-(0|0)+(-1|1)\}. \tag{9.131}$$

So, in some cases it is not possible to write a single determinantal wavefunction that has the proper symmetry properties, as in this case, with reagard to angular momentum.

The nitrogen atom has the ground-state electron configuration $(1s)^2(2s)^2(2p)^3$, in which the number of possible determinants is $\binom{6}{3} = 20$, which we can put in a table analogous to the one used for carbon

M_L/M_s	$\frac{3}{2}$	$\frac{1}{2}$	$-\frac{1}{2}$	$-\frac{3}{2}$
2		(10|1)	(1|01)	
1		(1 − 1|1) (10|0)	(1|1 − 1), (0|10)	
0	(10 − 1|	(10| − 1), (0 − 1|1), (1 − 1|0)	(−1|10), (1|0 − 1), (0|1 − 1)	|10 − 1)
−1		(1 − 1|−1), (−10|0)	(−1|1 − 1), (0|−10)	
−2		(−10|−1)	(−1|0 − 1)	

which can be decomposed as before

$$\begin{array}{cc} 1 & 1 \\ 1 & 1 \\ 1 & 1 \\ 1 & 1 \\ 1 & 1 \end{array} + \begin{array}{cc} 1 & 1 \\ 1 & 1 \\ 1 & 1 \end{array} + \begin{array}{cccc} 1 & 1 & 1 & 1 \end{array}, \tag{9.132}$$

yielding the possible terms ${}^2D, {}^2P, {}^4S$, and the distribution of determinantal wavefunctions (or combinations thereof) among those terms. For instance, the determinant

$$(10-1| = \det\{1s\alpha 1s\beta 2s\alpha 2s\beta 2p_+\alpha 2p_0\alpha 2p_-\alpha\} \tag{9.133}$$

belongs to the ground-state term ${}^4S$, which is the one lowest in energy according to Hund's rule.

Example 9.3 Using the angular momentum ladder operators, show that $|11\tfrac{1}{2}\tfrac{1}{2}\rangle = (1/\sqrt{2})[(1-1|0)-(10|0)]$.

The oxygen atom has the ground state electron configuration

$$(1s)^2(2s)^2(2p)^4, \tag{9.134}$$

which is missing two electrons in a complete $2p$ shell. One says that there is two "holes" in a complete p shell. One can proceed exactly as in the case of the carbon atom by listing the two p orbitals of the "holes" and the details of obtaining the atomic terms are the same. Thus, oxygen has the possible terms ${}^1D, {}^3P, {}^1S$, where the triplet P is the ground state. The fluorine atom has a single "hole" in a complete p shell, making the ground-state term a ${}^2P$.

We can now consider any atomic electron configuration and with more or less difficulty follow the same procedure to obtain the possible atomic terms. For instance, the ground-state electron term of the titanium atom is

$$(1s)^2(2s)^2(2p)^6(3s)^2(3p)^6(4s)^2(3d)^2, \tag{9.135}$$

in which the only open-shell part is the two electrons in the $3d$ spin orbitals. The number of possible determinantal wavefunctions are $\binom{10}{2} = 45$. The table of determinants can be obtained as

M_L/M_s	1	0	−1
4		$(2\|2)$	
3	$(21\|$	$(2\|1)$, $(1\|2)$	$\|21)$
2	$(20\|$	$(2\|0)$, $(0\|2)$, $(1\|1)$	$\|20)$
1	$(2-1\|$, $(10\|$	$(2\|-1)$, $(-1\|2)$, $(1\|0)$, $(0\|1)$	$\|2-1)$, $\|10)$
0	$(2-2\|$, $(1-1\|$	$(2\|-2)$, $(-2\|2)$, $(1\|-1)$, $(-1\|1)$, $(0\|0)$	$\|2-2)$, $\|1-1)$

where we have listed the determinants only for $M_L \geq 0$, since the rest follows by symmetry ($m \rightarrow -m$) The number of determinants for each combination of M_L and M_S

M_L/M_s	1	0	−1
4		1	
3	1	2	1
2	1	3	1
1	2	4	2
0	2	5	2
−1	2	4	2
−2	1	3	1
−3	1	2	1
−4		1	

which can be resolved as before into

$$\begin{array}{ccccccccccccc}
1 & & & & & & & & & & & & \\
1 & & 1 & 1 & 1 & & & & & & & & \\
1 & & 1 & 1 & 1 & & 1 & & & & & & \\
1 & & 1 & 1 & 1 & & 1 & & 1 & 1 & 1 & & \\
1 & + & 1 & 1 & 1 & + & 1 & + & 1 & 1 & 1 & + & 1\,, \\
1 & & 1 & 1 & 1 & & 1 & & 1 & 1 & 1 & & \\
1 & & 1 & 1 & 1 & & 1 & & & & & & \\
1 & & 1 & 1 & 1 & & & & & & & & \\
1 & & & & & & & & & & & &
\end{array} \tag{9.136}$$

yielding the terms ${}^1G, {}^3F, {}^1D, {}^3P,\ {}^1S$.

9.6 SIMPLE MOLECULAR ORBITALS FOR DIATOMICS

It is, of course, quite possible to discuss *molecular orbitals* as energy eigenfunctions for an electron with its kinetic energy, and its potential energy due to the presence of all the atomic nuclei and the charge density of all the other electrons without reference to atomic orbitals. However, it makes some sense that the electron as it moves in the vicinity of one atomic nucleus has a wavefunction that closely resembles an atomic orbital on that nucleus and when it moves in proximity to more than one nucleus its wavefunction is some linear combination of atomic orbitals centered on these nuclei. Molecular orbitals formed as linear combinations of atomic orbitals (MO-LCAO) actually provide the most common way to describe molecular electronic structure.

The atomic orbitals that we discussed in the previous section become a basis for the molecular orbitals. The best mixing coefficients are obtained from a secular problem.

The effective one-electron Hamiltonian for a diatomic molecule is

$$F = -\frac{1}{2}\vec{\nabla}^2 - \frac{Z_A}{r_A} - \frac{Z_B}{r_B} + V(\vec{r}), \tag{9.137}$$

where atomic units have been used and where Z_A and Z_A are the atomic numbers of atom A and atom B, respectively, r_A and r_B are the distances of the electron from nucleus A and nucleus B, respectively, and $V(\vec{r})$ is the average potential energy of the electron due to the presence of all the other electrons.

The eigenvalue problem

$$F\phi_i = \epsilon_i \phi_i \tag{9.138}$$

can be solved approximately by introducing a basis of atomic orbitals $\{\chi_\nu\}$ centered on the two participating nuclei. Thus expansion in a basis

$$\phi_i = \sum_{\nu=1}^{M} \chi_\nu c_{\nu i} \tag{9.139}$$

leads, as before, to the secular equations

$$\sum_{\nu=1}^{M} [(\chi_\mu | F \chi_\nu) - \epsilon_i (\chi_\mu | \chi_\nu)] c_{\nu i} = 0 \tag{9.140}$$

for $\mu = 1, 2, \ldots, M$.

For the purpose of this discussion we consider the Fock matrix elements $F_{\mu\nu} = (\chi_\mu | F\chi_\nu)$ and the metric matrix elements $\Delta_{\mu\nu} = (\chi_\mu | \chi_\nu)$ to be known integrals, so the molecular orbitals (MOs) of Eq. (9.139) and the corresponding MO energies ϵ_i can be determined from the secular equations in Eq. (9.140).

The kinetic-energy operator for an electron in a diatomic molecule AB can be expressed in terms of the spherical polar coordinates (see Fig. 9.3) with their center at, say, A. This means that we can write

$$-\frac{1}{2}\vec{\nabla}^2 = -\frac{1}{2}\left[\frac{\partial^2}{\partial r_A^2} + \frac{2}{r_A}\frac{\partial}{\partial r_A} + \frac{1}{r_A^2 \sin\theta_A}\frac{\partial}{\partial\theta}\left(\sin\theta_A \frac{\partial}{\partial\theta_A}\right) + \frac{1}{r_A^2 \sin^2\theta_A}\frac{\partial^2}{\partial\varphi_A^2}\right] \tag{9.141}$$

and since there is no longer spherical symmetry as in the case of an atom, but an axis (a C_∞ axis) of symmetry (the z axis), the total angular momentum of the electron, which can be expressed as above in terms of derivatives with respect to θ_A, φ_A, no longer commutes with F (since $r_B = [r_A^2 + R^2 - 2r_A R\cos\theta_A]^{1/2}$).

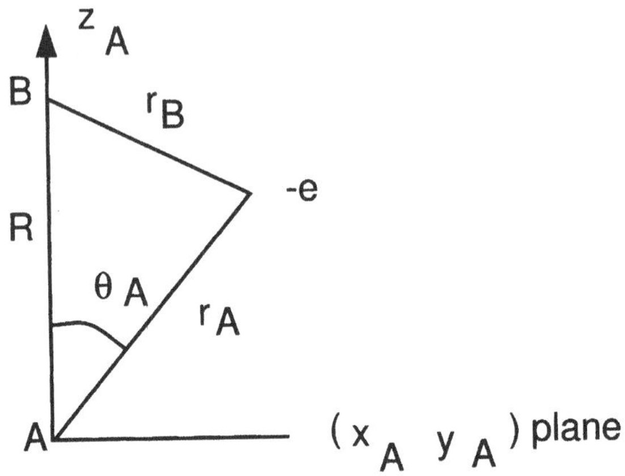

Figure 9.3 Spherical polar coordinates for an electron in a diatomic molecular AB.

However, the z component of angular momentum $l_z = -i(\partial/\partial\varphi_A)$ does commute with F, meaning that these two operators have common eigenfunctions. The diatomic MOs then must have a factor $\exp(im\varphi_A)$ and the MO energies depend on m^2, since

$$\frac{\partial^2}{\partial\varphi^2}\exp(im\varphi) = -m^2\exp(im\varphi). \tag{9.142}$$

This also means that MOs that are the same except for a factor $\exp(im\varphi)$ or $\exp(-im\varphi)$ are degenerate. Molecular orbitals for diatomics are labeled by σ for $|m| = 0$, π for $|m| = 1$, δ for $|m| = 2$, and ϕ for $|m| = 3$. Obviously, MO energy levels for $|m| \neq 0$ are doubly degenerate.

In addition to the $|m|$ quantum number, MOs of homonuclear diatomics are labeled according to their behavior under the inversion operation. The subscripts g (German *gerade* for even) and u (German *ungerade* for odd) are used for MOs that are invariant or change sign, respectively, under the inversion operation.

With this in mind we can express the very simplest MOs for homonuclear diatomics as combinations of the $1s$, $2s$, $2p$, and further AOs on the participating atoms. In unnormalized form we can write

$$1\sigma_g = 1s_A + 1s_B, \tag{9.143}$$

which can be given a graphical representation as shown in Figure 9.4.

Similarly

$$1\sigma_u = 1s_A - 1s_B, \tag{9.144}$$

see also Figure 9.5.

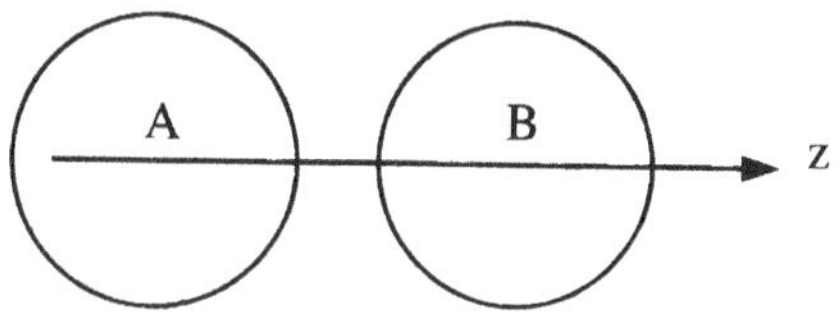

Figure 9.4 Graphical representation of a σ_g molecular orbital for the homonuclear diatomic AB.

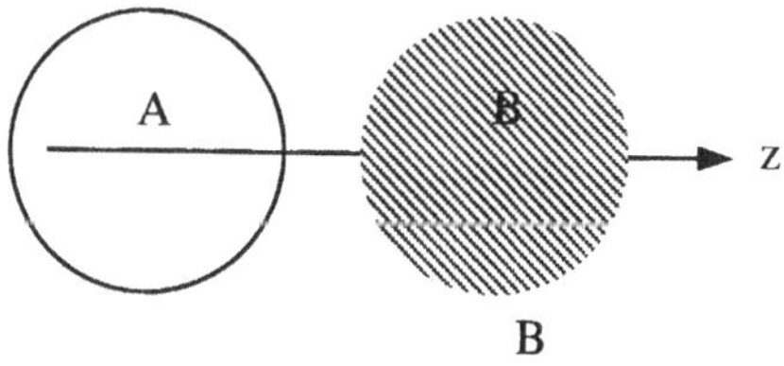

Figure 9.5 Graphical representation of a σ_u molecular orbital.

The other simple combinations of the AOs that are occupied in the ground state of the atoms of the elements in the first row of the periodic table are

$$2\sigma_g = 2s_{\mathrm{A}} + 2s_{\mathrm{B}}; \qquad 2\sigma_u = 2s_{\mathrm{A}} - 2s_{\mathrm{B}}; \qquad 3\sigma_g = 2p_{z\mathrm{A}} - 2p_{z\mathrm{B}};$$
$$3\sigma_u = 2p_{z\mathrm{A}} + 2p_{z\mathrm{B}}; \qquad 1\pi_{xu} = 2p_{x\mathrm{A}} + 2p_{x\mathrm{B}}; \qquad 1\pi_{yu} = 2p_{y\mathrm{A}} + 2p_{y\mathrm{B}};$$
$$1\pi_{xg} = 2p_{x\mathrm{A}} - 2p_{x\mathrm{B}}; \qquad 1\pi_{yu} = 2p_{y\mathrm{A}} - 2p_{y\mathrm{B}}; \qquad (9.145)$$

where the integer in front of σ or π, such as 2 in $2\sigma_u$, simply means the second sigma ungerade MO in order of energy from the lowest to the highest.

Graphical representations help in assigning the symmetry labels. The graphical representation of the $1\pi_u$ MO is is given in Figure 9.6, and the $1\pi_g$ is as shown in Figure 9.7.

In general, for heteronuclear diatomics, since there is no center of symmetry, the atomic orbitals combine with different weights on the participating AOs. Very approximate MOs $\phi = \chi_{\mathrm{A}} c_{\mathrm{A}} + \chi_{\mathrm{B}} c_{\mathrm{B}}$ for a general diatomic AB can be

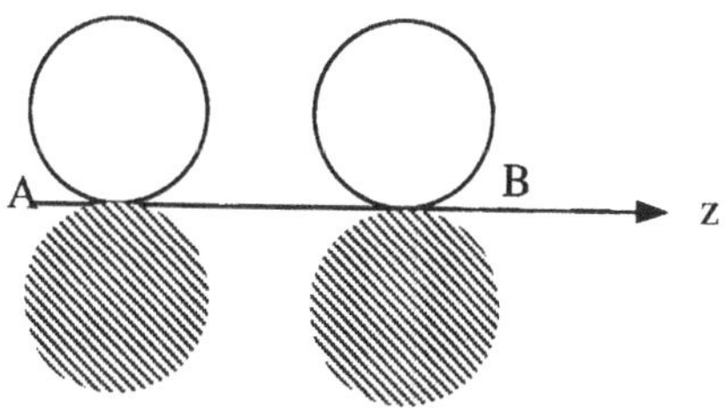

Figure 9.6 Graphical representation of a π_u molecular orbital.

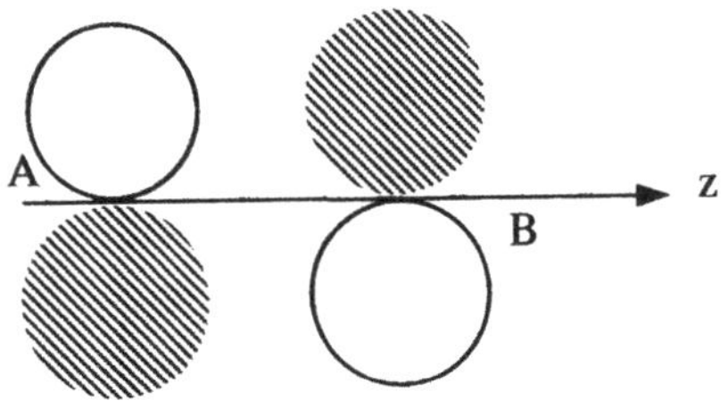

Figure 9.7 Graphical representation of a π_g molecular orbital.

formed as a linear combination of two atomic orbitals, one centerd on A and the other on B. The secular equations are

$$\begin{aligned}(F_{AA} - \epsilon)c_A + (F_{AB} - \epsilon S)c_B &= 0, \\ (F_{BA} - \epsilon S) + (F_{BB} - \epsilon)c_B &= 0,\end{aligned} \tag{9.146}$$

where the MO energy is denoted by ϵ and the atomic orbitals are normalized to unity and are assumed to be real so that

$$\begin{aligned}&F_{AA} = (\chi_A|F\chi_A); \qquad F_{BB} = (\chi_B|F\chi_B); \qquad F_{AB} = F_{BA} = (\chi_A|F\chi_B); \\ &(\chi_A|\chi_A) = (\chi_B|\chi_B) = 1; \qquad (\chi_A|\chi_B) = (\chi_B|\chi_A) = S;\end{aligned} \tag{9.147}$$

Nontrivial solutions to the system of linear (secular) equations require that the coefficient matrix have a vanishing determinant:

$$\begin{vmatrix}(F_{AA} - \epsilon) & (F_{AB} - \epsilon S) \\ (F_{AB} - \epsilon S) & (F_{BB} - \epsilon)\end{vmatrix} = 0. \tag{9.148}$$

When the two atoms are very far apart, it is possible to write

$$F_{AA} = (\chi_A|[-\frac{1}{2}\vec{\nabla}^2 - \frac{Z_A}{r_A} - \frac{Z_B}{r_B} + V_A(\vec{r}) + V_B(\vec{r})]\chi_A), \tag{9.149}$$

where the average potential energy of the electron due to the presence of all the other electrons has been divided into the potential energy due to the electrons on atom A and another term due to those on atom B. Obviously, as the nuclei are infinitely far apart and since the atomic orbital on A goes to zero at great distances from nucleus A, we can conclude that

$$F_{AA} \underset{R\to\infty}{\longrightarrow} (\chi_A|[-\frac{1}{2}\vec{\nabla}^2 - \frac{Z_A}{r_A} + V_A(\vec{r})]\chi_A), \tag{9.150}$$

which equals the atomic orbital energy of an electron on atom A in orbital χ_A.

A rough idea of what happens to the MO energies as the two atoms approach each other to ordinary bond distances can be obtained from studying two specific cases:

1. $F_{AA} \approx F_{BB}$, which, from Eq. (9.148), yields the two energy roots

$$\begin{aligned} \epsilon_1(\text{bonding}) &= \frac{F_{AA} + F_{AB}}{1 + S}, \\ \epsilon_2(\text{antibonding}) &= \frac{F_{AA} - F_{AB}}{1 - S}. \end{aligned} \tag{9.151}$$

   Normally $F_{AA} < 0$, $F_{AB} < 0$, $S > 0$ leading to the orbital energy-level diagram shown in Figure 9.8, where the "antibonding" orbital is more "antibonding" than the "bonding" is bonding, due to the *overlap integral S*.

2. $F_{AA} >> F_{BB}$ and $|F_{AA} - F_{BB}| >> |F_{AB}|$, while $S = 0$. With these conditions Eq. (9.148) yields

$$\begin{aligned} \epsilon_1 &= F_{BB} - \frac{|F_{AB}|^2}{|F_{AA} - F_{BB}|}, \\ \epsilon_2 &= F_{AA} + \frac{|F_{AB}|^2}{|F_{AA} - F_{BB}|}. \end{aligned} \tag{9.152}$$

   The MO energy-level diagram now resembles that shown in Figure 9.9.

These considerations make possible a building-up principle for molecules. The homonuclear diatomic MO energy levels in a generic diagram for

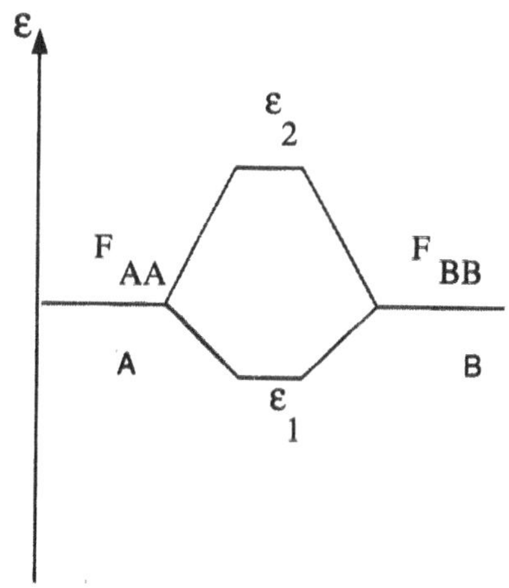

Figure 9.8 Schematic MO energy levels resulting from the combination of two atomic orbitals of comparable energies.

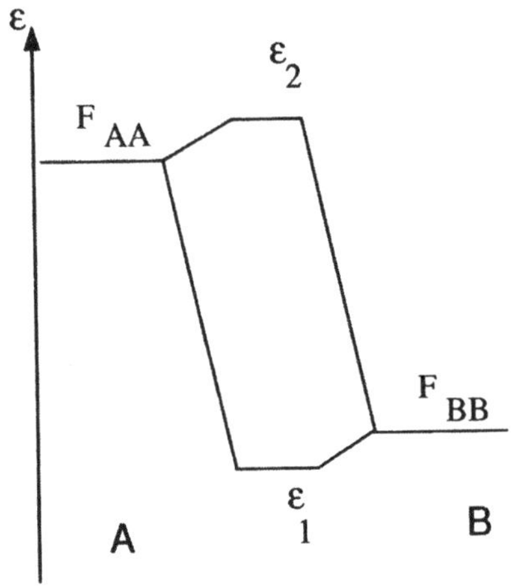

Figure 9.9 Schematic MO energy levels resulting from the combination of two atomic orbitals of unequal energies.

equilibrium bond distances are ordered as follows: $1\sigma_g < 1\sigma_u < 2\sigma_g < 2\sigma_u < 1\pi_u < 3\sigma_g < 1\pi_g < 3\sigma_u \ldots$; the only ambiguity is the relative order of $1\pi_u$ and $3\sigma_g$ that depends on level of theory and system. The ground-state electron configuration and the corresponding molecular term are given in the table below for some homonuclear systems. The molecular term $^{2S+1}\Lambda$ is decided by calculating $|\sum_{i=1}^{N} m_i| = \Lambda$. For $\Lambda = 0$ the symbol is Σ, for $\Lambda = 1$ the symbol is Π, and for $\Lambda = 2$ the symbol is Δ. Spin multiplicity is handled in the same way as for atoms. The subscript u or g refers to the symmetry of the total determinantal wavefunction under inversion. Only an odd number of u orbitals can lead to u symmetry.

Molecule	Ground-State Electron Configuration	Ground-State Term
H_2	$1\sigma_g^2$	$^1\Sigma_g$
He_2	$1\sigma_g^2 1\sigma_u^2$	$^1\Sigma_g$
Li_2	$1\sigma_g^2 1\sigma_u^2 2\sigma_g^2$	$^1\Sigma_g$
Be_2	$1\sigma_g^2 1\sigma_u^2 2\sigma_g^2 2\sigma_u^2$	$^1\Sigma_g$
B_2	$(Be_2) 1\pi_u^2$	$^3\Sigma_g$
C_2	$(Be_2) 1\pi_u^4$	$^1\Sigma_g$
N_2	$(Be_2) 1\pi_u^4 3\sigma_g^2$	$^1\Sigma_g$
O_2	$(Be_2) 3\sigma_g^2 1\pi_u^4 1\pi_g^2$	$^3\Sigma_g$
F_2	$(Be_2) 3\sigma_g^2 1\pi_u^4 1\pi_g^4$	$^1\Sigma_g$
Ne_2	$(Be_2) 3\sigma_g^2 1\pi_u^4 1\pi_g^4 3\sigma_u^2$	$^1\Sigma_g$

As an example we sketch a MO energy-level diagram for O_2 (see Fig. 9.10). Since only sigma orbitals combine with sigma, pi with pi, and so on and only atomic orbitals with comparable energies combine to any great extent, we can use the results discussed in this section to combine atomic orbital energies. The oxygen atomic orbital energies from calculations using good quality basis sets for the ground-state ^3P term are listed in the table below. These values can be used to get a rough idea about the MO energy-level diagram for O_2. The electrons occupy the orbitals as indicated with spin-up and spin-down arrows, and the ground state according to Hund's rule becomes a spin triplet as indicated.

^3P,O	AO energies in atomic units or hartrees (1 au = 27.209 eV)		
	$\epsilon_{1s} = -20.66866$	$\epsilon_{2s} = -1.24433$	$\epsilon_{2p} = -0.63192$

As an example of a heteronuclear diatomic molecule, we consider HF. The AO energies are listed in the table below for the fluorine atom ground state. The hydrogen atom 1*s* level lies at – 0.50000 au.

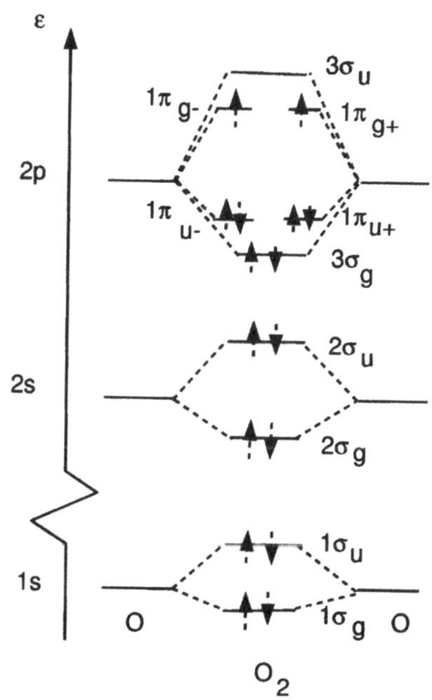

Figure 9.10 Sketched MO energy-level diagram for the oxygen molecule.

^2P, F	AO energies in atomic units or hartrees (1 au = 27.209 eV)		
	$\epsilon_{1s} = -26.38273$	$\epsilon_{2s} = -1.57254$	$\epsilon_{2p} = -0.73001$

Placing the 10 electrons in these MO energy levels yields the ground-state electron configuration

$$(1\sigma)^2(2\sigma)^2(3\sigma)^2(1\pi)^4. \tag{9.153}$$

The diagram in Figure 9.11 also shows that the 1σ and 2σ MOs are in this approximation simply the fluorine atomic $1s$ and $2s$ orbitals, respectively. The $3\sigma = c_1 2p_z + c_2 1s_{\mathrm{H}}$, assuming the z axis to be the molecular axis. The 1π orbitals are simply the $2p_x, 2p_y$ orbitals on fluorine in this—the simplest of—descriptions.

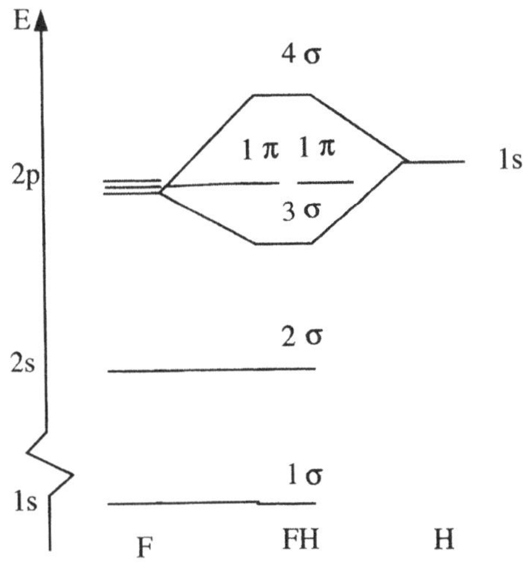

Figure 9.11 Sketched MO energy level diagram for the *FH* molecule.

10

THE SYMMETRIC GROUP

10.1 PRELIMINARIES

Several facts about permutation groups have already been discussed. We repeat some of the more important conclusions and we will discuss some new particulars. Most facts are illustrated by examples, and detailed proofs are omitted.

Every permutation of the symmetric group S_n can be expressed as a product of nonoverlapping cycles, and each cycle is a permutation in its own right. For instance, one can write

$$\begin{pmatrix} 12345678 \\ 25173468 \end{pmatrix} = (1253)\,(476)\,(8) \in S_8. \tag{10.1}$$

All permutations with the same cycle structure belong to the same conjugate class. For instance, all the permutations of S_8 that have one 1-cycle, one 3-cycle, and one 4-cycle as the one shown in Eq. (10.1) belong to the class $(1,3,4)$. In this way the cycle structure labels the class. Since each cycle can be expressed as a product of (overlapping) simple interchanges (2-cycles), the permutations are classified as even when they contain an even number of simple interchanges and odd when they contain an odd number. The subset of even permutations form an invariant subgroup $A_n \subset S_n$ called the *alternating group*.

The different possible cycle structures of S_n correspond to all possible integer partitions of the integer n. For example, S_4 has the cycle structures

$$(1^4),\ (1^2,2),\ (1,3),\ (2^2),\ (4) \tag{10.2}$$

corresponding to the following integer partitions of 4

$$\begin{aligned} &1 + 1 + 1 + 1 \\ &1 + 1 + 2 \\ &1 + 3 \\ &2 + 2 \\ &4. \end{aligned} \tag{10.3}$$

Since for any group the number of conjugate classes equals the number of irreps, one can prove that there is a meaningful one-to-one correspondence between the integer partitions of n and the irreps. The irreps of, say, S_4 can then be given the notations

$$[4], [1^4], [3, 1], [2^2], [2, 1^2]; \tag{10.4}$$

thus, the irreps are labeled by the partitions with the numbers in descending order inside square brackets. It will become clear that [4] and $[1^4]$ are one-dimensional, as are in general $[n]$ and $[1^n]$, called the *symmetric* and *antisymmetric irrep*, respectively.

Each partition and thus each irrep corresponds to a graph. For example, the irrep [4,2] of S_6 has the graph

$$\begin{array}{llll} \bullet & \bullet & \bullet & \bullet \\ \bullet & \bullet & & \end{array}\ , \tag{10.5}$$

namely, a top row with four dots and a second row with two dots, left-adjusted. Similarly, the irrep [3,2,1] has the graph

$$\begin{array}{lll} \bullet & \bullet & \bullet \\ \bullet & \bullet & \\ \bullet & & \end{array}\ . \tag{10.6}$$

These graphs have properties that can be used to develop a fast algorithm to compute the value of the character of its irrep at a given conjugate class.

10.2 CHARACTERS

A graph with nonincreasing row length from top to bottom is called a "proper" graph. A. J. Coleman, (Adv. in Quantum Chem. *4*, 83 (1968)) drawing on the work of Young and Robinson, devised a method for calculating the value of the character of any irrep of a symmetric group at a given class. This scheme can be described as follows:

1. *Hook.* A hook is defined as a "line" (with at most a 90° downward bend) that starts in a point of a proper graph with no other point to its right and ends in a point with no other point below. A single point can be a hook:

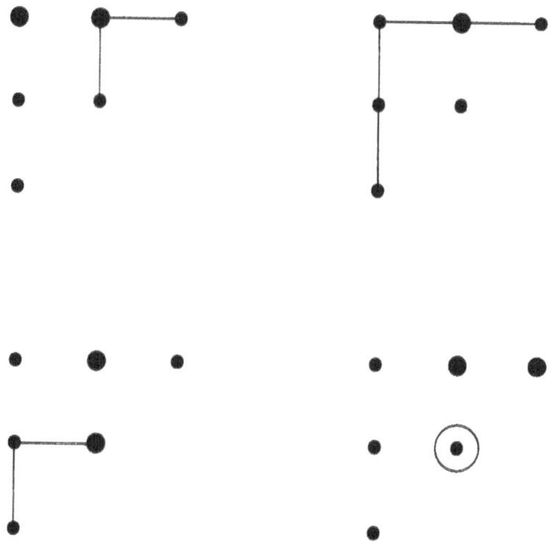

2. *Hook Length.* The length of a hook is simply the number of points in the graph that it covers.
3. *Hook Head.* The head of a hook is the point in the hook furthest to the top and farthest to the left. Every point in the graph is the head of a hook. The hook length of each hook is written at the head of the hook. For the [3,2,1] graph above this gives

$$\begin{matrix} 5 & 3 & 1 \\ 3 & 1 & \\ 1 & & \end{matrix} \tag{10.7}$$

The hook lengths of the leftmost column are called the *principal hook lengths.*

4. The principal hook lengths l_i are arranged in descending order into a symbol referred as the "*determinant*", which is appropriate since it has several properties of a determinant:

$$|l_1, l_2, \ldots, l_k| . \tag{10.8}$$

When the entries are such that $l_1 > l_2 > \cdots > l_k$, they are in *normal order.*

5. In the process of manipulating the entries of the determinant, new determinants will appear for which two entries might be identical or one entry negative. Such determinants are zero. A determinant might also appear with the entries not in normal order. Normal order can be restored resulting in at most a factor of (-1), when the permutation required to restore the entries to normal order is odd.

The actual use of these concepts and the operations on a determinant of a particular graph in order to obtain the value of the character of the corresponding irrep at a particular class are best introduced by examples.

Example 10.1 Consider irrep [3,2,1] of S_6. The principal hook lengths are 5, 3, and 1 as shown above. The determinant

$$|5,3,1| \tag{10.9}$$

will yield the character of this irrep at a given class, say, $(1^3,3)$, by using this class symbol as an operator on the determinant and applying it as follows:

$$(1^3,3)|5,3,1| = (1^3)\{|2,3,1| + |5,0,1| + |5,3,-2|\}; \tag{10.10}$$

thus, the rightmost number in the class symbol (in this case 3) is subtracted in turn from each of the entries in the determinant. This is a linear operation (like differentiation of a product). Of the resulting determinants, the last one is zero according to the rules given above. The application of the class symbol continues by subtracting 1 from each new determinant to obtain

$$(1^2)\{|1,3,1| + |2,2,1| + |2,3,0| + |4,0,1| + |5,-1,1| + |5,0,0|\}, \tag{10.11}$$

where, according to the rules, the first two and the last two determinants are zero. Continuing this procedure, the next step yields

$$(1)\{|1,3,0| + |3,0,1|\}, \tag{10.12}$$

where only the nonvanishing determinants are given. The last step yields

$$|1,2,0| + |2,0,1| = -2 \times |2,1,0|, \tag{10.13}$$

where we have brought the determinants to normal order. The final result is a single determinant of normal order times a number (in this case -2). This is a general result. The numerical factor is the value of the character. In summary

$$(1^3,3)\,|5,3,1| = -2 \times |2,1,0|, \tag{10.14}$$

and the charcater of the irrep has the value – 2 at the class $(1^3,3)$ or

$$\chi^{[3,2,1]}((1^3,3)) = -2. \tag{10.15}$$

Example 10.2 The character of the irrep $[2^2, 1^3]$ of S_7 at the class $(1^2, 2, 3)$ can be calculated in the same way, by first getting the hook lengths

$$\begin{matrix} 6 & 2 \\ 5 & 1 \\ 3 & \\ 2 & \\ 1 & \end{matrix} \quad , \tag{10.16}$$

forming the determinant of the principal hook lengths, and applying the class symbol

$$\begin{aligned} (1^2, 2, 3)|6, 5, 3, 2, 1| &= (1^2, 2)|6, 5, 0, 2, 1| \\ &= (1^2)\{|4, 5, 0, 2, 1| + |6, 3, 0, 2, 1|\} \\ &= (1)\{|3, 5, 0, 2, 1| + |5, 3, 0, 2, 1|\} = 0|5, 3, 2, 1, 0|\,, \end{aligned} \tag{10.17}$$

yielding

$$\chi^{[2^2,1^3]}((1^2, 2, 3)) = 0\,. \tag{10.18}$$

Because of this efficient algorithm for calculating the values of characters, there is no need to list the character tables of the permutation groups.

10.3 THE ORTHOGONAL IRREPS

The dimension of the irrep of S_n having graph with the hook lengths $\{h_j\}$ is

$$f = \frac{n!}{\prod\limits_j h_j} \tag{10.19}$$

Example 10.3 The irrep $[2^2, 1]$ of S_5 has the graph

$$\begin{matrix} \bullet & \bullet \\ \bullet & \bullet \\ \bullet & \end{matrix} \tag{10.20}$$

with the hook lengths

$$\begin{matrix} 4 & 2 \\ 3 & 1 \\ 1 & \end{matrix} \tag{10.21}$$

and the dimension

$$f = \frac{5!}{2 \cdot 3 \cdot 4} = 5. \tag{10.22}$$

Instead of representing the graph with dots, one may use squares or "boxes" and by putting the integers $1, 2, 3, \ldots, n$ in these boxes one obtains the Young tableaux. Restricting the entries of the integers to that of *standard order*, in which the integers increase from left to right and from top to bottom, one creates the *standard Young tableaux.*

Example 10.4 The five standard Young tableaux of the irrep $[2^2, 1]$ of S_5 are

$$\begin{array}{ccccc}
\text{I} & \text{II} & \text{III} & \text{IV} & \text{V} \\
\begin{array}{|c|c|}\hline 1 & 2 \\ \hline 3 & 4 \\ \hline 5 \\ \cline{1-1}\end{array} &
\begin{array}{|c|c|}\hline 1 & 2 \\ \hline 3 & 5 \\ \hline 4 \\ \cline{1-1}\end{array} &
\begin{array}{|c|c|}\hline 1 & 3 \\ \hline 2 & 4 \\ \hline 5 \\ \cline{1-1}\end{array} &
\begin{array}{|c|c|}\hline 1 & 3 \\ \hline 2 & 5 \\ \hline 4 \\ \cline{1-1}\end{array} &
\begin{array}{|c|c|}\hline 1 & 4 \\ \hline 2 & 5 \\ \hline 3 \\ \cline{1-1}\end{array}
\end{array}$$

It is no coincidence that the number of standard Young tableaux is the same as the dimension of the irrep. That is always so.

The standard Young tableaux are arranged in lexicographic order, and we have labeled them by uppercase roman numerals. The lexicographic order is established by forming the vector $M = (m_1 \quad m_2 \quad \cdots \quad m_n)$ for each tableau, in which the components are the entries from left to right row by row from the top. In this way tableau II above would have the vector

$$(1 \quad 2 \quad 3 \quad 5 \quad 4). \tag{10.23}$$

In order to establish the relative order of two Young tableaux with vectors M and M', the difference $(M - M')$ is formed, and if the first nonzero component $\Delta m_i = m_i - m_i'$ of the difference vector is positive, then M comes after M' in the lexicographic order. The standard Young tableaux (I – V) listed above are already in lexicographic order.

The standard Young tableaux of an irrep of S_n can serve as a basis for that irrep and generate orthogonal representation matrices. Such a set of matrices is called an *orthogonal irrep.* Before we enter into the details of the construction of the orthogonal representation matrices, we consider a few facts about the full symmetric group.

Theorem 10.1 The entire symmetric group S_n can be generated from the $(n - 1)$ simple transpositions of type $(i, i + 1)$.

Proof

1. All transpositions of type $(1, a)$ are generated from all the $(i, i+1)$:

$$
\begin{aligned}
&(23)(12)(23) = (13)\,; \\
&(34)(13)(34) = (14)\,; \\
&\vdots \\
&(n-1, n)(1, n-1)(n-1, n) = (1, n)\,.
\end{aligned}
\tag{10.24}
$$

2. From all the transpositions $(1, a)$, the general transpositions can be formed:

$$(1, a)(1, b)(1, a) = (a, b) \tag{10.25}$$

3. Any permutation of S_n can be expressed as a product of transpositions.

QED

This means that by generating the representation matrices for the transpositions $(i, i+1)$, one can generate the entire representation. For instance, the irreps of S_5 can be obtained by considering only the transpositions

$$(12),\ (23),\ (34),\ (45)\,. \tag{10.26}$$

The standard Young tableaux I–V are used as a basis of the irrep $[2^2, 1]$ and the elements of the representation matrices are determined by the locations of the numbers i and $(i+1)$ in the Young tableaux. For instance

$$
\Gamma(12) = \begin{array}{cccccc}
 & \mathrm{I} & \mathrm{II} & \mathrm{III} & \mathrm{IV} & \mathrm{V} \\
\mathrm{I} & 1 & 0 & 0 & 0 & 0 \\
\mathrm{II} & 0 & 1 & 0 & 0 & 0 \\
\mathrm{III} & 0 & 0 & -1 & 0 & 0 \\
\mathrm{IV} & 0 & 0 & 0 & -1 & 0 \\
\mathrm{V} & 0 & 0 & 0 & 0 & -1
\end{array}, \tag{10.27}
$$

where the following rules have been used:

1. The number +1 is assigned to the diagonal element if i and $(i+1)$ appear in the same row of the corresponding Young tableau, and –1 if they appear in the same column.

   If i and $(i+1)$ are not in the same row or column, they appear in two Young tableax (YT) interchanged (the other entries in the two YT are identical). For example, 2 and 3 appear in YTs I and III interchanged and similarly in YTs II and IV. When this is the case, the next rule (2) is used.

2. The two tableaux (e.g., YTs I and III) define a 2×2 submatrix

$$\begin{bmatrix} -\rho & -(1-\rho^2)^{1/2} \\ -(1-\rho^2)^{1/2} & \rho \end{bmatrix}, \tag{10.28}$$

where $\rho = [h(i, i+1) - 1]^{-1}$, in which $h(i, i+1)$ is the "city block distance" or partial hook length in the YT going from i to $(i+1)$.

Using rules 1 and 2 results in

$$\Gamma(23) = \begin{bmatrix} -\frac{1}{2} & 0 & -\frac{\sqrt{3}}{2} & 0 & 0 \\ 0 & -\frac{1}{2} & 0 & -\frac{\sqrt{3}}{2} & 0 \\ -\frac{\sqrt{3}}{2} & 0 & \frac{1}{2} & 0 & 0 \\ 0 & -\frac{\sqrt{3}}{2} & 0 & \frac{1}{2} & 0 \\ 0 & 0 & 0 & 0 & -1 \end{bmatrix} \tag{10.29}$$

since $h(2,3) = 3$ for YTs I and III and YTs II and IV.

Using the same rules, the representation matrices for (34) and (45) are found to be

$$\Gamma(34) = \begin{bmatrix} 1 & 0 & 0 & 0 & 0 \\ 0 & -1 & 0 & 0 & 0 \\ 0 & 0 & -1 & 0 & 0 \\ 0 & 0 & 0 & -\frac{1}{3} & -\frac{\sqrt{8}}{3} \\ 0 & 0 & 0 & -\frac{\sqrt{8}}{3} & \frac{1}{3} \end{bmatrix}, \tag{10.30}$$

and

$$\Gamma(45) = \begin{bmatrix} -\frac{1}{2} & -\frac{\sqrt{3}}{2} & 0 & 0 & 0 \\ -\frac{\sqrt{3}}{2} & \frac{1}{2} & 0 & 0 & 0 \\ 0 & 0 & -\frac{1}{2} & -\frac{\sqrt{3}}{2} & 0 \\ 0 & 0 & -\frac{\sqrt{3}}{2} & \frac{1}{2} & 0 \\ 0 & 0 & 0 & 0 & -1 \end{bmatrix}. \tag{10.31}$$

The rest of the representation matrices are obtained by simply invoking the combination law of the group.

Example 10.4 Find all the matrices of the irrep [2,1] of S_3. The standard Young tableaux are

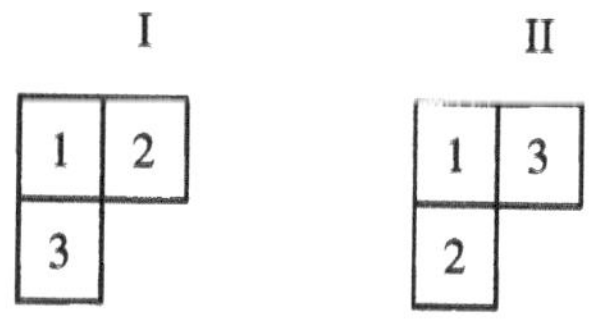

and the matrices are

$$\Gamma(12) = \begin{bmatrix} 1 & 0 \\ 0 & -1 \end{bmatrix}, \qquad \Gamma(23) = \begin{bmatrix} -\frac{1}{2} & -\frac{\sqrt{3}}{2} \\ -\frac{\sqrt{3}}{2} & \frac{1}{2} \end{bmatrix} \tag{10.32}$$

Then it follows readily from the combination law that

$$\Gamma(13) = \begin{bmatrix} -\frac{1}{2} & \sqrt{\frac{3}{2}} \\ \sqrt{\frac{3}{2}} & \frac{1}{2} \end{bmatrix} = \Gamma(23)\Gamma(12)\Gamma(23), \tag{10.33}$$

$$\Gamma(321) = \begin{bmatrix} -\frac{1}{2} & \sqrt{\frac{3}{2}} \\ -\sqrt{\frac{3}{2}} & -\frac{1}{2} \end{bmatrix} = \Gamma(23)\Gamma(12), \tag{10.34}$$

$$\Gamma(123) = \Gamma^{-1}(321) = \Gamma^{T}(321) = \begin{bmatrix} -\frac{1}{2} & -\sqrt{\frac{3}{2}} \\ \sqrt{\frac{3}{2}} & -\frac{1}{2} \end{bmatrix} \tag{10.35}$$

and, of course

$$\Gamma(e) = \begin{bmatrix} 1 & 0 \\ 0 & 1 \end{bmatrix}. \tag{10.36}$$

10.3.1 The Conjugate Representation

An irrep $[\lambda]$ of S_n is associated with another representation $[\tilde{\lambda}]$, the matrices of which are obtained from those of $[\lambda]$ by simply multiplying each representation matrix by the number $(-1)^p$, where p is the parity of the permutation, specifically, $p = 0$ for representation matrices of even permutations and $p = 1$ for those of odd permutations. The two representations $[\lambda]$ and $[\tilde{\lambda}]$ are said to be conjugate. The totally symmetric irrep $[n]$ and the antisymmetric irrep $[1^n]$ are examples of mutually conjugate representations. The graph of $[\tilde{\lambda}]$ is obtained from that of $[\lambda]$ by interchanging rows and columns, such that if, say

$$\begin{array}{lll} \bullet & \bullet & \bullet \\ \bullet & \bullet & \\ \bullet & \bullet & \\ \bullet & & \end{array} = |\lambda|, \tag{10.37}$$

then

$$\begin{array}{llll} \bullet & \bullet & \bullet & \bullet \\ \bullet & \bullet & \bullet & \\ \bullet & & & \end{array} = |\tilde{\lambda}|. \tag{10.38}$$

Conjugate representations have the same dimension, since their graphs contain the same hook lengths. Representations that have graphs that are invariant to this type of conjugation are said to be self-conjugate.

11

GROUPS OF CONSTANTS OF MOTION

The many-electron time-independent Schrödinger equation

$$H\Psi = E\Psi \tag{11.1}$$

can be solved approximately by invoking the variational principle in a basis. This means that the average value of the many-electron Hamiltonian

$$(\Phi|H\Phi)/(\Phi|\Phi) = E \tag{11.2}$$

is formed for some trial function

$$\Phi = \sum_{k=1}^{M} \Phi_k C_k , \tag{11.3}$$

where $\{\Phi_k\}$ is a basis set of many-electron functions. Such a basis can, for instance, be constructed as all possible determinantal wavefunctions in terms of a set of spin orbitals $\{\phi_l\}$. The resulting secular equations can be expressed in matrix form as

$$(\mathbf{H} - \mathbf{\Delta}E)\mathbf{C} = \mathbf{0}, \tag{11.4}$$

where the matrix element are $H_{kl} = (\Phi_k|H\Phi_l) = H_{lk}^*$ and $\Delta_{kl} = (\Phi_k|\Phi_l) = \Delta_{lk}^*$. The energy roots of the secular problem are found from the secular determinant

$$\det\{\mathbf{H} - \mathbf{\Delta}E\} = 0. \tag{11.5}$$

Similarly, the "best" (one-electron) orbitals to use in forming the determinantal basis set can be found as approximate eigenfunctions of a Fock operator F and the corresponding secular equations

$$(\mathbf{F} - \boldsymbol{\Delta}\epsilon)\mathbf{c} = \mathbf{0}. \tag{11.6}$$

We proceed by using the notation $\{\Phi_k\}$ for the basis in either the many-electron or the one-electron problem. The metric matrix is positive, that is, its eigenvalues $d_i \geq 0$ and the vanishing eigenvalues occur only in the case of linear dependencies among the M basis functions. This means that

$$\det \boldsymbol{\Delta} = \prod_{i=1}^{M} d_i \geq 0 \tag{11.7}$$

and is equal to 0 only for the case of linear dependencies. We consider the $\det \boldsymbol{\Delta}$ for some special cases:

- $M = 1$:
$$(\Phi_1|\Phi_1) \geq 0 \tag{11.8}$$
equality only for the case $\Phi_1 = 0$.
- $M = 1$:
$$\begin{vmatrix} (\Phi_1|\Phi_1) & (\Phi_1|\Phi_2) \\ (\Phi_2|\Phi_1) & (\Phi_2|\Phi_2) \end{vmatrix} \geq 0 \tag{11.9}$$
equality for $\Phi_2 = c\Phi_1$.

In general, there would be $n \leq M$ linearly independent functions $\{\Phi_k\}$ and the sums over all possible determinants for a given number of basis functions, such that

$$D_1 = \sum_{k=1}^{M} (\Phi_k|\Phi_k) \geq 0,$$

$$D_2 = \sum_{k=1}^{M}\sum_{l>k}^{M} \begin{vmatrix} (\Phi_k|\Phi_k) & (\Phi_k|\Phi_l) \\ (\Phi_l|\Phi_k) & (\Phi_l|\Phi_l) \end{vmatrix} \geq 0$$

$$\vdots \tag{11.10}$$

$$D_M = \begin{vmatrix} (\Phi_1|\Phi_1) & (\Phi_1|\Phi_2) & \cdots & (\Phi_1|\Phi_M) \\ (\Phi_2|\Phi_1) & (\Phi_2|\Phi_2) & \cdots & (\Phi_2|\Phi_M) \\ \vdots & \vdots & \cdots & \vdots \\ (\Phi_M|\Phi_1) & (\Phi_M|\Phi_2) & \cdots & (\Phi_M|\Phi_M) \end{vmatrix} \geq 0.$$

One would then find that $D_i > 0$ for $i = 1, 2, \ldots, n$ and $D_i = 0$ for $i = n+1, n+2, \ldots, M$.

11.1 GROUP ELEMENTS ACTING ON A FUNCTION

Consider a many-electron basis function $\Phi(X)$, with the coordinates in configuration space given as

$$X = (x_1, x_2, \ldots, x_N), \tag{11.11}$$

and let $g \in G$ be a coordinate transformation such that

$$g^{-1}X = X'. \tag{11.12}$$

Example 11.1 For $N = 2$, consider the operation of counterclockwise rotation of the coordinate axes by ϕ radians, which can be expressed as shows in Figure 11.1.

$$g^{-1}(x_1 \quad x_2) = (x_1' \quad x_2') = (x_1 \quad x_2)\begin{pmatrix} \cos\phi & -\sin\phi \\ \sin\phi & \cos\phi \end{pmatrix}. \tag{11.13}$$

The operation of a transformation (group operation) on a function is defined as

$$g\Phi(X) = \Phi(g^{-1}X). \tag{11.14}$$

This can be interpreted as follows. If $g \in G$ is a rotation about a symmetry axis, then g that operates on Φ is a rotation counterclockwise by ϕ radians of the contours of constant function value, while g^{-1} operating on X is a rotation of the coordinate axes clockwise by the same amount. This can be illustrated by the following example.

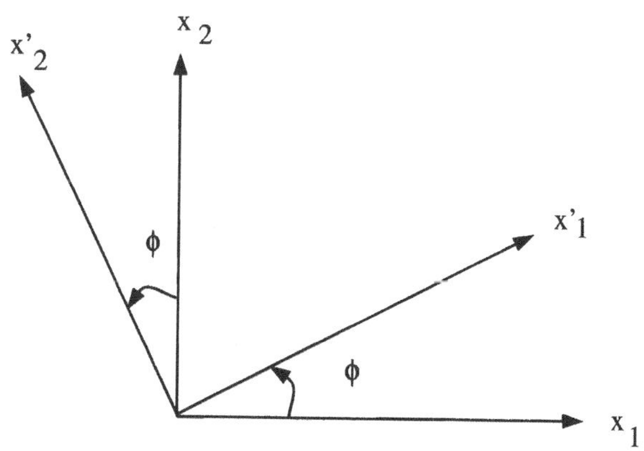

Figure 11.1 The rotation of two coordinate axes in their plane by an angle of ϕ radians.

Example 11.2 Let $N = 2$ and g be a rotation counterclockwise by an angle of $\pi/2$ radians. We choose the function

$$\Phi(X) = ax_1^2 + bx_2^2, \tag{11.15}$$

making the contours of constant value ellipses. This can be depicted as in Figure 11.2, illustrating that

$$g\Phi(X) = \Phi(g^{-1}X). \tag{11.16}$$

In particular, we can see how the point (P, Q) becomes the point $(-Q, P)$ under either operation.

This meaning of operation on a function means that

$$(gh)\Phi(X) = g\Phi(h^{-1}X) = \Phi(h^{-1}g^{-1}X). \tag{11.17}$$

In quantum mechanics all symmetry operators are unitary (the only exception is the time-reversal operator, which is antiunitary). For a unitary coordinate transformation $\mathrm{d}X = \mathrm{d}X'$ (the Jacobian is unity):

$$\begin{aligned}(g\Phi_1|g\Phi_2) = (\Phi_1|g^{\dagger}g\Phi_2) &= \int \Phi_1^*(g^{-1}X)\Phi_1(g^{-1}X)\,dX \\ &= \int \Phi_1^*(X')\Phi_1(X')\,\mathrm{d}X' = (\Phi_1|\Phi_2),\end{aligned} \tag{11.18}$$

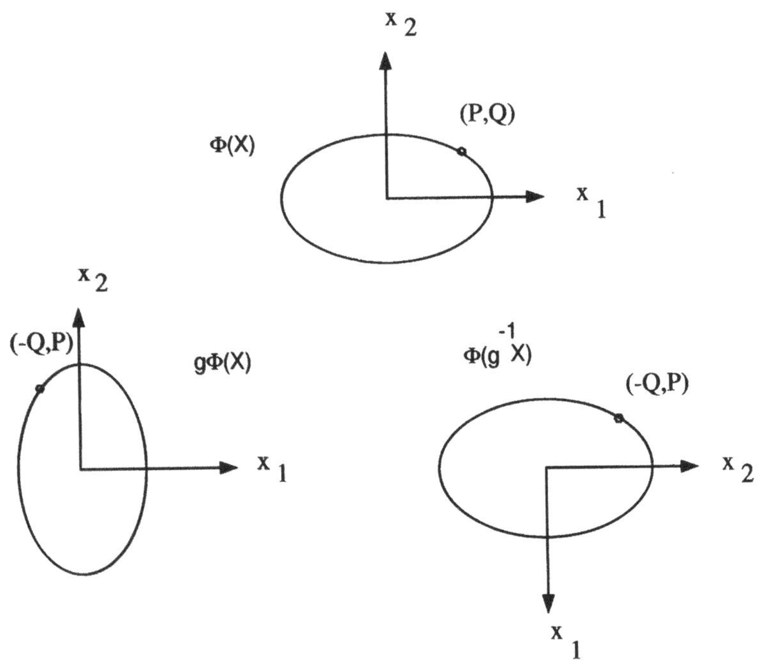

Figure 11.2 The action of a group element on a function and on the coordinate axes.

which means that

$$g^{\dagger}g = e \tag{11.19}$$

or that

$$g^{\dagger} = g^{-1}; \tag{11.20}$$

therefore, g is unitary.

When a set of operators G, combining under some multiplication rule satisfying the associative law, commutes with a Hamiltonian, the set forms a group. This is so since if

$$gH = Hg; \qquad hH = Hh; \quad \text{for} \quad g, h \in G \tag{11.21}$$

then

$$g^{-1}H = Hg^{-1}; \qquad eH = He; \qquad ghH = Hgh \tag{11.22}$$

such that the operators satisfy all four group postulates. Such a group is called a group of *constants of motion.*

Commuting operators have common eigenfunctions, so if a set of operators commute with each other and with the Hamiltonian, they characterize the exact eigenfunctions of the Hamiltonian. Even if only approximate eigenfunctions of the Hamiltonian can be found, it might be possible to make them exact eigenfunctions of a set of commuting symmetry operators, thus making the approximate solutions to the Schrödinger equation obtain some of the important characteristics of the unattainable exact solutions. For an Abelian symmetry group the elements themselves $\{g\}$ can be used for such a set of commuting operators. For a general group the character projectors $\{Q^{\alpha}\}$ can be used since they span the center of the group algebra and obviously also commute with the Hamiltonian. There are other possible choices as well as the interesting problem of finding the maximal set of constants of the motion.

11.2 INTEGRALS AND SYMMETRY

Symmetry properties of the integrand can be used to decide whether an integral is zero because of symmetry. A simple example is an intgral

$$\int_{-a}^{a} f(x)\, dx \tag{11.23}$$

in one dimension, which can be shown to be zero if the integrand is odd under the operation $x \to -x$:

$$f(-x) = -f(x)\,. \tag{11.24}$$

Using the results we have obtained for groups and their representations, we can look for a general statement about when an integral vanishes.

Consider an integral

$$I = \int I(X)\, dX \tag{11.25}$$

over all the configuration space of a physical system with a symmetry group G. Multiplying the integrand with the identity $e = \sum_\alpha \sum_{k=1}^{f^\alpha} P_{kk}^\alpha$ yields

$$I = \sum_\alpha \int \sum_{k=1}^{f^\alpha} P_{kk}^\alpha I(X)\, dX = \sum_\alpha \int I^\alpha(X)\, dX\,, \tag{11.26}$$

where

$$\begin{aligned} I^\alpha(X) &= \sum_{k=1}^{f^\alpha} f^\alpha \mathrm{M}_g \Gamma_{kk}^\alpha(g) I(gX) \\ &= f^\alpha \mathrm{M}_g \chi^\alpha(g) I(X')\,. \end{aligned} \tag{11.27}$$

The orthogonality relation for characters of irreps states that

$$(\chi^\alpha * \chi^\beta)_e = \mathrm{M}_g \chi^\alpha(g) \chi^\beta(g^{-1}) = \delta^{\alpha\beta} \tag{11.28}$$

so if, in particular $\beta = J$, the identity representation, or totally symmetric irrep, the character of which has the value 1 at each conjugate class, we can write

$$(\chi^\alpha * \chi^J) = \mathrm{M}_g \chi^\alpha(g) = \delta^{\alpha J}. \tag{11.29}$$

Thus

$$I = \int I(X) \mathrm{d}X \neq 0 \tag{11.30}$$

if and only if the integrand $I(X)$ has at least one component that transforms according to the totally symmetric irrep.

In general, if $\{\phi_i^\alpha\}$ transforms as the irrep Γ^α, $\{\varphi_j^\beta\}$ as Γ^β, and $\{\psi_k^\gamma\}$ as Γ^γ, then

$$\begin{aligned} (g,h)\phi_i^\alpha \varphi_j^\beta = g\phi_i^\alpha h \varphi_j^\beta &= \sum_{l=1}^{f^\alpha} \sum_{m=1}^{f^\beta} \phi_l^\alpha \varphi_m^\beta \Gamma_{li}^\alpha(g) \Gamma_{mj}^\beta(h) \\ &= \sum_{l,m} \phi_l^\alpha \varphi_m^\beta [\boldsymbol{\Gamma}^\alpha(g) \otimes \boldsymbol{\Gamma}^\beta(h)]_{lm;ij}, \end{aligned}$$

$$(g,h,s)\phi_i^\alpha \varphi_j^\beta \psi_k^\gamma = \sum_{l,m,n} \phi_l^\alpha \varphi_m^\beta \psi_n^\gamma [\boldsymbol{\Gamma}^\alpha(g) \otimes \boldsymbol{\Gamma}^\beta(h) \otimes \boldsymbol{\Gamma}^\gamma(s)]_{lmn;ijk}\,. \tag{11.31}$$

In this way the direct products of representations become important.

In spectroscopy the intensity of spectral lines is related to the magnitude of transition moment integrals, such as

$$(\Phi^{\alpha}|\hat{A}\Psi^{\beta}) = \int \Phi^{\alpha *}(X)\hat{A}\Psi^{\beta}(X)\,dX \tag{11.32}$$

where the operator $\hat{A}$ would be, for instance, the electric dipole moment operator of the molecule in question in the case of ordinary absorption spectroscopy. The electric dipole moment operator is simply the sum of the position vectors of the electrons and of the nuclei, each multiplied by the charge of the particle. This quantity transforms as the Cartesian coordinates x, y, z. For Raman spectroscopy the transition moment integral involves the components of the tensor

$$\begin{pmatrix} 3x^2 - r^2 & xy & xz \\ yx & 3y^2 - r^2 & yz \\ zx & zy & 3z^2 - r^2 \end{pmatrix} \tag{11.33}$$

This is one of the reasons why the point group character tables include information on the transformation properties of these quantities. Whether the transition moment integral vanishes and the transition is *forbidden* or is nonzero, making the transition *allowed* depends on the symmetry properties of the integrand, which transforms as the direct product representation

$$\Gamma^{\alpha} \otimes \Gamma^{A} \otimes \Gamma^{\beta}. \tag{11.34}$$

For the transition to be allowed, it must hold that

$$\Gamma^{J} \in \Gamma^{\alpha} \otimes \Gamma^{A} \otimes \Gamma^{\beta}, \tag{11.35}$$

that is, the identity irrep must be contained in the direct product representation.

In particular, in absorption spectra (excluding so-called hot bands) the initial state of the molecule is the ground state (electronic and vibrational), which often is totally symmetric. Say that $\Gamma^{\alpha} = \Gamma^{J}$; then the character of the direct product representation must satisfy

$$\begin{aligned} \mathrm{M}_g \chi^{\alpha}(g)\chi^{A}(g)\chi^{\beta}(g) &= \mathrm{M}_g \chi^{A}(g)\chi^{\beta}(g) \\ &= \mathrm{M}_g \chi^{A}(g)\chi^{\beta *}(g^{-1}) \\ &= \mathrm{M}_g \chi^{A}(g)\chi^{\beta}(g^{-1}) = \delta^{A\beta}, \end{aligned} \tag{11.36}$$

where in the last step a real character has been assumed. There are point groups for which the characters of irreps are complex as, for instance, Abelian groups such as some of the C_{nh} that have one-dimensional irreps. For such groups it is often common (see the character tables) to combine two irreps, whose characters are each other's complex conjugate, to form a reducible two-dimensional representation with a real character. It is then possible to state as a general truth that in absorption from a totally symmetric state, the final state symmetry has to match that of the operator.

11.3 SYMMETRY AND THE SECULAR PROBLEM

The construction of the gerade and ungerade sigma and pi molecular orbitals for the first-row homonuclear diatomics was based on symmetry considerations without invoking the full power of group theory. The construction of these orbitals is an example of the construction of *symmetry-adapted* functions. That is an important problem in the study of molecular systems. In this section we study this problem from the point of view of group algebra and, in particular, the consequences for the secular equations.

An element

$$P^{\alpha}_{km} = \mathrm{M}_g f^{\alpha} \Gamma^{\alpha}_{km}(g) g^{-1} \tag{11.37}$$

of the group algebra of constants of the motion acting on a given trial function $\Phi(X)$ yields[1]

$$P^{\alpha}_{km}\Phi(X) = \Phi^{\alpha}_{km} = \mathrm{M}_g f^{\alpha}\Gamma^{\alpha}_{km}(g)\Phi(gX)\,. \tag{11.38}$$

For a given $\Phi(X)$ not all functions $\{\Phi^{\alpha}_{km}\}$ are nonzero. A symmetry-adapted function Φ^{α}_{kk} belonging to a particular irrep Γ^{α} and a particular row k of that irrep can be obtained by applying the projector

$$P^{\alpha}_{kk}\Phi(X) = \Phi^{\alpha}_{kk}\,. \tag{11.39}$$

Two possibilities can occur. Either $\Phi^{\alpha}_{kk} = 0$, and then another trial function $\Phi(X)$ must be chosen, or $\Phi^{\alpha}_{kk} \neq 0$ when a set of f^{α} orthogonal (and thereby linearly independent) functions can be generated by

$$P^{\alpha}_{km}\Phi^{\alpha}_{kk} = P^{\alpha}_{km} \times P^{\alpha}_{kk}\Phi = P^{\alpha}_{km}\Phi = \Phi^{\alpha}_{km} \tag{11.40}$$

for $m = 1, 2, \ldots, f^{\alpha}$. Since the constants of the motion are all unitary and since the representation matrices can be assumed to be unitary, one can write

$$P^{\alpha\dagger}_{km} = \mathrm{M}_g f^{\alpha}\Gamma^{\alpha *}_{km}(g)(g^{-1})^{\dagger} = \mathrm{M}_g f^{\alpha}\Gamma^{\alpha}_{mk}(g^{-1})g = P^{\alpha}_{mk} \tag{11.41}$$

for the adjoint of a basis element of the linear space of the group algebra. It is interesting to note that

$$\begin{aligned}(\Phi^{\alpha}_{km}|\Phi^{\alpha}_{km}) &= (\Phi|P^{\alpha}_{mk} \times P^{\alpha}_{km}\Phi) = (\Phi|P^{\alpha}_{kk}\Phi) \\ &= (\Phi|P^{\alpha}_{kk} \times P^{\alpha}_{kk}\Phi) = (\Phi^{\alpha}_{kk}|\Phi^{\alpha}_{kk})\,,\end{aligned} \tag{11.42}$$

---

[1] Remember the fundamental relation $P^{\alpha}_{km} \times P^{\beta}_{nl} = \delta^{\alpha\beta}\delta_{kl}P^{\alpha}_{nm}$.

so all these functions for $m = 1, 2, \ldots, f^\alpha$ have the same norm. So if

$$(\Phi^\alpha_{kk}|\Phi^\alpha_{kk}) \neq 0, \tag{11.43}$$

all $\Phi^\alpha_{km} \neq 0$. At the same time all these functions are orthogonal, since

$$(\Phi^\alpha_{km}|\Phi^u_{kn}) = (\Phi|P^\alpha_{mk} \times P^\alpha_{kn}\Phi) = \delta_{mn}(\Phi|\Phi^\alpha_{kk}). \tag{11.44}$$

Thus, functions belonging to different columns of the same irrep are orthogonal, and given one nonzero function of a particular row of an irrep, one can generate f^α orthogonal functions in that row. The f^α functions of a particular row space of an irrep are sufficient to generate the entire irrep, since

$$\begin{aligned} g\Phi^u_{km} &= g \times P^\alpha_{km}\Phi = \mathrm{M}_s f^\alpha \Gamma^\alpha_{km}(s) g s^{-1}\Phi \\ &= \mathrm{M}_u f^\alpha \Gamma^\alpha_{km}(ug) u^{-1}\Phi = \sum_{l=1}^{f^\alpha} \mathrm{M}_u f^\alpha \Gamma^\alpha_{kl}(u) u^{-1}\Phi \Gamma^\alpha_{lm}(g) \\ &= \sum_{l=1}^{f^\alpha} \Phi^\alpha_{kl}\Gamma^\alpha_{lm}(g). \end{aligned} \tag{11.45}$$

Symmetry-adapted functions obtained from appropriate trial functions when used as a basis will lead to considerably simplified secular equations. In a secular problem one needs metric matrix elements, which would be

$$(\Phi^\alpha_{km}|\Psi^\beta_{nl}) \tag{11.46}$$

between two symmetry-adapted basis functions obtained from (in general) different trial functions and for different irreps. Because of the basic properties of the basis elements of the group algebra, we find that

$$(\Phi^\alpha_{km}|\Psi^\beta_{nl}) = (\Phi|P^\alpha_{mk} \times P^\beta_{nl}\Psi) = \delta^{\alpha\beta}\delta_{ml}(\Phi|\Psi^\alpha_{nk}), \tag{11.47}$$

which means that symmetry-adapted functions are orthogonal when belonging to different irreps or different columns of the same irrep. The secular determinant also requires the Hamiltonian matrix elements

$$\begin{aligned} (\Phi^\alpha_{km}|H\Psi^\beta_{nl}) &= (\Phi|P^\alpha_{mk}H\Psi^\beta_{nl}) = (\Phi|HP^\alpha_{mk} \times P^\beta_{nl}\Psi) \\ &= \delta^{\alpha\beta}\delta_{ml}(\Phi|H\Psi^\alpha_{nk}), \end{aligned} \tag{11.48}$$

where we have used the fact we are considering a group of constants of the motion. It is clear from this that the matrix

$$\mathbf{H} - \mathbf{\Delta}E \tag{11.49}$$

of the secular problem will be block diagonal. One should also note that the nonvanishing integrals do not depend on the column index of the basis functions. It should also be clear from the previous discussion of symmetry-adapted basis functions that if there are, say, n nonzero, linearly independent functions $\{\Phi^{\alpha}_{kk}\}$ $(k = 1, 2, \ldots, n)$ for a particular irrep α, then there are a total of nf^{α} functions $(0 \leq n \leq f^{\alpha})$ linearly independent basis functions belonging to that irrep. The secular problem then blocks into independent secular problems for each irrep and a block of a particular irrep has the following structure (where we have assumed only one trial function Φ)

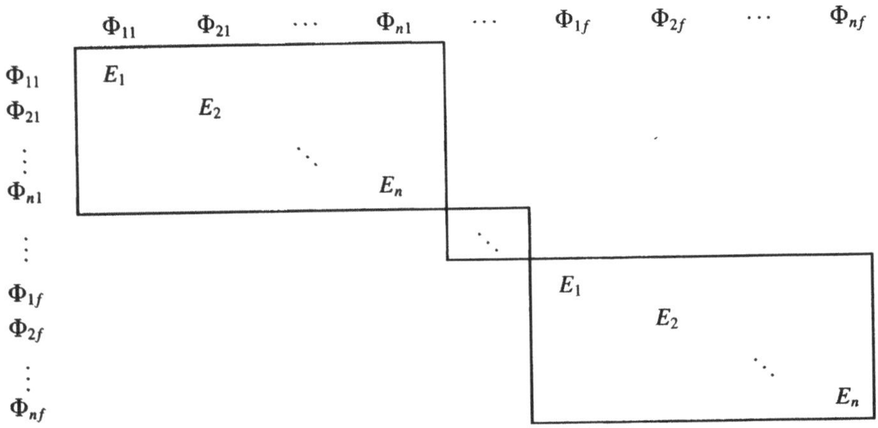

Thus, one obtains $f^{\alpha} = f$ identical blocks, and each root appears f^{α} times. In this way the degeneracy is completely specified and labeled by the irrep and the row and column indices of the irrep.

To determine the number n of linearly independent functions $\{\Phi^{\alpha}_{km}\}$ in a column space of an irrep obtainable from a single trial function [i.e., α and m fixed, and $k = 1, 2, \ldots, n$, $(0 \leq n \leq f^{\alpha})$], one can form sums over determinants such as

$$
\begin{aligned}
D_1 &= \sum_{k=1}^{f^{\alpha}} (\Phi^{\alpha}_{km} | \Phi^{\alpha}_{km}) = \sum_{k=1}^{f^{\alpha}} (\Phi | P^{\alpha}_{kk} \Phi) \\
&= f^{\alpha} \sum_{k=1}^{f^{\alpha}} \mathrm{M}_g \Gamma^{\alpha}_{kk}(g) (\Phi | g^{-1} \Phi) \\
&= f^{\alpha} \mathrm{M}_g \chi^{\alpha}(g) \Delta(g^{-1}) = f^{\alpha} (\chi^{\alpha} * \Delta)_e ,
\end{aligned}
\tag{11.50}
$$

where $\Delta(g) = (\Phi | g \Phi)$ is a function over the group determined by the trial function Φ. Taking the basis functions two at a time, we can form the sum of the determinants:

$$
\begin{aligned}
D_2 &= \sum_{k=1}^{f^\alpha}\sum_{l=1}^{f^\alpha}\frac{1}{2!}\begin{vmatrix}(\Phi^\alpha_{km}|\Phi^\alpha_{km}) & (\Phi^\alpha_{km}|\Phi^\alpha_{lm})\\ (\Phi^\alpha_{lm}|\Phi^\alpha_{km}) & (\Phi^\alpha_{lm}|\Phi^\alpha_{lm})\end{vmatrix}\\
&= \frac{(f^\alpha)^2}{2!}\sum_{k=1}^{f^\alpha}\sum_{l=1}^{f^\alpha}\mathrm{M}_g\mathrm{M}_h[\Gamma^\alpha_{kk}(g)\Gamma^\alpha_{ll}(h) - \Gamma^\alpha_{lk}(g)\Gamma^\alpha_{kl}(h)]\Delta(g^{-1})\Delta(h^{-1})\\
&= \frac{(f^\alpha)^2}{2!}\mathrm{M}_g\mathrm{M}_h[\chi^\alpha(g)\chi^\alpha(h) - \chi^\alpha(gh)]\Delta(g^{-1})\Delta(h^{-1})\\
&= \frac{(f^\alpha)^2}{2!}\mathrm{M}_g\mathrm{M}_h B^\alpha_2(g,h)\Delta(g^{-1})\Delta(h^{-1})\\
&= \frac{(f^\alpha)^2}{2!}[(\chi^\alpha * \Delta)^2_e - (\chi^\alpha * \Delta * \Delta)_e]. \qquad (11.51)
\end{aligned}
$$

It is rather straightforward to show that, in general, taking n functions at a time we get

$$D_n = \frac{(f^\alpha)^n}{n!}\mathrm{M}_{s_1}\mathrm{M}_{s_2}\cdots\mathrm{M}_{s_n}B^\alpha_n(s_1,s_2,\ldots,s_n)\Delta(s_1^{-1})\Delta(s_2^{-1})\cdots\Delta(s_n^{-1}), \qquad (11.52)$$

where

$$B^\alpha_n(s_1,s_2,\ldots,s_n) = \sum_{P\in S_n}(-1)^p\Pi^\alpha(P), \qquad (11.53)$$

specifically, a sum over all permutations P of the symmetric group of order $n!$ and with $\Pi^\alpha(P)$ a product of characters reflecting the cycle structure of P. For example, with $n = 5$ and $P = (1)(2)(354) \in S_5$, one finds

$$\Pi^\alpha(P) = \chi^\alpha(s_1)\chi^\alpha(s_2)\chi^\alpha(s_3s_5s_4). \qquad (11.54)$$

It is interesting, but not surprising, how the determinants introduce the permutations of the symmetric group. If there are exactly n linearly independent functions in a single column obtained from a particular trial function Φ, then $D_i > 0$ for $i \le n$, while $D_i = 0$ for $i \ge n + 1$, so the evaluation of these sums of determinants can, in principle, tell us how many linearly independent functions can be obtained from each trial function.

If in addition to the function $\Delta(g) = (\Phi|g\Phi)$ we calculate the function

$$H(g) = (\Phi|(H - E)g\Phi) \qquad (11.55)$$

as a function over the group for all $g \in G$, then the secular determinant for the block corresponding to irrep Γ^α can be expressed in an analogous way:

$$D_n(H) = \frac{(f^\alpha)^n}{n!} \mathrm{M}_{s_1} \mathrm{M}_{s_2} \cdots \mathrm{M}_{s_n} B_n^\alpha(s_1, s_2, \ldots, s_n) H(s_1^{-1}) H(s_2^{-1}) \cdots H(s_n^{-1}). \tag{11.56}$$

This shows explicitly how the secular problem and thus the energies of the system under consideration depend on the characters of the irreps and not on the details of the representation matrices. This is not too surprising since the characters are the invariants, but the explicit dependence is seldom discussed. Whether this formulation can be used with any advantage in actual applications is another matter.

The secular problem discussed above shows that the secular problem separates into individual problems for each irrep Γ^α and that since the functions belonging to different columns are orthogonal and noninteracting across the Hamiltonian, there are also f^α blocks of the secular determinant for a given irrep. Furthermore, since the nonvanishing integrals are independent of the column index, these f^α blocks are identical and of dimension $n \times n$ when there are n linearly independent functions belonging to a column space of that irrep. The end result is that one gets n energy roots labeled by the row indices, where each is degenerate of order f^α, and the substates of each degenerate level are completely labeled and characterized by the column indices.

Two energies belonging to different irreps or to different rows of the same irrep could be identical. Such a degeneracy is then not classified by the group structure of the chosen constants of the motion and is usually referred to as an "accidental degeneracy." The prevailing view of such degeneracies is that there is nothing accidental about them. One has simply not found the largest possible group of constants of the motion, but only a subgroup. Such statements are difficult to prove and one must usually proceed by trial and error to find additional symmetry operations that commute with the Hamiltonian.

Example 11.3 A famous example is the electronic states of the hydrogen atom Hamiltonian $-\frac{1}{2}\vec{\nabla}^2 - (1/r)$ and the obvious continuous symmetry group $SO(3)$ (the group of all rotations and rotation reflections by an arbitrary angle between 0 and 2π radians about any of the infinitely many symmetry axes of a sphere). This group has infinitely many irreps labeled $s, p, d, \ldots, l, \ldots$ and so on of dimensions $1, 3, 5, \ldots, (2l + 1), \ldots$. However, as we have seen, for the electron of the hydrogen atom the energies depend only on the principal quantum number n, such that, say, the $2s$ state and the three $2p$ states have the same energy as do the $3s$, the three $3p$, and the five $3d$ states, and so on. These accidental degeneracies that involve states belonging to different irreps of $SO(3)$ can be resolved by introducing the full group of all constants of the motion, namely, $SO(4)$ of which $SO(3)$ is a subgroup. $SO(4)$ contains additional symmetry operations that commute with the hydrogen atom Hamiltonian, but are less obvious, and the irreps of $SO(4)$ contain those of $SO(3)$ such that

s, $(s + p)$, $(s + p + d)$, and so on of dimensions $1, 4, 9, \ldots, n^2, \ldots$, which fully classify the hydrogen atom nonrelativistic energy spectrum.

11.4 ELECTRON SPIN AND THE SYMMETRIC GROUP

The fact that each electron has two possible spin states (α and β) limits the spatial wavefunctions of n-electron systems to those that transform according to irreps of S_n with graphs having at most two columns.[2] The proof of this fact is not difficult, but is omitted from this discussion.

For two-electron systems we have already seen that the requirement that the total wavefunction be antisymmetric under interchange of electron labels leads to that either the spin functions are symmetric, such as

$$\begin{aligned} &\alpha(1)\alpha(2) \\ &\alpha(1)\beta(2) + \beta(1)\alpha(2) \\ &\beta(1)\beta(2) \end{aligned} \tag{11.57}$$

and combined with an antisymmetric spatial wavefunction, such as

$$u_1(\vec{r}_1)u_2(\vec{r}_2) - u_2(\vec{r}_1)u_1(\vec{r}_2) \tag{11.58}$$

for the spin triplet states, or they are antisymmetric, such as

$$\alpha(1)\beta(2) - \beta(1)\alpha(2) \tag{11.59}$$

combined with a symmetric spatial function as

$$u_1(\vec{r}_1)u_2(\vec{r}_2) + u_2(\vec{r}_1)u_1(\vec{r}_2) \tag{11.60}$$

for the singlet states. The character table of the symmetric group S_2 is easily constructed from the rules discussed before as

S_2	(1^2)	(2)
[2]	1	1
$[1^2]$	1	-1

[2] One way to see that the two column irreps apply to electrons standard application of entries in the graphs to form the standard Young tableaux should increase from right to left and from top to bottom and there are two different spin functions for each orbital. For nucleai with dynamically equivalent protons and neutrons, irreps with at most four columns apply. This is so since, in addition to the half-integer spin quantum number, there is the so-called isospin that distinguishes the two kinds of particles.

and the symmetric irrep has the graph

$$\bullet \quad \bullet\,, \tag{11.61}$$

while the antisymmetric one has the graph

$$\begin{matrix}\bullet \\ \bullet\end{matrix}\,, \tag{11.62}$$

and obviously the direct product representation of the symmetric and the antisymmetric one-dimensional irreps is the antisymmetric one, as it should be.

For many-electron systems of more than two electrons the total wavefunction should still be antisymmetric but now the spatial functions by themselves are no longer either symmetric or antisymmetric in general, but transform as other irreps of S_n with graphs of at most two columns. It can be shown that if the total wavefunction should be antisymmetric, that is, transform according to the irrep $[1^n]$, the spin functions of the n-electron system must transform according to the conjugate representation of that of the spatial functions. This means that the graphs of the irreps of the n-electron spin functions must have two rows.

The connection between the graphs of the irreps and the total spin quantum number S of an n-electron system is such that the spatial functions must transform according to the irrep

$$[\lambda] = [2^{n/2-S},\ 1^{2S}], \tag{11.63}$$

while the spin functions span the conjugate irrep

$$[\tilde{\lambda}] = \left[\frac{n}{2} + S, \frac{n}{2} - S\right] \tag{11.64}$$

of S_n.

One possible construction of a totally antisymmetric wavefunction corresponding to a particular total spin quantum number S for a many electron wavefunction can (according to E. P. Wigner) be done as follows. A set of n-electron spatial functions $\{\Phi_k\}$ spanning the irrep $[\lambda]$ is constructed, as is a set of n-electron spin functions $\{\Theta_k\}$ spanning the conjugate irrep $[\tilde{\lambda}]$, such that

$$\begin{aligned} s\Phi_k &= \sum_l \Phi_l \Gamma_{lk}^{[\lambda]}(s)\,, \\ s\Theta_k &= \sum_m \Theta_m \Gamma_{mk}^{[\tilde{\lambda}]}(s)\,, \end{aligned} \tag{11.65}$$

for any permutation $s \in S_n$. Since the dimension of an irrep and its conjugate is the same, one can construct the wavefunction

$$\Psi = \sum_{k=1}^{f} \Phi_k \Theta_k \tag{11.66}$$

and find it to be antisymmetric. This follows from

$$
\begin{aligned}
(s,s)\Psi &= \sum_{k=1}^{f}\sum_{m=1}^{f}\sum_{l=1}^{f}\Phi_l\Theta_m\Gamma_{lk}^{[\lambda]}(s)\Gamma_{mk}^{[\tilde{\lambda}]}(s) \\
&= (-1)^s\sum_{k=1}^{f}\sum_{m=1}^{f}\sum_{l=1}^{f}\Phi_l\Theta_m\Gamma_{lk}^{[\lambda]}(s)\Gamma_{mk}^{[\lambda]}(s) \\
&= (-1)^s\sum_{m=1}^{f}\sum_{l=1}^{f}\Phi_l\Theta_m\sum_{k=1}^{f}\Gamma_{lk}^{[\lambda]}(s)\Gamma_{km}^{[\lambda]}(s^{-1}) \\
&= (-1)^s\sum_{l=1}^{f}\Phi_l\Theta_l = (-1)^s\Psi\,,
\end{aligned}
\tag{11.67}
$$

where we have used the inner direct product group with elements (s, s), which is isomorphic to S_n, and where the definition of a conjugate representation and the fact that the irreps are orthogonal have been used. Compare this general result to the special case of two-electron states, where one-dimensional irreps lead to a single product only in Eq. (11.66)!

In order to quickly determine which irrep to use for a particular number of electrons n and a given spin quantum number S, one can simply construct the graphs with at most two columns and simply place up-spins or α spin functions in the first column and β spin functions in the second column. The excess number of α spins times $\frac{1}{2}$ then yields the total spin quantum number of a particular graph.

Example 11.4 Consider $n = 6$ and the possible total spins

$S = 0$

α	β
α	β
α	β

$S = 1$

α	β
α	β
α	
α	

$S = 2$

α	β
α	
α	
α	
α	

$S = 3$

α
α
α
α
α
α

11.5 ORBITAL SYMMETRY

The symmetry group of homonuclear diatomics is $D_{\infty h}$ and for heteronuclear diatomics is $C_{\infty v}$. These are continuous groups with irreps Σ (one-dimensional), $\Pi, \Delta, \Phi, \ldots$ (two-dimensional). As one can see from the character tables the

Σ irreps have a superscript + or − that designate the behavior under, say a σ_v reflection. For instance, the two-electron determinants discussed for the oxygen molecule, that is, the $^3\Sigma_g$ state

$$|1\pi_{g+}(1)\alpha(1)1\pi_{g-}(2)\alpha(2)| \tag{11.68}$$

has a factor

$$e^{i\varphi_1}e^{-i\varphi_2} - e^{-i\varphi_1}e^{i\varphi_2}. \tag{11.69}$$

Under reflection in the xz plane $\varphi_j \to -\varphi_j$ and this factor changes sign, justifying the label $^3\Sigma_g^-$. Similarly, the $^1\Sigma_g$ state

$$|1\pi_{g+}(1)\alpha(1)1\pi_{g-}(2)\beta(2)| - |1\pi_{g+}(1)\beta(1)1\pi_{g-}(2)\alpha(2)| \tag{11.70}$$

has the factor

$$e^{i\varphi_1}e^{-i\varphi_2} + e^{-i\varphi_1}e^{i\varphi_2}, \tag{11.71}$$

justifying the symbol $^1\Sigma_g^+$. Diatomic molecular orbitals of σ type can have only + symmetry, making the use of this symbol for MOs superfluous.

We can use a procedure similar to that done for diatomics to obtain approximate MO energy-level diagrams for polyatomic molecules that exhibit some symmetry. Many polyatomic molecules exhibit point group symmetry in the sense that a rotation about a symmetry axis or reflection in a symmetry plane simply involves exchange of identical nuclei so that the nuclear framework of the molecule looks the same.

Example 11.5 The water molecule in its ground-state equilibrium geometry has two equal OH bonds and a bond angle of 104° (see Fig. 11.3). We place the molecule in the xz plane and use a *minimal basis* of AOs (atomic orbitals) consisting of $1s, 2s, 2p_x, 2p_y, 2p_z$ on oxygen and $1s$ on hydrogen. Obviously, this molecule transforms as the C_{2v} point group. The character table with the *symmetry-adapted* basis orbitals is

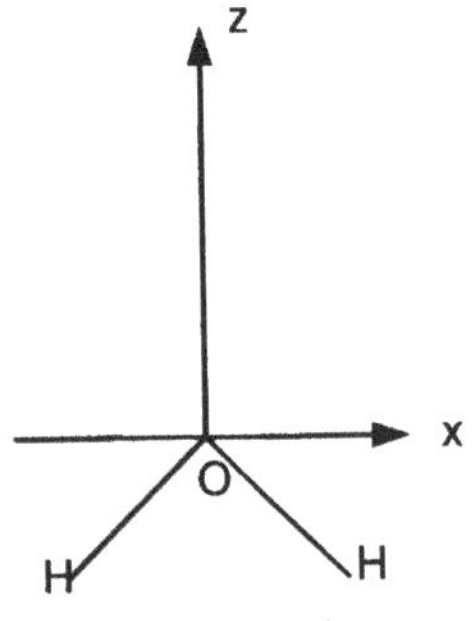

Figure 11.3 Coordinate system for the water molecule.

C_{2v}	E	C_2	$\sigma(xz)$	$\sigma(yz)$	Symmetry Orbitals
A_1	1	1	1	1	$1s, 2s, 2p_z, h_1 + h_2$
A_2	1	1	-1	-1	—
B_1	1	-1	1	-1	$2p_x, h_1 - h_2$
B_2	1	-1	-1	1	$2p_y$

where h_1, h_2 are the $1s$ orbitals on the two hydrogens. Using the atomic orbital energies for oxygen and hydrogen as before, we can construct the sketch shown in Figure 11.4.

The ground-state electron configuration is

$$(1a_1)^2(2a_1)^2(3a_1)^2(1b_1)^2(1b_2)^2, \tag{11.72}$$

and in this most approximate treatment the $1a_1$ and $2a_1$ MOs are simply the oxygen $1s$ and $2s$, respectively, while

$$3a_1 = c_1 2p_z + c_2(h_1 + h_2), \tag{11.73}$$

$$1b_1 = c_1 2p_x + c_2(h_1 - h_2), \tag{11.74}$$

and $1b_2 = 2p_y$.

The corresponding determinantal wave function is a closed-shell singlet, and the molecular term is

$$^1A_1 \,. \tag{11.75}$$

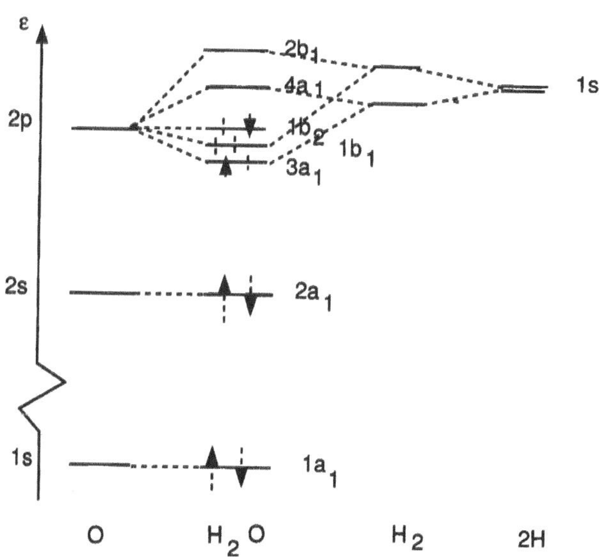

Figure 11.4 Sketched MO energy-level diagram for the water molecule.

Example 11.6 The ammonia molecule in its ground-state equilibrium configuration transforms as the C_{3v} point group (see Fig. 11.5).

The character table with the symmetry-adapted orbitals is

C_{3v}	E	C_3, C_3^2	$\sigma_1, \sigma_2, \sigma_3$	Symmetry Orbitals
A_1	1	1	1	$1s, 2s, 2p_z, h_1 + h_2 + h_3$
A_2	1	1	-1	—
E	2	-1	0	$(2p_x, 2p_y), (2h_1 - h_2 - h_3, h_2 - h_3)$

from a minimal basis of AOs consisting of $1s, 2s, 2p_x 2p_y, 2p_z$ on nitrogen, and $1s$ on the three hydrogens, which we denote by h_1, h_2, h_3, respectively.

The new feature is that the symmetry orbitals of type E transform in pairs, (e.g., $2p_x, 2p_y$), forming linear combinations among themselves (involving no other basis functions) under the geometric operations of the group. We can see how this comes about by representing these orbitals as the coordinate axis (x,y) (see Fig. 11.6).

From Fig. 11.6 above we can find the representation matrices, such as

$$C_3(x \quad y) = \left(-\frac{1}{2}x + \frac{\sqrt{3}}{2}y \quad -\frac{\sqrt{3}}{2}x - \frac{1}{2}y\right) = (x \quad y)\begin{bmatrix} -\frac{1}{2} & -\frac{\sqrt{3}}{2} \\ \frac{\sqrt{3}}{2} & -\frac{1}{2} \end{bmatrix} \tag{11.76}$$

and

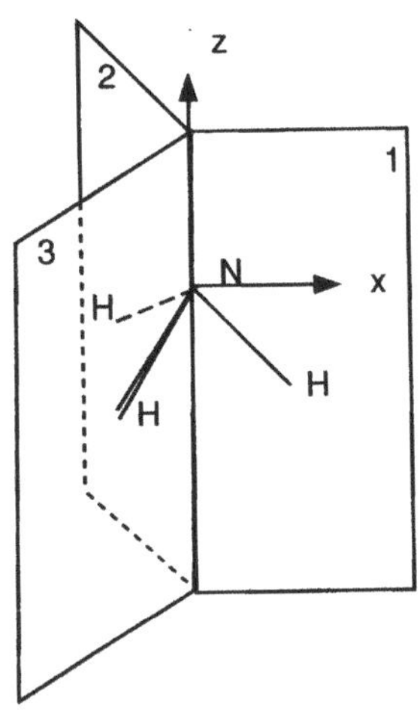

Figure 11.5 Symmetry planes and axis for the ammonia molecule in its ground-state equilibrium geometry.

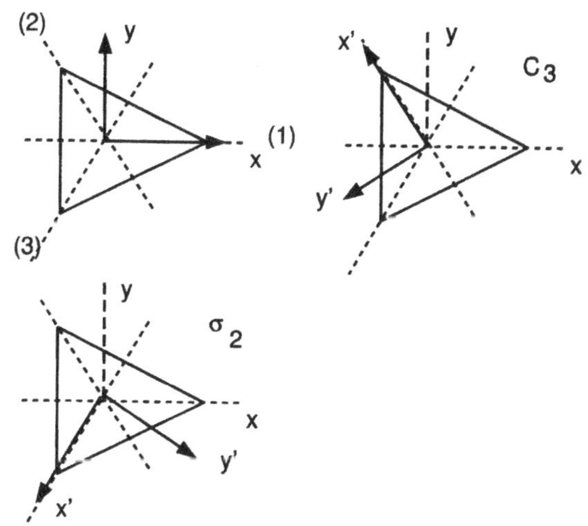

Figure 11.6 Symmetry operations on the p_x and p_y nitrogen atomic orbitals.

$$\sigma_2(x \quad y) = \left(-\frac{1}{2}x - \frac{\sqrt{3}}{2}y \quad -\frac{\sqrt{3}}{2}x + \frac{1}{2}y\right) = (x \quad y)\begin{bmatrix} -\frac{1}{2} & -\frac{\sqrt{3}}{2} \\ -\frac{\sqrt{3}}{2} & \frac{1}{2} \end{bmatrix}. \tag{11.77}$$

All the representation matrices are

$$\Gamma^E(E) = \begin{bmatrix} 1 & 0 \\ 0 & 1 \end{bmatrix}, \qquad \Gamma^E(C_3) = \begin{bmatrix} -\frac{1}{2} & -\frac{\sqrt{3}}{2} \\ \frac{\sqrt{3}}{2} & -\frac{1}{2} \end{bmatrix},$$
$$\Gamma^E(C_3^2) = \begin{bmatrix} -\frac{1}{2} & \frac{\sqrt{3}}{2} \\ -\frac{\sqrt{3}}{2} & -\frac{1}{2} \end{bmatrix}, \tag{11.78}$$

$$\Gamma^E(\sigma_1) = \begin{bmatrix} 1 & 0 \\ 0 & -1 \end{bmatrix}, \qquad \Gamma^E(\sigma_2) = \begin{bmatrix} -\frac{1}{2} & -\frac{\sqrt{3}}{2} \\ -\frac{\sqrt{3}}{2} & \frac{1}{2} \end{bmatrix},$$
$$\Gamma^E(\sigma_3) = \begin{bmatrix} -\frac{1}{2} & \frac{\sqrt{3}}{2} \\ \frac{\sqrt{3}}{2} & \frac{1}{2} \end{bmatrix}. \tag{11.79}$$

It is also obvious that the $1s$ orbitals on the hydrogens mix among themselves under the operations of the group, which means that they *span* a reducible representation with characters

C_{3v}	E	C_3, C_3^2	$\sigma_1, \sigma_2, \sigma_3$
Γ	3	0	1

The frequency numbers of the irreps in this reducible representation are

$$\begin{aligned} \nu^{A_1} &= \tfrac{1}{6}[1 \cdot 3 + 2 \cdot 1 \cdot 0 + 3 \cdot 1 \cdot 1] = 1\,, \\ \nu^{A_2} &= \tfrac{1}{6}[1 \cdot 3 + 2 \cdot 1 \cdot 0 + 3 \cdot (-1) \cdot 1] = 0\,, \\ \nu^{E} &= \tfrac{1}{6}[2 \cdot 3 + 2 \cdot (-1) \cdot 0 + 3 \cdot 0 \cdot 1] = 1\,. \end{aligned} \tag{11.80}$$

Thus, the three $1s$ orbitals on the hydrogens span the irreducible representations A_1 and E. Using the character projection operator

$$Q^{A_1} = \frac{1}{6}[1E + 1C_3 + 1C_3^2 + 1\sigma_1 + 1\sigma_2 + 1\sigma_3], \tag{11.81}$$

one obtains

$$\begin{aligned} Q^{A_1} h_1 &= \frac{1}{6}[h_1 + h_2 + h_3 + h_1 + h_3 + h_2] \\ &\propto [h_1 + h_2 + h_3]\,. \end{aligned} \tag{11.82}$$

The projector and step operator

$$\begin{aligned} P_{11}^{E} &= \frac{2}{6}\left[1E - \frac{1}{2}C_3 - \frac{1}{2}C_3^2 + 1\sigma_1 - \frac{1}{2}\sigma_2 - \frac{1}{2}\sigma_3\right], \\ P_{12}^{E} &= \frac{2}{6}\left[0E + \frac{\sqrt{3}}{2}C_3 - \frac{\sqrt{3}}{2}C_3^2 + 0\sigma_1 - \frac{\sqrt{3}}{2}\sigma_2 - \frac{\sqrt{3}}{2}\sigma_3\right] \end{aligned} \tag{11.83}$$

can be used to obtain the symmetry adapted orbitals for the E irrep, such that

$$\begin{aligned} P_{11}^{E} h_1 &= \frac{1}{3}[h_1 - \frac{1}{2}h_2 - \frac{1}{2}h_3 + h_1 - \frac{1}{2}h_3 - \frac{1}{2}h_2] \\ &\propto [2h_1 - h_2 - h_3], \end{aligned} \tag{11.84}$$

and

$$P^E_{12}h_1 = \frac{1}{3}\frac{\sqrt{3}}{2}[h_2 - h_3 - h_3 - h_2]$$
$$\propto [h_2 - h_3]. \tag{11.85}$$

This means that the ammonia MO energy-level diagram can be sketched as in Figure 11.7.

For a slightly more quantitative picture, we can note that the nitrogen atomic orbital energies for the ground-state ^4S term are

$$\epsilon_{1s} = -15.62909 \text{ au}, \qquad \epsilon_{2s} = -0.94531 \text{ au},$$
$$\epsilon_{2p} = -0.56758 \text{ au}. \tag{11.86}$$

The ground-state electron configuration is

$$(1a_1)^2(2a_1)^2(3a_1)^2(1e)^4, \tag{11.87}$$

which leads to the closed-shell determinant corresponding to the term $^1A_1$.

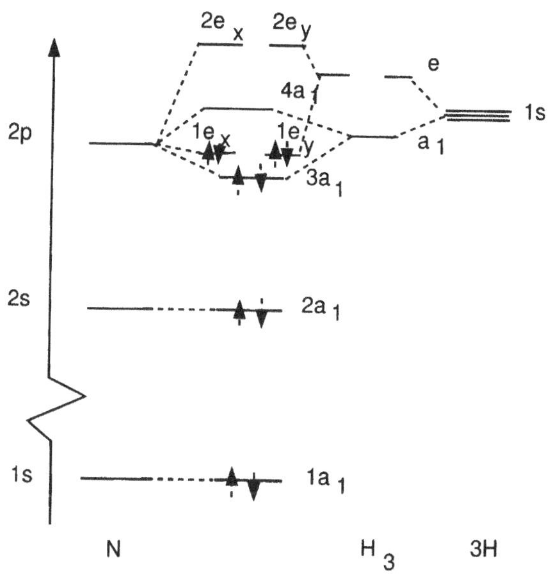

Figure 11.7 Sketched MO energy-level diagram for the ammonia molecule.

12

COUPLING OF IDENTICAL PARTICLES IN A FIELD OF GIVEN SYMMETRY

12.1 COMBINED PERMUTATION AND POINT GROUP SYMMETRY

In order to guarantee both the proper spin symmetry and the correct point group symmetry for an n-particle system one can consider the outer direct product of the symmetric group S_n and the appropriate point group G. One considers a set of f orthonormal orbitals

$$\{\psi_i^\alpha\} \tag{12.1}$$

that span an f-dimensional irrep γ^α of G. One can form a total of f^n products

$$\Phi^\alpha(\mu) = \psi_{\mu_1}^\alpha(1)\psi_{\mu_2}^\alpha(2)\cdots\psi_{\mu_n}^\alpha(n)\,, \tag{12.2}$$

where $1 \le \mu_1, \mu_2, \ldots, \mu_n \le f$. When $g \in G$ operates on such a product function, it is, strictly speaking, the element

$$g = (g, g, \ldots, g) \in \underbrace{G[\times]G[\times]\cdots[\times]G}_{n} \tag{12.3}$$

that acts, that is, an element of the n-fold inner direct product group. Thus

$$\begin{aligned} g\Phi^\alpha(\mu) &= \sum_{\mu_1'}\sum_{\mu_2'}\cdots\sum_{\mu_n'} \psi_{\mu_1'}^\alpha(1)\psi_{\mu_2'}^\alpha(2)\cdots\psi_{\mu_n'}^\alpha(n)\gamma_{\mu_1'\mu_1}^\alpha(g)\gamma_{\mu_2'\mu_2}^\alpha(g)\cdots\gamma_{\mu_n'\mu_n}^\alpha(g) \\ &= \sum_{\mu'} \Phi^\alpha(\mu')\Gamma_{\mu'\mu}^\alpha(g)\,, \end{aligned} \tag{12.4}$$

where

$$\Gamma^{\alpha} = \underbrace{\gamma^{\alpha}[\times]\gamma^{\alpha}[\times]\cdots[\times]\gamma^{\alpha}}_{n} \tag{12.5}$$

is the n-fold inner direct product representation.

For identical particles, such as electrons, one needs to consider the permutational symmetry. We consider the elements

$$P_{km}^{\sigma} = \frac{1}{n!} \sum_{s \in S_n} f^{\sigma} U_{km}^{\sigma}(s) s^{-1} \tag{12.6}$$

of the linear space of the group algebra of the full symmetric group S_n, where U^{σ} is an irrep of S_n. These elements can then be used to form the functions (fixed μ)

$$\Phi_{km}^{\sigma\alpha}(\mu) = P_{km}^{\sigma} \Phi^{\alpha}(\mu), \tag{12.7}$$

which span U^{σ}, specifically

$$s\Phi_{km}^{\sigma\alpha}(\mu) = \sum_{l=1}^{f^{\sigma}} \Phi_{kl}^{\sigma\alpha}(\mu) U_{lm}^{\sigma}(s), \tag{12.8}$$

for any $s \in S_n$. The set of $(f^{\sigma})^2$ functions

$$\{\Phi_{km}^{\sigma\alpha}(\mu)\} \tag{12.9}$$

for fixed μ is not, in general, linearly independent. However, if $\Phi^{\alpha}(\mu)$ has a nonzero component in a row space, that is, $P_{kk}^{\sigma}\Phi^{\alpha}(\mu) \neq 0$ for one k, then the shift operators P_{km}^{σ} for $m = 1, 2, \ldots, f^{\sigma}$ generates a basis for U^{σ}.

From the results in Chapter 5, one can conclude that properly chosen functions $\{\Phi_{km}^{\sigma\alpha}(\mu)\}$ span

$$U^{\sigma} \otimes \Gamma^{\alpha} \tag{12.10}$$

which is an irrep of the direct product group

$$S_n \otimes \{G[\times]G[\times]\cdots[\times]G\} \tag{12.11}$$

and a reducible representation of $G[\times]G[\times]\cdots[\times]G \simeq G$. The symmetry designation of all possible n-electron states with the total spin given by U^{σ} can then be obtained from the irreps of $G[\times]G[\times]\cdots[\times]G \simeq G$ contained in the representation

$$(U^{\sigma} \otimes \Gamma^{\alpha}) \downarrow \{G[\times]G[\times]\cdots[\times]G\}. \tag{12.12}$$

In order to obtain a basis $\{\Phi_{km}^{\sigma\alpha}(\mu)\}$ (linearly independent functions), one chooses k fixed, $m = 1, 2, \ldots, f^\sigma$ and considers only distinct ordered sequences

$$\mu_> = \{\mu_1 \geq \mu_2 \geq \cdots \geq \mu_n\}. \tag{12.13}$$

For each such ordered sequence there is a subgroup $\tau(\mu) \in S_n$ (improper or proper) consisting of all permutations that leave the sequence $\mu_>$ unchanged. For example, if the particular index μ_1 appears k_1 times, μ_2 appears k_2 times, and so on, then $t \in \tau(\mu)$ is any permutation of the k_1 $\psi_{\mu_1}^\alpha$ among themselves, and the k_2 $\psi_{\mu_2}^\alpha$ among themselves, and so on. The order of $\tau(\mu) \in S_n$ is $|\tau(\mu)| = k_1!k_2!\cdots k_n!$ with $k_i \leq 2$ for electrons, because of the Pauli principle. The functions $\{\Phi_{km}^{\sigma\alpha}(\mu_>)\}$ for k fixed, $m = 1, 2, \ldots, f^\sigma$, and ordered sequences $\mu_>$ are orthogonal, but not normalized. The norm is

$$\begin{aligned}(\Phi_{km}^{\sigma\alpha}(\mu_>)|\Phi_{km}^{\sigma\alpha}(\mu_>)) &= (P_{km}^\sigma\Phi^\alpha(\mu_>)|P_{km}^\sigma\Phi^\alpha(\mu_>))\\ &= (\Phi^\alpha(\mu_>)|P_{kk}^\sigma\Phi^\alpha(\mu_>))\\ &= \frac{f^\sigma}{n!}\sum_{t\in\tau(\mu)} U_{kk}^\sigma(t)(\Phi^\alpha(\mu_>)|t^{-1}\Phi^\alpha(\mu_>)) \qquad (12.14)\\ &= \frac{f^\sigma}{n!}\sum_{t\in\tau(\mu)} U_{kk}^\sigma(t),\end{aligned}$$

where $t^{-1}\Phi^\alpha(\mu_>) = \Phi^\alpha(\mu_>)$ and $(\Phi^\alpha(\mu_>)|\Phi^\alpha(\mu_>)) = 1$, while $(\Phi^\alpha(\mu_>)|s^{-1}\Phi^\alpha(\mu_>)) = 0$ for $s \notin \tau(\mu)$.

To evaluate the last sum in Eq. (12.14), one can consider the subduced representation $U^\sigma \downarrow \tau(\mu)$, which is, in general, reducible. In the basis that brings this representation to its fully reduced form, with, say, the one-dimensional irreps in the first diagonal positions, the U_{kk}^σ for such a k would simply be the character for a one-dimensional irrep. In such a case the sum

$$\frac{1}{|\tau(\mu)|}\sum_{t\in\tau(\mu)} U_{kk}^\sigma(t) \tag{12.15}$$

is simply the convolution product of such a character with the character of the identity irrep. Then, if, say, the identity representation appears in the first m diagonal positions

$$\frac{1}{|\tau(\mu)|}\sum_{t\in\tau(\mu)} U_{kk}^\sigma(t) = \begin{cases}1 & \text{for} \quad k \leq m\\ 0 & \text{otherwise.}\end{cases} \tag{12.16}$$

Thus, the normalization $(\Phi_{km}^{\sigma\alpha}(\mu_>)|\Phi_{km}^{\sigma\alpha}(\mu_>)) = f^\sigma|\tau(\mu)|/n! = C$ can be used to define the orthonormal basis $\{\Psi_{km}^{\sigma\alpha}(\mu_>)\}$ with

$$\Psi_{km}^{\sigma\alpha}(\mu_>) = C^{-1/2}\Phi_{km}^{\sigma\alpha}(\mu_>). \tag{12.17}$$

We use the fact that if an orthonormal basis $\{\psi_k\}$ spans a representation Γ of G, that is, $g\psi_l = \sum_k \psi_k \Gamma_{kl}(g)$, the character $\chi(g) = \mathrm{Tr}\{\mathbf{\Gamma}(g)\}$ at any $g \in G$ can simply be calculated as

$$\sum_k (\psi_k | g\psi_k) = \sum_k \Gamma_{kk}(g) = \chi(g). \tag{12.18}$$

We now use this approach to calculate the character of

$$U^\sigma \otimes \Gamma^\alpha \tag{12.19}$$

at

$$(e,g) \in \{G[\times]G[\times]\cdots[\times]G\} \subset S_n \otimes \{G[\times]G[\times]\cdots[\times]G\} \tag{12.20}$$

as

$$\begin{aligned}\chi^{\sigma\alpha}((e,g)) &= \sum_{m=1}^{f^\sigma} \sum_{\mu_>} (\Psi_{km}^{\sigma\alpha}(\mu_>)|(e,g)\Psi_{km}^{\sigma\alpha}(\mu_>)) \\ &= \sum_{m=1}^{f^\sigma} \sum_{\mu_>} \sum_{\mu'} (\Psi_{km}^{\sigma\alpha}(\mu_>)|\Psi_{km}^{\sigma\alpha}(\mu'))\Gamma_{\mu'\mu}^{\alpha}(g).\end{aligned} \tag{12.21}$$

The integrals in this sum will vanish unless μ' matches μ. This means that the sums over μ' can be limited to those sets of orbital indices that can be obtained from $\mu_> = \{\mu_1, \mu_2, \ldots, \mu_n\}$ by permutations $q \in S_n$ of the orbital subscripts μ_i. Such sequences are labeled $q\mu_>$, and the sum over μ' can be replaced by a sum over all permutations $q \in S_n$. Depending on the actual form of the compound index $\mu_> = \{\mu_1, \mu_2, \ldots, \mu_n\}$, not all such permutations give distinct results. All permutations qt with $t \in \tau(\mu)$ [i.e., those in the left coset of S_n by $\tau(\mu)$] give identical results. In order to only count distinct results, one divides by $|\tau(\mu)|$. The result of all this is that

$$\begin{aligned}\chi^{\sigma\alpha}((e,g)) = \chi^{\sigma\alpha}(g) &= \sum_{m=1}^{f^\sigma} \sum_{\mu_>} \sum_{q \in S_n} |\tau(\mu)|^{-1} (\Psi_{km}^{\sigma\alpha}(\mu_>)|q\Psi_{km}^{\sigma\alpha}(\mu_>))\Gamma_{q\mu,\mu}^{\alpha} \\ &= \sum_{m=1}^{f^\sigma} \sum_{l=1}^{f^\sigma} \sum_{\mu_>} \sum_{q \in S_n} |\tau(\mu)|^{-1} (\Psi_{km}^{\sigma\alpha}(\mu_>)|\Psi_{kl}^{\sigma\alpha}(\mu_>))U_{lm}^{\sigma}(q)\Gamma_{q\mu,\mu}^{\alpha} \\ &= \sum_{\mu_>} \sum_{q \in S_n} |\tau(\mu)|^{-1} u^\sigma(q)\Gamma_{q\mu,\mu}^{\alpha},\end{aligned} \tag{12.22}$$

where the orthogonality of the basis functions has been used and where

$$u^\sigma(q) = \sum_{m=1}^{f^\sigma} U_{mm}^{\sigma}(q) \tag{12.23}$$

is the character of U^σ at q. Relaxing the summation over ordered sequences $\mu_>$ to run over all sequences, one should note that all arrangements of a particular sequence $\{\mu_1, \mu_2, \ldots, \mu_n\}$ give the same result. Since there are $n!/|\tau(\mu)|$ distinct such arrangements, the character can be expressed as

$$\chi^{\sigma\alpha}(g) = \sum_{\mu} \frac{1}{n!} \sum_{q\in S_n} u^{\sigma}(q)\Gamma^{\alpha}_{q\mu,\mu}(g)$$

$$= \frac{1}{n!} \sum_{\mu_1=1}^{f} \sum_{\mu_2=1}^{f} \cdots \sum_{\mu_n=1}^{f} \sum_{q\in S_n} u^{\sigma}(q)\gamma^{\alpha}_{q\mu_1,\mu_1}(g)\gamma^{\alpha}_{q\mu_2,\mu_2}(g)\cdots\gamma^{\alpha}_{q\mu_n,\mu_n}(g). \tag{12.24}$$

The summation over S_n can be written as summations over classes, since the character u^{σ} is a class constant. A class C_q is labeled by its cycle structure, say

$$(1)^{i_1(q)}(2)^{i_2(q)}\cdots(n)^{i_n(q)}, \tag{12.25}$$

where $i_k(q)$ is the number of k-cycles and where, as we recall

$$i_1(q) + 2i_2(q) + \cdots + ni_n(q) = n. \tag{12.26}$$

This results in the final expression for the character as

$$\chi^{\sigma\alpha}(g) = \frac{1}{n!} \sum_{C_q\in S_n} |C_q|u^{\sigma}(C_q)[\chi^{\alpha}(g)]^{i_1(q)}[\chi^{\alpha}(g^2)]^{i_2(q)}\cdots[\chi^{\alpha}(g^n)]^{i_n(q)}, \tag{12.27}$$

where $|C_q|$ is the order of the class C_q. Since this is the character of the reducible representation $U^{\sigma} \otimes \Gamma^{\alpha} \downarrow \{G[\times]G[\times]\cdots[\times]G\}$ one can easily decompose this representations into its irreducibles of $\{G[\times]G[\times]\cdots[\times]G\} \simeq G$. This result was presented by M. Kotani [*J. Phys. Soc.* (Japan) **19**, 215 (1964)] and further elaborated by O. Goscinski and Y. Öhrn [*Int. J. Quantum Chem.* **II**, 845 (1968)].

12.2 TERM SYMBOLS AND SYMMETRY

The result of the previous section can be used in a variety of ways to determine the symmetry of many-particle states. We consider an illustrative example.

Example 12.1 The tetrahedral $CoCl_4^{2-}$ molecular ion has a dominant t_2^3 electron configuration. The possible electronic terms corresponding to this electron configuration can be obtained by using the character table of the tetrahedral point group T_d and the characters of the two-column irreps of the symmetric group S_3.

Using the rules for calculating characters for irreps of permutation groups, one can easily generate the necessary characters:

S_3	(1^3)	3, (1,2)	2, (3)
$[1^3]$	1	−1	1
[2,1]	2	0	−1

The [2,1] irrep corresponds to spin doublet states and yields the character

$$\chi^{[2,1]\alpha}(g) = \frac{1}{3}[(\chi^\alpha(g))^3 - \chi^\alpha(g^3)], \tag{12.28}$$

while the $[1^3]$ irrep corresponds to spin quartet states and yields the character

$$\chi^{[1^3]\alpha}(g) = \frac{1}{6}[(\chi^\alpha(g))^3 - 3\chi^\alpha(g)\chi^\alpha(g^2) + 2\chi^\alpha(g^3)], \tag{12.29}$$

where we have applied the formula in Eq. (12.27).

The character table of the tetrahedral group T_d is used below to complete the analysis:

T_dx	E	$8C_3$	$3C_2$	$6S_4$	$6\sigma_d$
A_1	1	1	1	1	1
A_2	1	1	1	−1	−1
E	2	−1	2	0	0
T_1	3	0	−1	1	−1
T_2	3	0	−1	−1	1
g^2	E	C_3	E	C_2	E
g^3	E	E	C_2	S_4	σ_d
$\chi^{T_2}(g^2)$	3	0	3	−1	3
$\chi^{T_2}(g^3)$	3	3	−1	−1	1
$\chi^{[2,1]T_2}(g)$	8	−1	0	0	0
$\chi^{[1^3]T_2}(g)$	1	1	1	−1	−1

Using the relation

$$(\chi * \chi^\alpha)_e = \nu^\alpha \tag{12.30}$$

for the frequency of an irrep in a general representation with the character $\chi = \chi^{[2,1]T_2}$ one calculates the nonzero frequencies to be

$$\nu^E = \nu^{T_1} = \nu^{T_2} = 1 \tag{12.31}$$

and for $\chi = \chi^{[1^3]T_2}$ only

$$\nu^{A_2} = 1 \tag{12.32}$$

is nonzero. This means that the t_2^3 electron configuration corresponds to the spin doublet terms

$$^2E, {}^2T_1, {}^2T_2 \tag{12.33}$$

and the spin quartet term

$$^4A_2 \,. \tag{12.34}$$

12.3 POINT GROUPS AS SUBGROUPS OF $SO(3)$

The symmetry group of a sphere consists of the elements $C(\alpha)$, a rotation by an angle α about any axis through the center of the sphere, and $S(\alpha)$, a rotation reflection or improper rotation, that is, a rotation about a symmetry axis by an angle α followed by a reflection in the symmetry plane perpendicular to the rotation axis. This group is called $SO(3)$, and its irreps Γ^l are spanned by the spherical harmonics $\{Y_{lm}(\theta, \varphi)\}$ and are of dimension $(2l + 1)$. All rotations about any symmetry axis by a fixed angle α belong to the same class. The same holds for the improper rotations. This can easily be seen in Figure 12.1, where the conjugation operation $C_{\mathbf{n}}(\alpha) = C(-\beta)C_{\mathbf{n}'}(\alpha)C(\beta)$ is depicted.

The rotations by β and by $-\beta$ is about the symmetry axis that is perpendicular to the plane of the two rotation axes $\mathbf{n}$ and $\mathbf{n}'$. And, of course, the relation $C^{-1}(\beta) = C(-\beta)$ holds.

We recall the spherical harmonics

$$Y_{lm}(\theta, \varphi) \propto P_l^{|m|}(\cos\theta)e^{im\varphi} \,, \tag{12.35}$$

where the associated Legendre polynomials satisfy the generating relation

$$P_l^{|m|}(\omega) \propto (1 - \omega^2)^{|m|/2} \frac{d^{m+l}}{d\omega^{m+l}}(\omega^2 - 1)^l \,. \tag{12.36}$$

Since the rotations by a fixed angle about any symmetry axis are in the same class, we are, for the purposes of determining characters, free to choose any

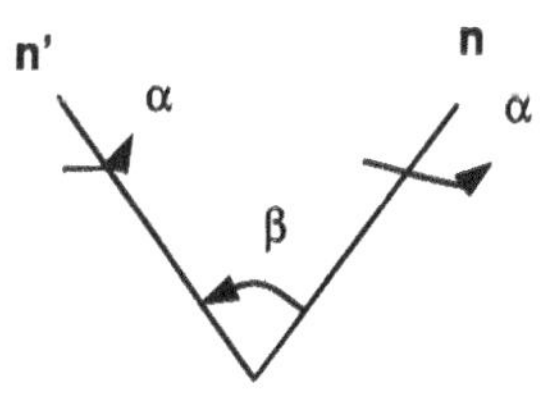

Figure 12.1 Illustration of the conjugation $C_{\mathbf{n}}(\alpha) = C(-\beta)C_{\mathbf{n}'}(\alpha)C(\beta)$.

axis. Therefore we choose the z axis for which it is readily determined that the angular coordinates under a rotation $C(\alpha)$ change as follows:

$$\theta \to \theta; \qquad \varphi \to \varphi + \alpha \tag{12.37}$$

and

$$C(\alpha)Y_{lm}(\theta,\varphi) = Y_{lm}(\theta,\varphi)e^{im\alpha} = \sum_{m'=-l}^{l} Y_{lm'}(\theta,\varphi)\Gamma^{l}_{m'm}(C(\alpha)), \tag{12.38}$$

which means that

$$\Gamma^{l}(C(\alpha)) = \begin{bmatrix} e^{il\alpha} & 0 & \cdots & 0 \\ 0 & e^{i(l-1)\alpha} & \cdots & 0 \\ \vdots & \vdots & \ddots & \vdots \\ 0 & 0 & \cdots & e^{-il\alpha} \end{bmatrix} \tag{12.39}$$

and the character is

$$\chi^{l}(C(\alpha)) = \sum_{m=-l}^{l} e^{im\alpha} = 1 + 2\sum_{m=1}^{l} \cos m\alpha\,. \tag{12.40}$$

Similarly, $S(\alpha)$ changes the angular coordinates such that

$$\theta \to \pi - \theta; \qquad \varphi \to \varphi + \alpha \tag{12.41}$$

and since $\cos(\pi - \theta) = -\cos\theta = -\omega$, it follows that

$$P_l^{|m|}(\omega) \to P_l^{|m|}(-\omega) = (-1)^{l+m}P_l^{|m|}(\omega)\,. \tag{12.42}$$

Then, since

$$S(\alpha)Y_{lm}(\theta,\varphi) = (-1)^{l+m}Y_{lm}(\theta,\varphi)e^{im\alpha}\,, \tag{12.43}$$

the character at $S(\alpha)$ is

$$\chi^{l}(S(\alpha)) = (-1)^{l}\sum_{m=-l}^{l}(-1)^{m}e^{im\alpha} = (-1)^{l}[1 + 2\sum_{m=1}^{l}(-1)^{m}\cos m\alpha]. \tag{12.44}$$

The group $SO(3)$ will be discussed in terms of more formal theory in Chapter 17. From the simple discussion here it should be clear that the spherical symmetry with its infinite number of symmetry axes and planes contains as special cases the point groups whose elements involve at most a few symmetry axes and a few symmetry planes with particular angles and directions. This means that all the point groups are subgroups of $SO(3)$. This also means that the irreps of $SO(3)$ in general are reducible representations of any point group, a fact that can be used to obtain the subduced representations

$$\Gamma^{l} \downarrow G \tag{12.45}$$

for any point group G.

Example 12.2 We consider a transition metal octahedral complex AB_6 and attempt to determine which symmetry orbitals have $3d$ components on the central transition metal atom. The subduced representation

$$\Gamma^2 \downarrow O_h \tag{12.46}$$

should give the answer. The symmetry elements of $SO(3)$ are restricted to finite angles such that

$$\begin{aligned} &C(2\pi) = E, \qquad C\frac{\pi}{2} = C_4; \qquad C(\pi) = C_2; \qquad \left(C\frac{2\pi}{3}\right) = C_3; \\ &S(2\pi) = \sigma; \qquad S(\pi) = i; \qquad S\frac{\pi}{2} = S_4; \qquad S\frac{\pi}{3} = S_6 . \end{aligned} \tag{12.47}$$

The character table of O_h can then be used in the following manner:

O_h	E	$8C_3$	$6C_2$	$6C_4$	$3C_2$	i	$6S_4$	$8S_6$	$3\sigma_h$	$6\sigma_d$
A_{1g}	1	1	1	1	1	1	1	1	1	1
A_{2g}	1	1	−1	−1	1	1	−1	1	1	−1
E_g	2	−1	0	0	2	2	0	−1	2	0
T_{1g}	3	0	−1	1	−1	3	1	0	−1	−1
T_{2g}	3	0	1	−1	−1	3	−1	0	−1	1
A_{1u}	1	1	1	1	1	−1	−1	−1	−1	−1
A_{2u}	1	1	−1	−1	1	−1	1	−1	−1	1
E_u	2	−1	0	0	2	−2	0	1	−2	0
T_{1u}	3	0	−1	1	−1	−3	−1	0	1	1
T_{2u}	3	0	1	−1	−1	−3	1	0	1	−1
Γ^2	5	−1	1	−1	1	5	−1	−1	1	1

The nonzero frequency numbers of the various irreps of O_h can readily be found as

$$\nu^{E_g} = 1; \qquad \nu^{T_{2g}} = 1, \tag{12.48}$$

which shows what we could expect, namely, that no *ungerade* irreps can appear, since $l = 2$ orbitals are even under inversion. Thus the $3d$ atomic orbitals on the central atom split into two (degenerate) e_g orbitals and three (degenerate) t_{2g} orbitals. From the section on point group character tables it can be seen that it is the pair

$$d_{3z^2-r^2}, d_{x^2-y^2} \tag{12.49}$$

that forms the e_g orbitals, while the d_{xy}, d_{xz}, d_{yz} form the t_{2g} orbitals. For an octahedral molecule or ion with a transition metal central atom and typical ligand atoms, the t_{2g} orbitals are as a rule lower in energy than the e_g. A tetrahedral transition metal complex also splits the central atom d orbitals. In this case it is the e orbitals that as a rule are lower in energy than the t_2.

The result of this chapter for the combined permutation and space group symmetry can also be used to yield the appropriate atomic terms that correspond to a particular electron configuration.

Example 12.3 We consider a d^3 transition metal atom such as the ground state of a vanadium atom. The possible spin doublet terms can be calculated from the characters[1]

$$\chi^{[2,1]l}(C(\alpha)) = \tfrac{1}{3}[(\chi^l(C(\alpha)))^3 - \chi^l(C(3\alpha))], \tag{12.50}$$

where $l = 2$, and

$$\chi^2(C(\alpha)) = 1 + 2\cos\alpha + 2\cos 2\alpha\,. \tag{12.51}$$

Using some elementary trigonometry, we find that

$$\begin{aligned}\chi^{[2,1],2}(C(\alpha)) &= 6 + 12\cos\alpha + 10\cos 2\alpha + 6\cos 3\alpha \\ &\quad + 4\cos 4\alpha + 2\cos 5\alpha\,,\end{aligned} \tag{12.52}$$

which can be decomposed as

$$\begin{aligned}\chi^{[2,1],2}(C(\alpha)) &= \left(1 + 2\sum_{m=1}^{5}\cos m\alpha\right) + \left(1 + 2\sum_{m=1}^{4}\cos m\alpha\right) \\ &\quad + \left(1 + 2\sum_{m=1}^{3}\cos m\alpha\right) + 2\left(1 + 2\sum_{m=1}^{2}\cos m\alpha\right) \\ &\quad + (1 + 2\cos\alpha)\,.\end{aligned} \tag{12.53}$$

This means that the d^3 configuration corresponds to the following doublet terms

$${}^2H, {}^2G, {}^2F, 2{}^2D, {}^2P \tag{12.54}$$

corresponding to angular momentum quantum numbers $L = 5, 4, 3, 2$, and 1, respectively. Similarly, the spin quartet terms can be determined from

$$\begin{aligned}\chi^{[1^3],2}(C(\alpha)) &= \tfrac{1}{6}[(\chi^2(C(\alpha)))^3 - 3\chi^2(C(\alpha))\chi^2(C(2\alpha)) + 2\chi^2(C(3\alpha))] \\ &= 2 + 4\cos\alpha + 2\cos 2\alpha + 2\cos 3\alpha\,,\end{aligned} \tag{12.55}$$

leading to the terms

$${}^4F, {}^4P\,. \tag{12.56}$$

A check that can be performed is to count the possible number of substates, which should equal the number of ways three electrons can be distributed among 10 d spin orbitals, or

$$\binom{10}{3} = 120\,. \tag{12.57}$$

The ${}^2H$ term has $2\times 11 = 22$ substates, the ${}^2G$ term has $2\times 9 = 18$ substates, and so on, and ${}^4F$ has $4\times 7 = 28$, and so on, which add up to 120 as they should.

[1] Note that $(C(\alpha))^3 = C(3\alpha)$.

12.4 SPIN–ORBIT COUPLING

Qualitative arguments can often be discussed in terms of simple perturbation theory. We do not treat this very powerful methodology here but only comment on the typical scenario for such treatments.

The Hamiltonian

$$H = H_0 + \delta H, \tag{12.58}$$

is partitioned into an unperturbed part, H_0, whose eigenfunctions and eigenvalues are assumed known, at least in principle, and a perturbation δH. If the effect of the perturbation is small, then the correction of the energy of an unperturbed (normalized) state ψ is simply $\langle\psi|\delta H\psi\rangle$. This is called the *first-order correction* in the language of perturbation theory. An excellent treatise of perturbation theory can be found in the text by P. Jørgensen and J. Simons, *Second Quantization-Based Methods in Quantum Chemistry*, Academic Press, NewYork, 1981.

When the interaction of the electron spin magnetic moment with the electric field form the nucleus can be neglected, the symbols given in the last section for a d^3 atom provide a good way to label the many-electron states. This is the so-called Russell–Saunders coupling scheme, which couples all the electron spins to a total electron spin angular momentum with quantum numbers S, M_S, and all the electron orbital angular momenta to a total orbital angular momentum with quantum numbers L, M_L. However, this interaction, called the *spin–orbit interaction* is present, and although small for the lighter elements, it causes the atomic energy levels to be perturbed. For the heavier elements it can become dominant and more important than the electron–electron Coulomb interaction. The extra term in the Hamiltonian due to this interaction between the electron spin magnetic moments and the electric field is called the *spin–orbit term* and can be obtained from relativistic considerations (the Dirac equation). For a single electron this term can be expressed as

$$\delta H = \frac{e}{2m^2c^2r}\frac{dV}{dr}\vec{l}\cdot\vec{s} = \xi(r)\vec{l}\cdot\vec{s}, \tag{12.59}$$

where V is the electrostatic potential (for a central field $V = -Ze/r$). This makes the coupling constant $\xi(r)$ positive. For many electrons in a central field, one would write

$$\delta H = \sum_{\kappa=1}^{N} \xi(r_\kappa)\vec{l}_\kappa \cdot \vec{s}_\kappa \tag{12.60}$$

for the spin–orbit term to be added to the purely Coulombic Hamiltonian. This is a sum of one-electron operators. In a more ambitious treatment of the spin–orbit coupling, one would also consider coupling between the orbital angular momentum (electric field) and the spin of different electrons. In molecular systems the electric field is a sum of central fields from the various nuclei and the term in the Hamiltonian would simply be a sum of terms like the one in Eq. (12.60) when only the one-electron contributions are considered.

We repeat here some of the discussion on angular momentum. All the orbital angular momenta add up to a total orbital angular momentum

$$\vec{L} = \sum_{\kappa=1}^{N} \vec{l}_\kappa , \tag{12.61}$$

and all the spin angular momenta of individual electrons add up to a total spin angular momentum

$$\vec{S} = \sum_{\kappa=1}^{N} \vec{s}_\kappa . \tag{12.62}$$

We recall from the previous discussion that a general angular momentum operator has Cartesian components that satisfy the commutation relations

$$\begin{aligned} [M_x, M_y] &= i\hbar M_z , \\ [M_y, M_z] &= i\hbar M_x , \\ [M_z, M_x] &= i\hbar M_y , \end{aligned} \tag{12.63}$$

where $\hbar$, which has the dimension of angular momentum, is commonly put equal to unity, so that angular momenta are measured in units of $\hbar$. If we do that, we can write

$$\vec{M}^2 = M_x^2 + M_y^2 + M_z^2 = M_+M_- + M_z^2 - M_z \tag{12.64}$$

in terms of the so-called ladder operators $M_\pm = M_x \pm iM_y$. When two angular momenta $\vec{M}_1$ and $\vec{M}_2$ commute their sum it is also an angular momentum. Two angular momenta commute; that is, they can act in any order with the same final result, or equivalently stated, their commutator is zero, when they refer to different variables. This is the case, for instance, for the orbital angular momenta of individual electrons that add to a total orbital angular momentum in Eq. (12.61). Similarly, the orbital angular momentum and the spin angular momentum of a single electron act on different variables and, thus, add up to a total angular momentum

$$\vec{j} = \vec{l} + \vec{s} \tag{12.65}$$

or for many electrons

$$\vec{J} = \vec{L} + \vec{S} = \sum_{\kappa=1}^{N} (\vec{l}_\kappa + \vec{s}_\kappa) = \sum_{\kappa=1}^{N} \vec{j}_\kappa . \tag{12.66}$$

It is straightforward to show that

$$[\vec{l}^2, \vec{l} \cdot \vec{s}] = 0,$$
$$[\vec{s}^2, \vec{l} \cdot \vec{s}] = 0, \tag{12.67}$$

but that $[l_z, \vec{l} \cdot \vec{s}] \neq 0$, and $[s_z, \vec{l} \cdot \vec{s}] \neq 0$, so the total spin and orbital angular momenta are "good quantum numbers" even when spin–orbit coupling is introduced, but their z components are not. On the other hand, the total angular momentum

$$\vec{j}^2 = (\vec{l} + \vec{s})^2 \tag{12.68}$$

and its z component do commute with the spin–orbit term and thus with the entire Hamiltonian. The corresponding quantum numbers are j and $m_j = -j, -j + 1, \ldots, j$. For a many-electron system, $\vec{J}^2$ and J_z do commute with the Hamiltonian, including the spin–orbit coupling. Obviously

$$\vec{L} \cdot \vec{S} = \frac{1}{2}[\vec{J}^2 - \vec{L}^2 - \vec{S}^2], \tag{12.69}$$

a fact that can be used when manipulating the angular momentum operators.

In general, the angular momentum operators have eigenstates $|km\rangle$ such that

$$\vec{M}^2|km\rangle = k(k+1)|km\rangle,$$
$$M_z|km\rangle = m|km\rangle. \tag{12.70}$$

Thus, for a fixed quantum number k the quantum number $m = -k, -k + 1, \ldots, k$ so that there are $(2k + 1)$ degenerate eigenstates of $\vec{M}^2$ corresponding to the eigenvalue $k(k + 1)$.

As was shown earlier in the discussion of angular momenta, when $\vec{M} = \vec{M}_1 + \vec{M}_2$, $|k_1 m_1\rangle$ and $|k_2 m_2\rangle$ eigenstates of $\vec{M}_1^2, M_{1z}$ and $\vec{M}_2^2, M_{2z}$, respectively, the quantum number k can assume the values

$$k = k_1 + k_2, k_1 + k_2 - 1, \ldots, |k_1 - k_2|. \tag{12.71}$$

This can be seen to be consistent with the fact that the space of eigenstates of $\vec{M}^2, M_z$ is the space of product states $|k_1 m_1\rangle|k_2 m_2\rangle$ of dimension $(2k_1 + 1)(2k_2 + 1)$ and the expression

$$\sum_{k=|k_1-k_2|}^{k_1+k_2} (2k + 1) = (2k_1 + 1)(2k_2 + 1) \tag{12.72}$$

for the sum of dimensions of the various subspaces.

The quantitative determination of the effect of spin–orbit coupling on the spectra of general molecular systems is still a developing area of research.

However, some simple observations can be made for specific situations. In terms of a single determinantal wavefunction in terms of orthonormal spin orbitals for an atom

$$D = \frac{1}{\sqrt{N!}} \det\{\phi_1 \phi_2 \cdots \phi_N\}, \tag{12.73}$$

the spin–orbit energy can be expressed as

$$\langle D|\delta H D\rangle = \sum_{\kappa=1}^{N} (\phi_\kappa|\xi(r)\vec{l}_\kappa \cdot \vec{s}_\kappa|\phi_\kappa). \tag{12.74}$$

When the atomic spin orbitals are products of a radial, an angular, and a spin factor,

$$R_{nl}(r) Y_{lm}(\theta, \varphi) \begin{cases} \alpha \\ \beta \end{cases}, \tag{12.75}$$

then the integrals over angular and spin factors can easily be made in any particular case and one is left with the integrals

$$\int_0^\infty R_{nl}^2(r)\xi(r) r^2 dr = \frac{\zeta_{nl}}{\hbar^2}, \tag{12.76}$$

where we have explicitly again introduced the $\hbar$. The quantity ζ_{nl} is called the *spin–orbit coupling constant* for an *nl* orbital.

In the following examples, a single electron in a p orbital has the following Russell–Saunders substates of its ^2P term with $l = 1$, and $s = \frac{1}{2}$, where the states are labeled by the various m_l and m_s values:

$$\begin{array}{ll} |1, \frac{1}{2}\rangle & |1, -\frac{1}{2}\rangle \\ |0, \frac{1}{2}\rangle & |0, -\frac{1}{2}\rangle \\ |-1, \frac{1}{2}\rangle & |-1, -\frac{1}{2}\rangle \end{array}$$

In this case there are only two possible j values $j = 1 + s = \frac{3}{2}$, and $j = l - s = \frac{1}{2}$, yielding the two terms $^2P_{3/2}$ and $^2P_{1/2}$. We can calculate the spin–orbit energies

$$\begin{aligned} E(^2P_{3/2}) &= \langle 1, \tfrac{1}{2}|\xi(r)\vec{l} \cdot \vec{s}|1, \tfrac{1}{2}\rangle \\ &= \langle 1, \tfrac{1}{2}|\xi(r)[\tfrac{1}{2}\, l_+ s_- + \tfrac{1}{2}\, l_- s_+ + l_z s_z]|1, \tfrac{1}{2}\rangle \\ &= \langle 1, \tfrac{1}{2}|\xi(r)\, \tfrac{1}{2}[\vec{j}^2 - \vec{l}^2 - \vec{s}^2]|1, \tfrac{1}{2}\rangle \\ &= \tfrac{1}{2}[\tfrac{3}{2}(\tfrac{3}{2} + 1) - 1(1 + 1) - \tfrac{1}{2}(\tfrac{1}{2} + 1)]\zeta_p = \tfrac{1}{2}\zeta_p. \end{aligned} \tag{12.77}$$

In order to calculate $E(^2P_{1/2})$, we recognize that the states

$$|j,m\rangle = |\tfrac{1}{2},\tfrac{1}{2}\rangle, \qquad |j,m\rangle = |\tfrac{3}{2},\tfrac{1}{2}\rangle \tag{12.78}$$

can be expressed as linear combinations of the states $|m_l, m_s\rangle = |1, -\frac{1}{2}\rangle$, and $|m_l, m_s\rangle = |0, \frac{1}{2}\rangle$, which means that the trace of the 2 × 2 spin–orbit energy matrix is

$$E(^2P_{3/2}) + E(^2P_{1/2})$$
$$= \langle 0, \tfrac{1}{2}|\xi(r)\vec{l}\cdot\vec{s}|0,\tfrac{1}{2}\rangle + \langle 1, -\tfrac{1}{2}|\xi(r)\vec{l}\cdot\vec{s}|1,-\tfrac{1}{2}\rangle = 0 - \tfrac{1}{2}\zeta_p . \tag{12.79}$$

Thus $E(^2P_{1/2}) = -\zeta_p$, that is, lower than $E(^2P_{3/2})$, which is typical for less than half-filled shells. For example, for the excited-state electron configuration $(1s)^2 2p$ of the doubly positive boron ion (B^{+2}), the $^2P_{1/2}$ term lies 34.1 cm^{-1} below the $^2P_{3/2}$, making $\zeta_p = 22.7$ cm^{-1}. There are, however, exceptions to this rule. In this way one can obtain estimates of spin–orbit energies by neglecting closed shells, as long as one stays within one term and only relative energies are considered.

Similarly, for the ^3P term of the electron configuration $(2p)^2$, one obtains[2]

$$|J, M_J\rangle = |2,2\rangle = |m_{l_1}, m_{s_1}, m_{l_2}, m_{s_2}\rangle = |1, \tfrac{1}{2}, 0, \tfrac{1}{2}\rangle, \tag{12.80}$$

and the corresponding spin–orbit energy is

$$E(^3P_2) = \langle 1, \tfrac{1}{2}, 0, \tfrac{1}{2}|\xi(r)\textstyle\sum_\kappa\{\frac{1}{2}[l_{\kappa+}s_{\kappa-} + l_{\kappa-}s_{\kappa+}]$$
$$+ l_{\kappa z}s_{\kappa z}\}|1, \tfrac{1}{2}, 0, \tfrac{1}{2}\rangle \tag{12.81}$$
$$= \zeta_p .$$

Using the invariance of the trace to a basis set transformation and the two obvious basis states

$$|m_{l_1}, m_{s_1}, m_{l_2}, m_{s_2}\rangle = |1, \tfrac{1}{2}, 0, -\tfrac{1}{2}\rangle,$$
$$|m_{l_1}, m_{s_1}, m_{l_2}, m_{s_2}\rangle = |1, -\tfrac{1}{2}, 0, \tfrac{1}{2}\rangle, \tag{12.82}$$

one finds

$$E(^3P_2) + E(^3P_1) = \langle 1, \tfrac{1}{2}, 0, -\tfrac{1}{2}|\xi(r)\textstyle\sum_\kappa\{\frac{1}{2}[l_{\kappa+}s_{\kappa-}$$
$$+ l_{\kappa-}s_{\kappa+}] + l_{\kappa z}s_{\kappa z}\}|1, \tfrac{1}{2}, 0, -\tfrac{1}{2}\rangle$$
$$+ \langle 1, -\tfrac{1}{2}, 0, \tfrac{1}{2}|\xi(r)\textstyle\sum_\kappa\{\frac{1}{2}[l_{\kappa+}s_{\kappa-} + l_{\kappa-}s_{\kappa+}]$$
$$+ l_{\kappa z}s_{\kappa z}\}|1, -\tfrac{1}{2}, 0, \tfrac{1}{2}\rangle$$
$$= 0 \tag{12.83}$$

and thus $E(^3P_1) = -\zeta_p$. Similarly, the appropriate basis states yields $E(^3P_0) = -2\zeta_p$.[3]

[2] We used the notation (1, 0 | for the determinant $|1, \frac{1}{2}, 0, \frac{1}{2}\rangle$ earlier in the discussion of atomic terms.

[3] The basis states $|1, -\frac{1}{2}, 0, -\frac{1}{2}\rangle$, and $|-1, \frac{1}{2}, 0, \frac{1}{2}\rangle$ yield nonzero contributions.

For the carbon doubly positive ion $C^{+2}(1s)^2(2p)^2)$ the spin–orbit splittings are -52, -29, and 28 cm^{-1}, in reference to the center of gravity of the term.

The examples above [following Eq. (12.76)] represent the case where the electrostatic energy dominates over the spin–orbit coupling and one best diagonalizes the Coulombic Hamiltonian first and then considers the spin–orbit coupling as a perturbation term. This is called *LS coupling*. In these cases the spin–orbit energy obviously is quite a bit smaller than the electron–electron repulsion. This is not always the case. The opposite may be true. First diagonalizing the spin–orbit energy and then treating the electron–electron repulsion as a perturbation is referred to as *jj* coupling. Some configurations of a given atom might obey *jj* coupling, while others obey *LS* coupling (Russell–Saunders coupling). Obviously, first diagonalizing one or the other part of the energy determines the basis, and a complete calculation in any of these two bases gives the same result. The use of a particular basis is referred to as using the corresponding coupling scheme. When the first-order treatment in a particular basis agrees well with the experimental result, one says that the atomic system obeys the corresponding coupling scheme. Some of the material on spin–orbit presented above coupling follows closely material in the text by J. S. Griffith, *The Theory of Transition-Metal Ions*, Cambridge University Press, 1961.

13

MOLECULAR ORBITAL THEORY

Molecular orbitals as solutions to an effective single-electron Schrödinger equation were discussed already without giving the specifics of the effective single-particle Hamiltonian or Fock operator. We focused on the questions of point group symmetry and the MO-LCAO approximation, in which the molecular orbitals are represented in a basis of atomic orbitals. We studied the resulting secular problem and arrived at rules for constructing molecular orbital energy-level diagrams for simple systems. One strict rule was that only symmetry-adapted orbitals of the same symmetry combine to form molecular orbitals.

This comes from the fact that the Fock operator

$$F = -\frac{1}{2}\vec{\nabla}^2 - \sum_A \frac{Z_A}{r_A} + V(\vec{r}) \tag{13.1}$$

can be assumed to be totally symmetric, that is, to transform as the identity representation of the appropriate point group. The kinetic-energy operator is the same as used for atomic systems and has spherical symmetry and thus commutes with all point group operations. The potential energy $V(\vec{r})$ can be constructed in such a way that it also is totally symmetric. This will become clear in this chapter. The sum over electron–nuclear attraction terms transforms into itself under point group operations, since such operations simply result in permutations of identical nuclei. This means that the typical matrix elements in the secular ptoblem

$$F_{ij} = (\phi_i^\alpha | F \phi_j^\beta), \qquad \Delta_{ij} = (\phi_i^\alpha | \phi_j^\beta), \tag{13.2}$$

where ϕ_i^α and ϕ_j^β are any two symmetry-adapted orbitals, will be zero unless the integrands have a component that transform as the identity irrep. Since F is totally symmetric, this means that $\Gamma^\alpha \otimes \Gamma^\beta \ni \Gamma^J$, that is, $\alpha = \beta$ for the matrix elements to be nonzero. So, as we saw in the discussion about symmetry and the secular equations, we can conclude that the secular problem will block into smaller problems for each symmetry type or irrep of the appropriate point group.

In this chapter we first discuss approximate molecular orbital theory and so-called molecular orbital correlation diagrams as applied to predictions about chemical reactions when an element of symmetry is preserved throughout the reaction. Then we go on to derive the full details about the Fock operator and the Hartree–Fock approximation to molecular orbitals. Other sections deal with various forms of approximate molecular orbital theory and in particular parametrized forms or so-called semiempirical theory.

13.1 FRONTIER ORBITALS AND THE NONCROSSING RULE

The fact that low-lying excited states of many molecules involve primarily changes in the highest occupied molecular orbital (HOMO) and lowest unoccupied molecular orbital (LUMO) mean that in low-energy processes as in ordinary chemical reactions involving the breaking or formation of chemical bonds, these molecular orbitals and the ones closest to them in energy are the ones most affected and in many ways steer the process in a qualitative manner. This is the primary idea in the so called frontier orbital theory, which manages to make useful predictions about simple chemical reactions and about the relative stability of high-symmetry molecular conformations from considerations involving only the HOMO and the LUMO and other orbitals closest to them in energy, the so-called frontier orbitals.

Molecular orbital theory can in this and related ways be used to study reaction mechanisms for many concerted chemical reactions, such as certain thermal addition reactions or electrocyclic reactions. One normally studies various possible reaction paths, but limits the considerations to paths that preserve at least one symmetry element in going from reactants to products. One reason for why this works is that such paths tend to lead to more MO energy-level crossings—and thus state energy-level crossings—than do arbitrary paths. Such crossings are necessary for reactions to take place, and the true reaction path is often likely to be closely approximated by such symmetry-preserving paths, at least at low energies.

In order to make qualitative arguments about what happens to molecular orbitals and their energies during a chemical transformation, one relies on symmetry properties and the *noncrossing rule*. This addresses the question of whether two MO energy levels can or cannot cross during the path from reagents to products.

Suppose that two MOs ϕ_1 and ϕ_2 have energies that come close together at some nuclear geometry R_0 of the total system of reacting species. At this geometry the matrix elements of the Fock operator with respect to these eigenfunctions are $F_{11} = (\phi_1|F\phi_1)$, $F_{22} = (\phi_2|F\phi_2)$, with $F_{11} \approx F_{22}$ and $F_{12} = (\phi_1|F\phi_2) = 0$, $(\phi_1|\phi_2) = 0$. When the molecular system is distorted away from R_0, preserving at least an element of symmetry (such as a plane or an axis) to a nearby geometry R, the new molecular orbitals can to a good approximation be expressed as linear combinations of the orbitals at R_0. The new energy levels (at R) are then solutions to the secular problem

$$\begin{vmatrix} F_{11} - \epsilon & F_{12} \\ F_{21} & F_{22} - \epsilon \end{vmatrix} = 0, \tag{13.3}$$

that is

$$\epsilon = F_{11} \pm |F_{12}|, \tag{13.4}$$

where we have used the assumption that $F_{11} \approx F_{22}$ and the fact that $F_{12} = F_{21}^*$.

If $F_{12} = 0$ the two energy levels are the same and they will cross at R. This is a question of the symmetry type of the two MOs. The integral $(\phi_1|F\phi_2) = F_{12}$ is different from zero only if the integrand contains a part that is totally symmetric with respect to the symmetry operation preserved during the distortion. This result follows from our discussion of integrals and symmetry. It can also be seen by elementary reasoning. Suppose that the preserved symmetry operation is a reflection in a symmetry plane; then, since F is totally symmetric by construction, the integrand of F_{12} would be totally symmetric or contain such a part only if ϕ_1 and ϕ_2 are of the same symmetry type, either both even or both odd under reflection in the symmetry plane. It follows that when $F_{12} \neq 0$, the two energy levels would not cross. If this is not the case, since the symmetry plane divides the space in two equal parts, the integrand will be positive in one half and negative of the same magnitude in the other half, causing the integral over all of space to vanish. This is the so-called noncrossing rule, which states that *energy levels of two orbitals of the same symmetry cannot cross*, while those of orbitals of different symmetry may cross.

As an example we consider the ring closure of the pentadienyl cation (see Fig. 13.1) to form the cyclopentenyl cation (see Fig. 13.2). In Figures 13.1 and 13.2 the highest occupied and lowest unoccupied orbitals are depicted. Considerations are limited to the four least bound electrons. A useful form of the π orbitals can be found from the simple π Hückel theory by pencil and paper.

Two reaction paths are considered: (1) conrotatory ring closure (see Fig. 13.3), which preserves the C_2 symmetry axis, and (2) the disrotatory ring closure (see Fig. 13.4), which preserves the σ_v symmetry plane. The orbitals are classified as symmetric (S) or antisymmetric (A) with respect to these symmetry operations, and displayed in a schematic energy-level diagram

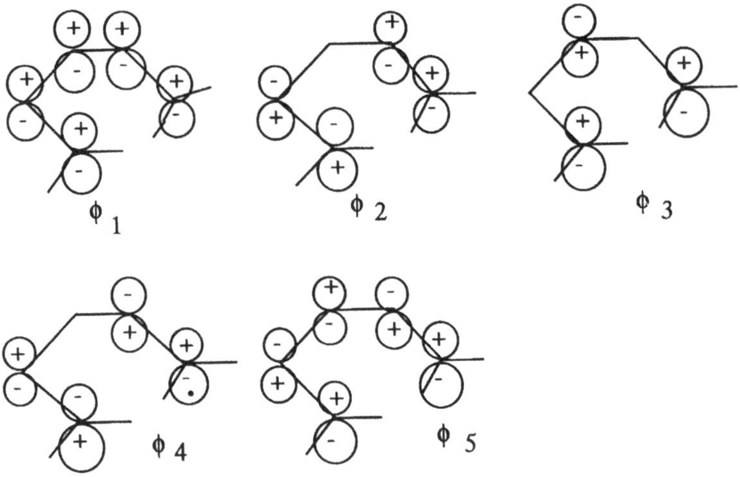

Figure 13.1 Pentadienyl cation π orbitals.

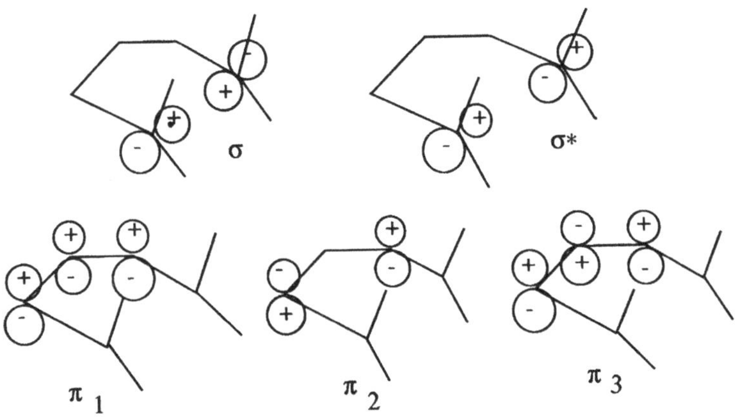

Figure 13.2 Cyclopentenyl cation orbitals.

for the reactant on the left and the product on the right. The energy levels of the reactant are joined with the energy levels of the product by straight lines according to the noncrossing rule. Actual calculations for various geometries along the reaction path would yield more irregular variations of the orbital energy levels. The straight-line construction, which results in what is known as a *molecular orbital energy-level correlation diagram*, is the only thing one can do with complete ignorance of the orbital energies at intermediate geometries.

The four least bound electrons can now be placed in the lowest two orbitals on the reactant side to form an electron configuration for the ground state, which then correlates to an excited product state in the disrotatory case and to the ground state for the conrotatory path. On the other hand, the doubly

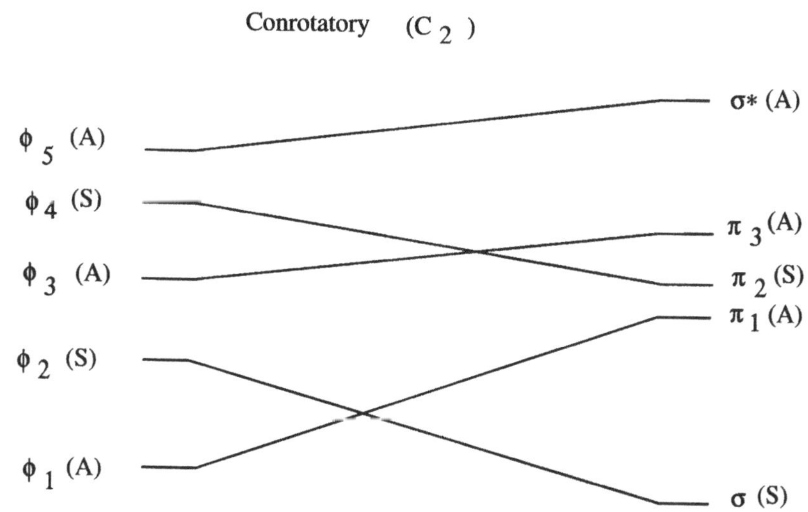

Figure 13.3 Molecular energy-level correlation diagram for the conrotatory reaction path.

excited reactant configuration $(\phi_1)^2(\phi_3)^2$ in the disrotatory case correlates to the product ground state (see Fig. 13.4). Simple considerations of this kind together with the calculated energies make possible the predictions that the thermal electrocyclic reaction is conrotatory, while the disrotatory reaction is photochemically allowed. The same predictions follow in this case from corresponding correlation diagram between many-electron states.

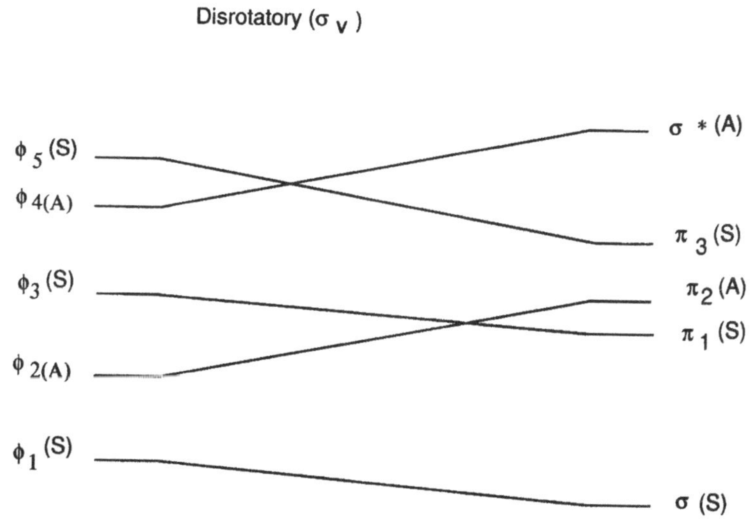

Figure 13.4 Molecular energy-level correlation diagram for the disrotatory reaction path.

13.2 THE HARTREE–FOCK APPROXIMATION

The many-electron Hamiltonian of a molecule may be expressed as

$$H = \sum_{i=1}^{N} h_i + \frac{1}{2}\sum_{i \neq j} g_{ij}, \tag{13.5}$$

where

$$\begin{aligned} h_i &= -\frac{1}{2}\vec{\nabla}_i^2 - \sum_{k=1}^{Q} \frac{Z_k}{|\vec{r}_i - \vec{R}_k|}, \\ g_{ij} &= \frac{1}{|\vec{r}_i - \vec{r}_j|}. \end{aligned} \tag{13.6}$$

Atomic units are used, and $\vec{r}_i$ is the position vector of one of the N electrons and $\vec{R}_k$ is the position vector of one of the Q nuclei. The atomic number of nucleus k is Z_k. This assumes clamped nuclei, and the solution of the corresponding electronic Schrödinger equation for a variety of nuclear geometries constitutes the traditional electronic structure problem in quantum chemistry. Its solutions have proved useful for the understanding of molecular structure, spectra, and energetics. To obtain accurate solutions, quite elaborate theory and computations have to be performed. Many of the more popular such accurate treatments start from a basic approach referred to as the *Hartree–Fock approximation*, or sometimes as the *independent particle approximation.*

A determinantal wavefunction built from orthonormal molecular spin orbitals $\{\phi_i(x)\}$ can be expressed as

$$\begin{aligned} &(N!)^{-1/2}\begin{vmatrix} \phi_1(x_1) & \phi_1(x_2) & \cdots & \phi_1(x_N) \\ \phi_2(x_1) & \phi_2(x_2) & \cdots & \phi_2(x_N) \\ \vdots & \vdots & \cdots & \vdots \\ \phi_N(x_1) & \phi_N(x_2) & \cdots & \phi_N(x_N) \end{vmatrix} \\ &= (N!)^{-1/2}\det\{\phi_1(x_1)\phi_2(x_2)\cdots\phi_N(x_N)\} \\ &= (N!)^{1/2} Q^{[1^N]} \prod_{i=1}^{N} \phi_i(x_i) = D, \end{aligned} \tag{13.7}$$

where $Q^{[1^N]} = (N!)^{-1}\sum_{P \in S_N}(-1)^p P = Q$ is the character projection operator of the antisymmetric irrep of the symmetric group S_N. Obviously it satisfies the relations

$$Q^2 = Q, \qquad Q^\dagger = Q \tag{13.8}$$

and

$$QH = HQ. \tag{13.9}$$

The norm of D can be calculated as an illustration of how a determinantal wavefunction of orthonormal spin orbitals can be manipulated[1]:

$$\begin{aligned}(D|D) &= N!\left(Q\prod_{i=1}^{N}\phi_i(x_i)|Q\prod_{i=1}^{N}\phi_i(x_i)\right)\\ &= N!\left(\prod_{i=1}^{N}\phi_i(x_i)|Q^\dagger Q\prod_{i=1}^{N}\phi_i(x_i)\right)\\ &= N!\left(\prod_{i=1}^{N}\phi_i(x_i)|Q\prod_{i=1}^{N}\phi_i(x_i)\right)\\ &= \sum_{P\in S_N}(-1)^p\left(\prod_{i=1}^{N}\phi_i(x_i)|P\prod_{i=1}^{N}\phi_i(x_i)\right) = 1\,,\end{aligned} \tag{13.10}$$

and because of the orthonormality of the spin orbitals,

$$(\phi_i|\phi_j) = \delta_{ij}\,. \tag{13.11}$$

The permutations P may be interpreted as rearranging the electron labels among the spin orbitals or as rearranging the spin orbital labels among the electrons. Both ways work, but should, of course, not be mixed in a particular problem. Similarly, the integral over the Hamiltonian can be expressed as

$$\begin{aligned}(D|HD) &= (N!)\left(Q\prod_{i=1}^{N}\phi_i(x_i)|HQ\prod_{i=1}^{N}\phi_i(x_i)\right)\\ &= (N!)\left(\prod_{i=1}^{N}\phi_i(x_i)|Q^\dagger HQ\prod_{i=1}^{N}\phi_i(x_i)\right)\\ &= (N!)\left(\prod_{i=1}^{N}\phi_i(x_i)|HQ\prod_{i=1}^{N}\phi_i(x_i)\right)\\ &= \sum_{P\in S_N}(-1)^p\left(\prod_{i=1}^{N}\phi_i(x_i)|HP\prod_{i=1}^{N}\phi_i(x_i)\right).\end{aligned} \tag{13.12}$$

[1] Sometimes the notation $(\psi(x)|\phi(x))$ for an integral $\int \psi^*(x)\phi(x)\,dx$ is used for clarity rather than the more common $(\psi|\phi)$ with the variables of integration suppressed.

The orthonormality of the spin orbitals limits the sum over permutations to those that yield

$$
\begin{aligned}
(D|HD) &= \sum_{i=1}^{N}(\phi_i(x_1)|h_1\phi_i(x_1)) \\
&\quad + \frac{1}{2}\sum_{i=1}^{N}\sum_{j=1}^{N}(\phi_i(x_1)\phi_j(x_2)|g_{12}(1 - P_{12})\phi_i(x_1)\phi_j(x_2)) \qquad (13.13)\\
&= \sum_{i=1}^{N}(i|h|i) + \frac{1}{2}\sum_{i=1}^{N}\sum_{j=1}^{N}\langle ij||ij\rangle,
\end{aligned}
$$

where the one-electron integrals $(i|h|i)$ and the antisymmetric (Coulomb minus exchange) two-electron integrals $\langle ij||ij\rangle = \langle ij|ij\rangle - \langle ij|ji\rangle$, with

$$
\langle ab|cd\rangle = \int dx_1 \int dx_2 a^*(x_1)b^*(x_2)g_{12}c(x_1)d(x_2) \qquad (13.14)
$$

are the natural building blocks usually obtained from integrals over basis sets.

The average total electronic energy is

$$
\begin{aligned}
\langle H\rangle &= \frac{(D|HD)}{(D|D)} \\
&= \sum_{i=1}^{N}(\phi_i(1)|h_1\phi_i(1)) \qquad (13.15)\\
&\quad + \tfrac{1}{2}\sum_{i=1}^{N}\sum_{j=1}^{N}(\phi_i(1)\phi_j(2)|g_{12}(1 - P_{12})\phi_i(1)\phi_j(2)).
\end{aligned}
$$

The variation of this functional with respect to the spin orbitals is

$$
\begin{aligned}
\delta\langle H\rangle &= \sum_{i=1}^{N}[(\delta\phi_i(1)|h_1\phi_i(1)) \\
&\quad + \sum_{j=1}^{N}(\delta\phi_i(1)\phi_j(2)|g_{12}(1 - P_{12})\phi_i(1)\phi_j(2)) \qquad (13.16)\\
&\quad + \text{complex conjugate}.
\end{aligned}
$$

In this expression we are using the shorthand notation 1 for the compound position-spin coordinate $x_1 = \{\vec{r}_1, m_{s_1}\}$. We have also used the integral symmetry $\langle ij||ij\rangle = \langle ji||ji\rangle$ to equate the two expressions

$$
\begin{aligned}
&\sum_{i=1}^{N}\sum_{j=1}^{N}(\delta\phi_i(1)\phi_j(2)|g_{12}(1 - P_{12})\phi_i(1)\phi_j(2)) \\
&= \sum_{i=1}^{N}\sum_{j=1}^{N}(\phi_i(1)\delta\phi_j(2)|g_{12}(1 - P_{12})\phi_i(1)\phi_j(2)) \qquad (13.17)
\end{aligned}
$$

resulting from the variation.

The variations $\delta\phi_i$ of the spin orbitals are subject to the constraint that $(D|D) = 1$, which means that the N^2 orthonormality conditions

$$(\phi_i|\phi_j) = \delta_{ij} \tag{13.18}$$

must remain valid during the variations. The problem then is to find an extremum value with constraints that can be handled with the use of undetermined Lagrangian multipliers. We construct the functional

$$I = \langle H \rangle - \sum_{i=1}^{M}\sum_{j=1}^{N}[(\phi_i|\phi_j) - \delta_{ij}]\lambda_{ji} \tag{13.19}$$

and consider the stationarity of this functional under variations of the spin orbitals, namely, $\delta I = 0$, which, of course, also means that $\delta I^* = 0$, and

$$\delta I - \delta I^* = 0. \tag{13.20}$$

Since $\delta\langle H \rangle$, as shown above, is a sum of a term and its complex conjugate, it must be real, so

$$\delta I - \delta I^* = 0 = \sum_{i=1}^{N}\sum_{j=1}^{N}[\lambda_{ji} - \lambda_{ij}^*]\delta(\phi_i|\phi_j). \tag{13.21}$$

In the last expression we have used the fact that $(\phi_i|\phi_j) = (\phi_j|\phi_i)^*$ and renamed (interchanged) the summation indexes in the second term. This relation obviously must hold for arbitrary variations of the spin orbitals, which forces the relation $\lambda_{ji} = \lambda_{ij}^*$ to hold. This means that the matrix of Lagrangian undetermined multipliers $\boldsymbol{\lambda}$ is Hermitian and then can be brought to diagonal form by a unitary transformation $\mathbf{U}$, such that

$$\mathbf{U}^\dagger \boldsymbol{\lambda} \mathbf{U} = \boldsymbol{\epsilon}, \tag{13.22}$$

where $\boldsymbol{\epsilon}$ is a diagonal matrix with elements ϵ_k. It is then possible to write

$$\lambda_{ji} = \sum_{k=1}^{N} U_{jk}\epsilon_k U_{ik}^*, \tag{13.23}$$

which, when inserted into

$$\sum_{i=1}^{N}\sum_{j=1}^{N}(\phi_i|\phi_j)\lambda_{ji}, \tag{13.24}$$

leads to

$$\sum_{k=1}^{N}\epsilon_k\left(\sum_{i=1}^{N}\phi_i U_{ik}\Big|\sum_{j=1}^{N}\phi_j U_{jk}\right) = \sum_{k=1}^{N}\epsilon_k(\psi_k|\psi_k), \tag{13.25}$$

and the spin orbitals have undergone a unitary transformation such that

$$\psi_k = \sum_{i=1}^{N} \phi_i U_{ik} \,. \tag{13.26}$$

The determinantal wavefunction D is invariant (up to an unimportant phase factor) to a unitary transformation of the spin orbitals:

$$\det\{\psi_1 \psi_2 \cdots \psi_N\} = \det \mathbf{U} \det\{\phi_1 \phi_2 \cdots \phi_N\} \,. \tag{13.27}$$

This means that the Lagrangian functional I can be written in terms of the spin orbitals $\{\psi_k\}$ with a diagonal multiplier matrix, such that

$$\delta I = \sum_{k=1}^{N} \left(\delta\psi_k(1) \left| \left[h_1 + \sum_{j=1}^{N} (\psi_j(2)|g_{12}(1 - P_{12})\psi_j(2)) - \epsilon_k \right] \right| \psi_k(1) \right)$$
$$+ \text{complex conjugate} = 0. \tag{13.28}$$

Since $\delta\psi_k$ in general is complex, there are two real degrees of freedom of variation, the real and the imaginary parts of $\delta\psi_k$, or equivalently, $\delta\psi_k$ and $\delta\psi_k^*$ can be considered as independent variations. This means that the first part of the right-hand side (RHS) of Eq. (13.28) and its complex conjugate vanish separately. Furthermore, since $\delta\psi_k$ is arbitrary, it must hold that

$$\left[h_1 + \sum_{j=1}^{N} (\psi_j(2)|g_{12}(1 - P_{12})\psi_j(2)) \right] \psi_k(1) = \epsilon_k \psi_k(1) \,, \tag{13.29}$$

which can be written as

$$F\psi_k = \epsilon_k \psi_k, \tag{13.30}$$

with the effective one-electron Hamiltonian or Fock operator F. In more detail this pseudoeigenvalue problem can be expressed as

$$F\psi_k(1) \equiv -\frac{1}{2}\vec{\nabla}^2\psi_k(1) - \sum_{m=1}^{Q} \frac{Z_m}{r_{1m}} \psi_k(1)$$
$$+ \sum_{j=1}^{N} \int r_{12}^{-1} \psi_j^*(2)\psi_j(2)\mathrm{d}(2)\psi_k(1) \tag{13.31}$$
$$- \sum_{j=1}^{N} \int r_{12}^{-1} \psi_j^*(2)\psi_k(2)\mathrm{d}(2)\psi_j(1) = \epsilon_k \psi_k(1) \,.$$

The last term of the detailed action of the Fock operator F is a result of the so-called exchange potential acting on $\psi_k(1)$ and comes from the permutation operator P_{12}, which, in turn, comes from the antisymmetric form of the determinantal wavefunction D.

Introducing the quantity

$$\rho(x, x') = \sum_{j-1}^{N} \psi_j(x)\psi_j^*(x') \tag{13.32}$$

for the Fock–Dirac fundamental invariant, or the *single-particle reduced density matrix*, corresponding to the N-electron wavefunction D, we can express the Fock operator as

$$F(1) = h_1 + \int r_{12}^{-1}(1 - P_{12})\rho(x_2, x_2')\, dx_2 \,. \tag{13.33}$$

This notation warrants a comment. The convention has been adopted that the operator [in this case $r_{12}^{-1}(1 - P_{12})$] acts only on unprimed coordinates, and after the operator has acted, one puts $x_2 = x_2'$ before integration.

13.3 INTERPRETATION OF THE INVARIANT ρ

The probability interpretation of the wavefunction means that the probability per unit volume of finding an electron at position $\vec{r}_1$ with spin m_{s_1} can be written as

$$\int D^* D \, dx_2 \, dx_3 \cdots dx_N \,, \tag{13.34}$$

where integration over the position coordinates and summation over the spin coordinates of $N - 1$ electrons has been performed. The determinant D expanded with respect to its first column reads

$$D = N^{-1/2}[\psi_1(x_1)D_1 + \psi_2(x_2)D_2 + \cdots + \psi_N(x_N)D_N], \tag{13.35}$$

where D_i is the minor of D obtained by omitting the first column and the ith row of D and giving it the sign factor $(-1)^{i+1}$. In addition the normalization factor $(N!)^{-1/2}$ has been factored such that the factor $((N - 1)!)^{-1/2}$ is included in the notation D_i. In this way each D_i is a normalized $N - 1$ electron determinant made up of spin orbitals with the arguments $x_2, x_3, \ldots, x_N$, and all the D_i are mutually orthogonal and normalized to unity. When this expansion is inserted into Eq. (13.34), we obtain

$$\begin{aligned} \int D^* D \, dx_2 \, dx_3 \cdots dx_N &= N^{-1} \sum_{j=1}^{N} \psi_j(x_1)\psi_j^*(x_1) \\ &= \frac{\rho(x_1, x_1)}{N} \,. \end{aligned} \tag{13.36}$$

It follows from this result that if we sum over spin, then

$$\sum_{m_s} \rho(x,x) = \rho(\vec{r},\vec{r}) \tag{13.37}$$

is the probability per unit volume of finding an electron at $\vec{r}$ (regardless of its spin) times the number of electrons. This means that $\rho(\vec{r},\vec{r})$ times the electronic charge can be interpreted as the electron charge density of the system.

The second term on the RHS of Eq. (13.33) is the precise definition of the term $V(\vec{r})$ used earlier as the average potential energy of an electron due to the presence of all the other electrons. The first part

$$\int r_{12}^{-1}\rho(x_2,x_2)\,dx_2 = \int r_{12}^{-1}\rho(\vec{r}_2,\vec{r}_2)\,d\vec{r}_2 \tag{13.38}$$

is simply the Coulomb repulsion potential energy due to the charge distribution of all the electrons, and the second part

$$-\int r_{12}^{-1}P_{12}\rho(x_2,x_2')\,dx_2 = K \tag{13.39}$$

is the so-called exchange potential energy, which is a *nonlocal* potential energy in the sense that it is not just multiplicative like a *local* potential energy, but acts like an integral operator, specifically

$$K\psi_k(x_1) = \int K(x_1,x_2)\psi_k(x_2)\,dx_2\,, \tag{13.40}$$

where $K(x_1,x_2)$ is called the "kernel" of the integral operator. Thus

$$\begin{aligned} K\psi_k(x_1) &= -\int r_{12}^{-1}P_{12}\rho(x_2,x_2')\,dx_2\psi_k(x_1) \\ &= -\int r_{12}^{-1}\rho(x_1,x_2)\psi_k(x_2)\,dx_2 \end{aligned} \tag{13.41}$$

and the kernel is $r_{12}^{-1}\rho(x_1,x_2) = K(x_1,x_2)$. The term in Eq. (13.38), sometimes referred to as the "direct" term, accounts for the nonphysical interaction of the electron with itself, and one effect of the exchange potential energy in Eq. (13.39) is that it subtracts out this nonphysical self-energy term.

It is clear that Eq. (13.31) is an *integrodifferential equation* and that the potential-energy terms depend on the solutions, through $\rho(x,x')$, which depends on all the (occupied) spin orbitals. The solution of Eq. (13.31), referred to as the *Hartree–Fock equation*, must then be accomplished by an iterative procedure. This iterative procedure must be brought to convergence or *self-consistency*, and the procedure of solution is termed the self-consistent field (SCF) method. We will return to this after we have made some further simplifying assumptions.

13.4 CLOSED-SHELL SYSTEMS

A closed-shell system has an even number of electrons $N = 2n$, and the spin orbitals may be restricted to appear in pairs with the same orbital part differing only in spin function. The Hartree–Fock equation with this form of determinantal wavefunction is called the *restricted Hartree–Fock* (RHF) problem. The Fock–Dirac fundamental invariant can then be expressed as

$$\rho(x,x') = \sum_{i=1}^{n} \psi_i(\vec{r})\psi_i^*(\vec{r})[\alpha(m_s)\alpha(m_s') + \beta(m_s)\beta(m_s')]. \qquad (13.42)$$

The introduction of a basis of atomic orbitals $\{\chi_\mu\}$, so that the molecular orbitals are expressed as linear combinations of atomic orbitals (the MO-LCAO approximation), means that the molecular orbitals are

$$\psi_i(x) = \sum_{\mu=1}^{M} \chi_\mu(x)c_{\mu i}. \qquad (13.43)$$

The unknown coefficients $\{c_{\mu i}\}$ are obtained from a secular problem, but since the Fock operator depends on these coefficients, the problem is a bit more complex than what we have indicated so far.

Should the Fock operator *not* be dependent on the solutions through ρ, then, obviously, the variational principle will yield the secular equations

$$\sum_{\nu=1}^{M}[F_{\mu\nu} - \epsilon_i S_{\mu\nu}]c_{\mu i} = 0 \quad (\mu = 1, 2, \cdots, M), \qquad (13.44)$$

with $F_{\mu\nu} = (\chi_\mu|F\chi_\nu)$, $S_{\mu\nu} = (\chi_\mu|\chi_\nu)$. It is noteworthy that for χ_μ and χ_ν of different spin the matrix elements in Eq. (13.44) vanish, while for χ_μ and χ_ν having the same spin (either α or β), the result obtained after summation over spin is

$$\begin{aligned} F_{\mu\nu} &= (\chi_\mu\alpha|F\chi_\nu\alpha) \\ &= (\chi_\mu(\vec{r}_1)|h_1\chi_\nu(\vec{r}_1)) + 2\sum_{j=1}^{n}\int r_{12}^{-1}\chi_\mu^*(\vec{r}_1)\psi_j(\vec{r}_2)\psi_j^*(\vec{r}_2)\chi_\nu(\vec{r}_1)\,d\vec{r}_1\,d\vec{r}_2 \\ &\quad - \sum_{j=1}^{n}\int r_{12}^{-1}\chi_\mu^*(\vec{r}_1)\psi_j(\vec{r}_1)\psi_j^*(\vec{r}_2)\chi_\nu(\vec{r}_2)\,d\vec{r}_1\,d\vec{r}_2. \end{aligned} \qquad (13.45)$$

In this expression we have used the result of Eq. (13.42). When the expansion in the atomic orbital basis as expressed in Eq. (13.43) is used in the above expression for the Fock matrix elements [Eq. (13.45)] one readily obtains the expression

$$F_{\mu\nu} = (\mu|h|\nu) + \sum_{\kappa,\lambda} P_{\kappa\lambda}[\langle\mu\kappa|\nu\lambda\rangle - \tfrac{1}{2}\langle\mu\kappa|\lambda\nu\rangle], \qquad (13.46)$$

with

$$(\mu|h|\nu) = (\chi_\mu|h\chi_\nu) = \int \chi_\mu^*(\vec{r}_1)h\chi_\nu(\vec{r}_1)\,d\vec{r}_1,$$
$$\langle\mu\kappa|\nu\lambda\rangle = \int r_{12}^{-1}\chi_\mu^*(\vec{r}_1)\chi_\kappa^*(\vec{r}_2)\chi_\nu(\vec{r}_1)\chi_\lambda(\vec{r}_2)\,d\vec{r}_1\,d\vec{r}_2, \tag{13.47}$$

and

$$P_{\lambda\kappa} = 2\sum_{j=1}^{n} c_{\lambda j}c_{\kappa j}^*. \tag{13.48}$$

Since the atomic orbital basis is fixed, all the one- and two-electron integrals are usually calculated once and stored for a given nuclear geometry. An iterative procedure is then used to determine the molecular orbitals. The Fock matrix depends on the coefficients $\{c_{\mu i}\}$ through the *charge–bond-order* matrix

$$\mathbf{P} = \{P_{\lambda\kappa}\}. \tag{13.49}$$

Such an iterative procedure can in general be carried out as follows. An initial estimate of the coefficients of the occupied molecular orbitals or rather all the $\{P_{\lambda\kappa}\}$ is made (this can be done by performing an approximate calculation, namely, solving the secular problem with an approximate Fock matrix, say, from Hückel theory). Such an initial estimate of **P** together with the integrals over the atomic orbital basis yields an initial Fock matrix **F** and permits a solution of the secular problem

$$[\mathbf{F} - \epsilon_i\mathbf{S}]\mathbf{c}_i = \mathbf{0}, \tag{13.50}$$

yielding a set of molecular orbital energies ϵ_i and associated molecular orbital coefficients $\mathbf{c}_i$. These eigenvectors then give a new charge–bond-order matrix **P** and also a new Fock matrix. This iterative procedure can be visualized as a loop in flowchart:

$$\begin{array}{ccccccc}
\text{Start} \rightarrow & \mathbf{P} & \rightarrow & \mathbf{F} & \rightarrow & \{\mathbf{c}_i, \epsilon_i\} & \\
& \uparrow & & & & \downarrow & \\
& \text{“no”} & \leftarrow & & & |c_{\mu j}^{(n)} - c_{\mu j}^{(n-1)}| < \text{tolerance} & \\
& & & & & \downarrow & \\
& & & & & \text{“yes”} & \\
& & & & & \downarrow & \\
& & & & & \text{end} &
\end{array} \tag{13.51}$$

The end of the iterative process is reached when the eigenvector components from the nth pass through the loop differ by less than a preset tolerance[2] from

[2] The tolerance can have different values depending on the size of the system and how reliable an answer one wants. Typically $10^{-5}-10^{-7}$ is used.

those of the previous pass. Then self-consistency has been reached. In practice, convergence may be difficult to reach, and often one resorts to some form of interpolation procedure to get a new **P** from two or more previous passes. The final molecular orbitals and associated energies are the SCF solutions.

The total electronic energy of a molecule is

$$E = \langle H \rangle = \sum_{i=1}^{n} 2(i|h|i) + \sum_{i,j=1}^{n} [2\langle ij|ij\rangle - \langle ij|ji\rangle], \tag{13.52}$$

where the integrals are over molecular orbitals and where we have summed over spin. The MO-LCAO approximation yields the expression

$$E = \sum_{\mu,\nu=1}^{M} P_{\nu\mu}(\mu|h|\nu) + \frac{1}{2} \sum_{\mu,\kappa,\nu,\lambda=1}^{M} P_{\nu\mu}P_{\lambda\kappa}[\langle\mu\kappa|\nu\lambda\rangle - \tfrac{1}{2}\langle\mu\kappa|\lambda\nu\rangle]. \tag{13.53}$$

Furthermore, since the orbital energy can be expressed as the average value of the Fock operator with respect to one of the normalized SCF orbitals, namely

$$\epsilon_k = (\psi_k|F\psi_k) = (k|h|k) + \sum_{j=1}^{n} [2\langle kj|kj\rangle - \langle kj|jk\rangle], \tag{13.54}$$

the total electronic energy can also be expressed as

$$E = 2\sum_{k=1}^{n} \epsilon_k - \sum_{k,j=1}^{n} [2\langle kj|kj\rangle - \langle kj|jk\rangle] \tag{13.55}$$

or

$$E = \sum_{k=1}^{n} [\epsilon_k + (k|h|k)]. \tag{13.56}$$

To obtain the correct total energy of the molecule (with stationary nuclei) the repulsion energy of the nuclei

$$\frac{1}{2} \sum_{p \neq q} \frac{Z_p Z_q}{|\vec{R}_p - \vec{R}_q|} \tag{13.57}$$

must be added.

For a closed-shell system the Fock–Dirac fundamental invariant is totally symmetric, leading to a totally symmetric Fock operator, with the orbitals labeled by the irreps of the appropriate point group for a molecule or $SO(3)$ for an atom.

13.5 OPEN-SHELL SYSTEMS

An open-shell system is one for which the pairing constraint is relaxed. This could be an odd-electron system or an even-electron system. The details can be dealt with in different ways. For instance a certain number of orbitals may be chosen to be doubly occupied as in the closed-shell case, and the rest can be given the same spin, say, α, such that the correct total spin quantum number is obtained. This divides the orbitals into a closed-shell part and an open-shell part. This approach is not discussed further here. Another possibility is to construct a fixed combination of determinants that correspond to a particular substate of the desired molecular or atomic term. Application of the variational principle then leads to the so-called Roothaan equations. This procedure introduces extra constraints on the orbitals resulting in a more complicated Lagrangian multiplier matrix, which in general cannot be diagonalized simultaneously with the Fock matrix. We do not consider this approach further.

A conceptually simpler scheme is to remove all double occupancy of orbitals and to associate the α spins and the β spins with independent orbitals $\phi^\alpha(\vec{r})$ and $\phi^\beta(\vec{r})$, respectively. This leads to two separate but interconnected secular problems, one for α spin orbitals and another for β spin orbitals. The Fock operator differs from the one in the closed-shell case, since there are p α spins and q β spins with, in general, $p \neq q$, and $p + q = N$. The Fock–Dirac fundamental invariant can be expressed as

$$\rho(x,x') = \sum_{i=1}^{p} \phi_i^\alpha(\vec{r})\phi_i^\alpha(\vec{r}')\alpha(m_s)\alpha(m_s') + \sum_{i=1}^{q} \phi_i^\beta(\vec{r})\phi_i^\beta(\vec{r}')\beta(m_s)\beta(m_s'). \tag{13.58}$$

Normally, both sets of molecular orbitals are expanded in terms of the same basis:

$$\begin{aligned} \phi_i^\alpha(\vec{r}) &= \sum_{\mu=1}^{M} \chi_\mu(\vec{r})c_{\mu i}^\alpha, \\ \phi_i^\beta(\vec{r}) &= \sum_{\mu=1}^{M} \chi_\mu(\vec{r})c_{\mu i}^\beta. \end{aligned} \tag{13.59}$$

The Fock operator matrix elements then become

$$\begin{aligned} F_{\mu\nu}^{\alpha\alpha} &= (\chi_\mu\alpha|F\chi_\nu\alpha) \\ &= (\mu|h|\nu) + \sum_{\kappa,\lambda=1}^{M} \{[P_{\lambda\kappa}^\alpha + P_{\lambda\kappa}^\beta]\langle\mu\kappa|\nu\lambda\rangle - P_{\lambda\kappa}^\alpha\langle\mu\kappa|\lambda\nu\rangle\}, \end{aligned} \tag{13.60}$$

and

$$\begin{aligned} F_{\mu\nu}^{\beta\beta} &= (\chi_\mu\beta|F\chi_\nu\beta) \\ &= (\mu|h|\nu) + \sum_{\kappa,\lambda=1}^{M} \{[P_{\lambda\kappa}^\alpha + P_{\lambda\kappa}^\beta]\langle\mu\kappa|\nu\lambda\rangle - P_{\lambda\kappa}^\beta\langle\mu\kappa|\lambda\nu\rangle\}, \end{aligned} \tag{13.61}$$

generating an $\alpha\alpha$ block and a $\beta\beta$ block, but, obviously, the $\alpha\beta$ and $\beta\alpha$ blocks are zero. The secular problem becomes

$$
\begin{aligned}
[\mathbf{F}^{\alpha\alpha} - \epsilon^{\alpha}\mathbf{S}]\mathbf{c}^{\alpha} &= \mathbf{0}, \\
[\mathbf{F}^{\beta\beta} - \epsilon^{\beta}\mathbf{S}]\mathbf{c}^{\beta} &= \mathbf{0}.
\end{aligned}
\tag{13.62}
$$

The solutions of the first secular problem yield $P^{\alpha}_{\lambda\kappa} = \sum_j^p c^{\alpha}_{\lambda j} c^{\alpha *}_{\kappa j}$, which are then used in the construction of both $\mathbf{F}^{\alpha\alpha}$ and $\mathbf{F}^{\beta\beta}$, while the solutions of the second secular problem yield $P^{\beta}_{\lambda\kappa} = \sum_j^q c^{\beta}_{\lambda j} c^{\beta *}_{\kappa j}$ also used for the construction of both $\mathbf{F}^{\alpha\alpha}$ and $\mathbf{F}^{\beta\beta}$.

The total electronic energy for such a *spin unrestricted* determinantal wavefunction can be expressed as

$$
E = \tfrac{1}{2}\sum_{j=1}^{p}[\epsilon_j^{\alpha} + (j^{\alpha}|h|j^{\alpha})] + \tfrac{1}{2}\sum_{j=1}^{q}[\epsilon_j^{\beta} + (j^{\beta}|h|j^{\beta})]. \tag{13.63}
$$

Although the molecular spin orbitals are orthogonal (and normalized) as assumed in the derivation of the Hartree–Fock equations, the orbitals are not, when the summation over spin is left out. It holds that

$$
(\phi_i^{\alpha}|\phi_j^{\alpha}) = (\phi_i^{\beta}|\phi_j^{\beta}) = \delta_{ij}, \tag{13.64}
$$

but

$$
(\phi_i^{\alpha}|\phi_j^{\beta}) \neq \delta_{ij}, \tag{13.65}
$$

since in general $\mathbf{F}^{\alpha\alpha} \neq \mathbf{F}^{\beta\beta}$.

For an open-shell system the Fock–Dirac fundamental invariant is not in general totally symmetric, leading to a Fock operator that is not totally symmetric. However, often one chooses a symmetrized form of ρ, that is, it is expressed as a symmetric sum of degenerate orbitals of each partially occupied subshell, each of which is fractionally occupied so as to yield the correct average number of electrons in that subshell. This means that the orbitals at self-consistency may not be the ones that yield the determinantal wavefunction with lowest total energy. A totally symmetric Fock operator has the advantage that the resulting orbitals are labeled by irreps of the full molecular point group or the full $SO(3)$ for an atom. In fact, ρ constructed in this manner does not correspond to a single determinant, but rather to some form of statistical ensemble of degenerate determinants. Should the Fock–Dirac fundamental invariant be constructed with only the particular choice of open-shell orbitals that occur in the determinantal wavefunction, then one gets, in general, a Fock operator of a symmetry lower than that represented by the many-electron Hamiltonian of the system, and the orbitals are labeled by the irreps of a subgroup of the full symmetry group of the system. For instance, the carbon atom with real p orbitals would be treated in the D_{2h} subgroup rather than in the full $SO(3)$.

When the SCF calculation are the zeroth-order treatment of a more elaborate, so-called correlated approach, the choice of these different starting points for open-shell systems are usually not that crucial. The SCF part of the calculation plays the role of an orbital generator, and as long as the raw orbital basis is the same and the treatment is at a sufficiently high level, the final correlated results starting from different SCF approximations tend to differ very little. However, the question of symmetry labels still remains.

13.6 MOLECULAR PROPERTIES

Electronic properties of a molecular system are related to quantum-mechanical operators. For instance, the electric dipole moment due to the electrons is represented by the operator

$$\vec{d} = -\sum_{j=1}^{N} \vec{r}_j , \tag{13.66}$$

that is, a sum of all the electron position coordinates multiplied by the electron charge (-1 in atomic units). The average value of the electric dipole moment due to the electrons with respect to a determinantal wavefunction is

$$\langle \vec{d} \rangle = \sum_{j=1}^{N} (j|\vec{r}|j) = -\int \vec{r}\rho(x,x')\,dx . \tag{13.67}$$

In general, for an operator that is a symmetric sum of one-electron operators, such as

$$A = \sum_{j=1}^{N} a_j , \tag{13.68}$$

the average value is simply

$$\langle A \rangle = \sum_{j=1}^{N} (j|a|j) = \int a\rho(x,x')\,dx , \tag{13.69}$$

where the one-electron operator acts on the coordinate x and, after it has operated, the coordinate x' is put equal to x before integration. For instance, the total electronic kinetic energy is

$$\langle T \rangle = \sum_{j=1}^{N} (j|-\tfrac{1}{2}\vec{\nabla}^2|j) = \int (-\tfrac{1}{2}\vec{\nabla}^2)\rho(x,x')\,dx . \tag{13.70}$$

The electron density for a spin-unrestricted determinantal wavefunction is

$$\sum_{m_s} \rho(x,x') = \rho^{\alpha}(\vec{r},\vec{r}) + \rho^{\beta}(\vec{r},\vec{r})$$
$$= \sum_{j=1}^{p} \phi_j^{\alpha}(\vec{r})\phi_j^{\alpha}(\vec{r}) + \sum_{j=1}^{q} \phi_j^{\beta}(\vec{r})\phi_j^{\beta}(\vec{r}). \tag{13.71}$$

The average value of the z component of total electron spin angular momentum

$$S_z = \sum_{j=1}^{N} s_z(j), \tag{13.72}$$

which is zero for a closed-shell system, has the following average value for an open-shell spin unrestricted determinant:

$$\langle S_z \rangle = \sum_{j=1}^{N} (j|s_z|j) = \int s_z \rho(x,x')\, dx$$
$$= \tfrac{1}{2} \int [\rho^{\alpha}(\vec{r},\vec{r}) - \rho^{\beta}(\vec{r},\vec{r})]\, d\vec{r}. \tag{13.73}$$

The difference density

$$s(\vec{r},\vec{r}) = [\rho^{\alpha}(\vec{r},\vec{r}) - \rho^{\beta}(\vec{r},\vec{r})] \tag{13.74}$$

is called the *total spin density.* Its value at each nucleus is of importance in electron spin resonance studies.

13.7 SEPARATION OF CORE AND VALENCE STATES

The full MO-LCAO-SCF problem

$$[\mathbf{F} - \epsilon \mathbf{S}]\mathbf{c} = \mathbf{0} \tag{13.75}$$

can be solved for quite large systems even without using any symmetry. The size of systems that can be treated depends on methodology and computer software, and architecture. "Direct SCF," which computes the Fock matrix directly without intermediate storage of integrals and parallel computer architecture, uses distribution of tasks over (often a large number of) processors permitting SCF calculations of very large systems, treating all electrons explicitly. The capabilities for handling large molecular systems are always increasing. Any statement of the size of systems possible to study in this way is a moving target and strongly time-dependent. Routine calculations can now be performed for systems with up to 100 first-row atoms. When heavy elements are included, such as transition metals, rare earths, and actinides, the number of atoms possible to handle with adequate basis sets and by full MO-LCAO procedures quickly reduces to a few. For systems exhibiting high symmetry, larger calculations can be attempted. Fullerenes, and stereoregular polyenes with carbon and silicon backbones, which are infinite systems treated by cyclic boundary conditions, have been studied. There will, however, always be larger systems of interest that are beyond present capabilities for a full MO-LCAO treatment. Therefore there is always a need for simplified methods that can be used to make predictions about such systems.

It is a recognized fact that inner orbitals do not change drastically in going from free atoms to molecules, at least if we limit our interest to low-lying states. Observations such as these inspire methods that treat core orbitals as being essentially fixed and to focus on the valence orbitals, which undergo more dramatic changes during bond formation and bond breaking.

The core atomic orbitals are localized near the nucleus, and interactions between core electrons on distant and even on adjacent atoms can be neglected in the first approximation. The overlap integrals beween core orbitals are negligible. The atomic core orbitals are good approximations to molecular core orbitals. On the other hand, valence atomic orbitals, which are more extended in space, may have considerable overlap between adjacent centers and interact strongly with one another. One-center core–valence interaction is also generally small, when the atomic orbitals are SCF orbitals, which are orthogonal and also noninteracting across the atomic Fock operator. Two-center core–valence interaction, in contrast, is expected to be important, since overlap can be nonnegligible.

To exploit these observations, one seeks to separate the core (c) and valence (v) part of the molecular orbitals. To begin with this separation is completely arbitrary, and must be specified in each application. The notation

$$\mathbf{M} = \mathbf{F} - \epsilon\mathbf{S} \tag{13.76}$$

is introduced and the secular problem gets a partitioned form

$$\begin{bmatrix} \mathbf{M}_{vv} & \mathbf{M}_{vc} \\ \mathbf{M}_{cv} & \mathbf{M}_{cc} \end{bmatrix} \begin{bmatrix} \mathbf{c}_v \\ \mathbf{c}_c \end{bmatrix} = \begin{bmatrix} \mathbf{0} \\ \mathbf{0} \end{bmatrix}, \tag{13.77}$$

or

$$\begin{aligned} \mathbf{M}_{vv}\mathbf{c}_v + \mathbf{M}_{vc}\mathbf{c}_c &= \mathbf{0}, \\ \mathbf{M}_{cv}\mathbf{c}_v + \mathbf{M}_{cc}\mathbf{c}_c &= \mathbf{0}. \end{aligned} \tag{13.78}$$

We solve formally for $\mathbf{c}_c$ from the second equation to obtain

$$\mathbf{c}_c = -\mathbf{M}_{cc}^{-1}\mathbf{M}_{cv}\mathbf{c}_v \tag{13.79}$$

and substitute it into the first equation to yield

$$[\mathbf{M}_{vv} - \mathbf{M}_{vc}\mathbf{M}_{cc}^{-1}\mathbf{M}_{cv}]\mathbf{c}_v = \mathbf{0}, \tag{13.80}$$

which is a matrix problem of the dimension of the valence orbital space. Introducing the notation

$$\mathbf{G}(\epsilon) = -\mathbf{M}_{vc}\mathbf{M}_{cc}^{-1}\mathbf{M}_{cv}, \tag{13.81}$$

we write the secular problem as

$$[\mathbf{F}_{vv} + \mathbf{G}(\epsilon) - \epsilon\mathbf{S}_{vv}]\mathbf{c}_v = \mathbf{0}. \tag{13.82}$$

So far there are no approximations. This problem is entirely equivalent to the original one and is only a reformulation with a rather complicated term containing the effect of the core part on the valence part. However, this formulation is a useful starting point for approximations.

In the following Greek subscripts $(\mu, \nu, \kappa, \lambda, \ldots)$ denote valence atomic orbitals and the subscripts $(m, n, k, l, \ldots)$ denote core atomic orbitals. Explicitly, one can write

$$G_{\mu\nu}(\epsilon) = -\sum_{k,l}[F_{\mu k} - \epsilon S_{\mu k}][(\mathbf{F}_{cc} - \epsilon\mathbf{S}_{cc})^{-1}]_{kl}[F_{l\nu} - \epsilon S_{l\nu}], \tag{13.83}$$

which can be simplified, since the matrix

$$[\mathbf{F}_{cc} - \epsilon\mathbf{S}_{cc}] = \mathbf{A} + \mathbf{B} \tag{13.84}$$

is diagonally dominant because of the assumed slight interaction between core orbitals. The diagonal part **A** and the off-diagonal part **B** can be used in an expansion of the inverse matrix by iterating the identity

$$(\mathbf{A} + \mathbf{B})^{-1} = \mathbf{A}^{-1} - \mathbf{A}^{-1}\mathbf{B}(\mathbf{A} + \mathbf{B})^{-1}. \tag{13.85}$$

In this way different approximations can be obtained defined by the powers of the "small" matrix **B** that are kept. For instance, keeping only the first term $(\mathbf{A}^{-1})$ on the right yields the approximation

$$\begin{aligned} G_{\mu\nu}^{(0)} &= -[\mathbf{M}_{vc}\mathbf{A}^{-1}\mathbf{M}_{cv}]_{\mu\nu} \\ &= -\sum_k [F_{\mu k} - \epsilon S_{\mu k}][F_{kk} - \epsilon]^{-1}[F_{k\nu} - \epsilon S_{k\nu}] \\ &\approx -\sum_k S_{\mu k}[F_{kk} - \epsilon]S_{k\nu}, \end{aligned} \tag{13.86}$$

where the last expression is obtained by putting

$$F_{\mu k} = S_{\mu k}F_{kk}, \tag{13.87}$$

which is not exactly true, but would be if

$$F\chi_k = F_{kk}\chi_k, \tag{13.88}$$

that is, the atomic core orbital χ_k is an eigenfunction of the molecular Fock operator F. This is a good approximation for the deep core orbitals and less so for the higher ones.

The second term on the expansion of the inverse in Eq. (13.85) yields

$$\begin{aligned} G_{\mu\nu}^{(1)} &= [\mathbf{M}_{vc}\mathbf{A}^{-1}\mathbf{BA}^{-1}\mathbf{M}_{cv}]_{\mu\nu} \\ &\approx \sum_{k\neq l} S_{\mu k}[F_{kl} - \epsilon S_{kl}]S_{l\nu}, \end{aligned} \tag{13.89}$$

where the same assumptions as before have been used. The elements of the matrix

$$\mathbf{G} \approx \mathbf{G}^{(0)} + \mathbf{G}^{(1)} \tag{13.90}$$

are still dependent on the energy parameter ϵ, which introduces complications, which can be eliminated by some further considerations. The energy values of interest are those of a typical valence orbital eigenvalue, and thus numerically smaller, by a factor of at least 0.1–0.01, than F_{kk}, which is essentially a core orbital energy. It also follows straightforwardly from the assumption that the core atomic orbitals are approximate eigenfunctions to the molecular Fock operator, that $F_{kl} \approx S_{kl}(F_{kk} + F_{ll})/2$. This again shows that ϵS_{kl} is one to two orders of magnitude smaller than F_{kl}. Neglect of these smaller energy dependent terms is a more severe approximation than necessary in order to achieve a convenient form of the secular problem, but the arguments presented thus far illustrate the kind of reasoning that permits a formulation of a secular problem for valence orbitals only. Pursuing this approach to the end yields

$$\begin{aligned} F_{\mu\nu} &= (\mu|h|\nu) - \sum_{k} S_{\mu k}F_{kk}S_{k\nu} + \sum_{k\neq l} S_{\mu k}F_{kl}S_{l\nu} + \cdots \\ &\quad + \sum_{\lambda,\kappa} P_{\lambda\kappa}[\langle\mu\kappa|\nu\lambda\rangle - \tfrac{1}{2}\langle\mu\kappa|\lambda\nu\rangle] \\ &= H_{\mu\nu} + \sum_{\lambda,\kappa} P_{\lambda\kappa}[\langle\mu\kappa|\nu\lambda\rangle - \tfrac{1}{2}\langle\mu\kappa|\lambda\nu\rangle]. \end{aligned} \tag{13.91}$$

The one-electron term $H_{\mu\nu}$ contains contributions from the core, the core–valence overlap, and so on and in the common theoretical schemes is not calculated from the integrals but parametrized in some manner from experiment or from ab initio calculations on prototypical molecules.

13.8 ZERO-DIFFERENTIAL OVERLAP

Even with the rather severe approximations introduced in the previous section, further simplifications are often invoked, sometimes with weak justifications. A widely used assumption is that of zero-differential overlap (ZDO). The product of two atomic orbitals of the same coordinate

$$\chi_\mu^*(\vec{r})\chi_\nu(\vec{r}) \tag{13.92}$$

is called their differential overlap and approximated with

$$\chi_\mu^*(\vec{r})\chi_\nu(\vec{r}) = \delta_{\mu\nu}\chi_\mu(\vec{r})\chi_\mu(\vec{r}) \tag{13.93}$$

in all integrals, thereby considerably reducing the number of two-electron integrals that need to be considered. This seemingly drastic assumption can be rationalized by introducing a set of symmetrically orthogonalized[3] atomic orbitals, which yield results equivalent to the ZDO assumption. In some contexts the ZDO approximation yields realistic molecular orbitals; however, the justification is questionable for general molecular systems.

Introduction of the ZDO means that, for instance

$$\begin{aligned} \langle\mu\kappa|\nu\lambda\rangle &= \delta_{\mu\nu}\delta_{\kappa\lambda}\langle\mu\kappa|\mu\kappa\rangle, \\ S_{\mu\nu} &= \delta_{\mu\nu}, \end{aligned} \tag{13.94}$$

and leading to the Fock matrix with elements

$$\begin{aligned} F_{\mu\mu} &= H_{\mu\mu} - \tfrac{1}{2}P_{\mu\mu}\langle\mu\mu|\mu\mu\rangle + \sum_\kappa P_{\kappa\kappa}\langle\mu\kappa|\mu\kappa\rangle, \\ F_{\mu\nu} &= H_{\mu\nu} - \tfrac{1}{2}P_{\mu\nu}\langle\mu\nu|\mu\nu\rangle, \quad \text{for} \quad \mu \neq \nu. \end{aligned} \tag{13.95}$$

13.8.1 The Pariser–Parr–Pople model

This approach is typically used for planar unsaturated hydrocarbon molecules and can be readily extended to include molecules with first-row heteroatoms such as nitrogen and oxygen. It has been used for the study of polymeric systems including nuclear motion.

The basic assumption of the model can be summarized as follows:

- Only the $2p_z$ atomic orbitals on the participating carbon atoms are in the valence set. These atomic orbitals are odd under reflection in the molecular plane (the xy plane), and the molecular orbitals formed as linear combinations of them are called *π orbitals*. The other molecular orbitals formed from the $s, 2p_x$, and $2p_y$ atomic orbitals are not treated explicitly and are referred to as *σ orbitals*. They are symmetric under reflection in the molecular plane. This means that each carbon atom (and each nitrogen

[3] Symmetric orthogonalization means that the inverse square root of the metric matrix is used as a transformation matrix, that is, the orthonormal basis is $\phi_\kappa = \sum_\lambda^M \chi_\lambda(\mathbf{\Delta}^{-(1/2)})_{\lambda\kappa}$.

or oxygen heteroatom) contributes one $2p_z$ atomic orbital to the basis, and thus, there is a one-to-one correspondence between the label of a basis function and that of its atomic center.

•

$$
\begin{aligned}
H_{\mu\mu} &= (\mu|h|\mu) + G_{\mu\mu} \\
&= \left(\chi_\mu \middle\| \left[-\tfrac{1}{2}\vec{\nabla}^2 - \frac{Z_\mu}{r_\mu}\right] \middle\| \chi_\mu\right) + G_{\mu\mu} \\
&\quad - \sum_{\kappa\neq\mu} \left(\chi_\mu \middle\| \left[\frac{Z_\kappa}{r_\kappa}\right] \middle\| \chi_\mu\right) = U_{\mu\mu} - \sum_{\kappa\neq\mu} V_{\kappa\mu}.
\end{aligned}
\tag{13.96}
$$

- The two-electron integrals$\langle\mu\kappa|\mu\kappa\rangle = \gamma_{\mu\kappa}$ are parametrized in various ways; the simplest is $\gamma_{\mu\kappa} = 1/R_{\mu\kappa}$, where $R_{\mu\kappa}$ is the distance between atoms μ and κ.
- $V_{\kappa\mu} = Z_\kappa\gamma_{\kappa\mu}$, where Z_κ for carbon atoms is usually chosen to be 1. $H_{\mu\nu} = \beta_{\mu\nu}$ is a parameter, which is zero unless atoms μ and ν are directly bonded to each other. This parameter can be determined from calculations on small molecules by comparisons with known spectra. One usually chooses $\beta_{\mu\nu} = \beta$ for each C—C bond and determines its value from comparisons with spectra (ethene $\beta = -2.37$ eV) for a single molecule and uses this throughout the calculations on several molecules in a series of related systems.
- For hydrocarbons, $U_{\mu\mu}$ can be considered a constant, the same for each carbon atom, and its value is then of no consequence. In the same way $\gamma_{\mu\mu}$ need not be specified. However, when heteroatoms are involved, there is a need to assign values to these parameters. They can be estimated from the ionization potentials I_μ and electron affinities A_μ of the individual atoms, which can be approximated as[4]

$$
\begin{aligned}
-I_\mu &= U_{\mu\mu} + (Z_\mu - 1)\gamma_{\mu\mu}, \\
-A_\mu &= U_{\mu\mu} + Z_\mu\gamma_{\mu\mu}.
\end{aligned}
\tag{13.97}
$$

This yields

[4] The energy required to remove an electron from atom μ is I_μ, which means that one has to overcome its kinetic energy plus its attraction energy from the screened nucleus plus the electron repulsion from thr other electrons. Similarly, the electron affinity is the energy gained when an electron is added to the atom.

$$U_{\mu\mu} = -\tfrac{1}{2}(I_\mu + A_\mu) - \left(Z_\mu - \tfrac{1}{2}\right)\gamma_{\mu\mu},$$
$$\gamma_{\mu\mu} = (I_\mu - A_\mu), \tag{13.98}$$

- The Fock matrix, which determines the molecular orbitals in the Pariser–Parr–Pople (PPP) scheme is then

$$F_{\mu\mu} = U_{\mu\mu} + \tfrac{1}{2}P_{\mu\mu}\gamma_{\mu\mu} + \sum_{\kappa\neq\mu}(P_{\kappa\kappa} - 1)\gamma_{\kappa\mu},$$
$$F_{\mu\nu} = \beta_{\mu\nu} - \tfrac{1}{2}P_{\mu\nu}\gamma_{\mu\nu}. \tag{13.99}$$

13.8.2 The Hückel π-Orbital Model

This model, although historically quite a bit older than the PPP model, can be considered a simplification of the latter. The Fock matrix of the Hückel π-orbital method is obtained from the PPP Fock matrix by simply incorporating all terms in a diagonal element into a single parameter

$$F_{\mu\mu} = \alpha_\mu \tag{13.100}$$

(referred to as a *Coulomb integral*) and the terms in an off-diagonal element into a single parameter

$$F_{\mu\nu} = \beta_{\mu\nu} \tag{13.101}$$

(referred to as a *resonance integral*), which is zero unless μ and ν are bonded neighbors.

The parameters α_μ are different for each kind of atom. Thus, for the carbon atom framework in a hydrocarbon, there would only be one parameter value, $\alpha_\mu = \alpha$, which need not be specified, since it serves only as a reference level of the orbital energies. Similarly, there would only be one $\beta_{\mu\nu} = \beta$ for all C—C bonds. Thus, except for the value of β, it is only the topology that matters and that is input.

Parametrization schemes for heteroatoms exist, and some choices are listed in the table below.

Atom (X)	Sample Compound	Coulomb Integral (a)	Resonance Integral (X–Y)	Coulomb Integral (Y)
C	Benzene	α	β(carbon)	
N(1)	Pyridine	$\alpha + 0.5\beta$	β(pyridine N)	$a + 0.05\beta$
			0.8β(pyrrole N)	
			0.8β(furan O)	
N(2)	Pyrrole	$\alpha + 1.5\beta$	0.8β(carbon)	$a + 0.075\beta$
			0.8β(pyridine N)	
			0.6β(pyrrole N)	
			0.6β(furan O)	
N(3)	Nitrile	$\alpha + 0.5\beta$	β(carbon)	$a + 0.05\beta$
			β(carbon)	
N(4)	Nitro	$\alpha + 1.5\beta$	0.8β(oxygen)	$a + 0.075\beta$
O(1)	Acrolein	$\alpha + \beta$	β(carbon)	$a + 0.1\beta$
			β(pyridine N)	
			0.8β(pyrrole N)	
O(2)	Furan	$\alpha + 2\beta$	0.8β(carbon)	$a + 0.1\beta$
			0.6β(furan O)	
			0.8β(pyridine N)	
O(3)	Nitro	$\alpha + \beta$	0.8β(nitrogen)	$a + 0.1\beta$
F	Fluorobenzene	$\alpha + 3\beta$	0.7β(carbon)	$a + 0.3\beta$
Cl	Chlorobenzene	$\alpha + 2\beta$	0.4β(carbon)	$a + 0.2\beta$
Br	Bromobenzene	$\alpha + 1.5\beta$	0.3β(carbon)	$a + 0.15\beta$
CH_3	Toluene	$\alpha - 0.5\beta$	0β(carbon)	

The Hückel molecular orbital secular problem can be used as an example of how point group symmetry leads to simplifications, even when only a small subgroup of the full point group is used. For instance, the π MO treatment of the benzene molecule takes the form

$$\begin{bmatrix} \alpha-\epsilon & \beta & 0 & 0 & 0 & \beta \\ \beta & \alpha-\epsilon & \beta & 0 & 0 & 0 \\ 0 & \beta & \alpha-\epsilon & \beta & 0 & 0 \\ 0 & 0 & \beta & \alpha-\epsilon & \beta & 0 \\ 0 & 0 & 0 & \beta & \alpha-\epsilon & \beta \\ \beta & 0 & 0 & 0 & \beta & \alpha-\epsilon \end{bmatrix} \begin{bmatrix} c_1 \\ c_2 \\ c_3 \\ c_4 \\ c_5 \\ c_6 \end{bmatrix} = \begin{bmatrix} 0 \\ 0 \\ 0 \\ 0 \\ 0 \\ 0 \end{bmatrix}. \tag{13.102}$$

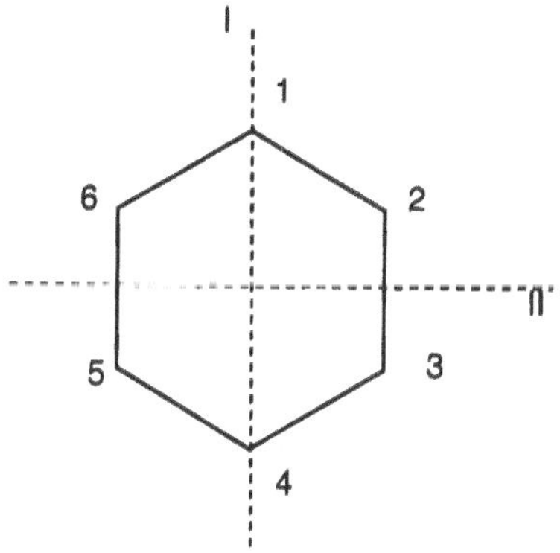

Figure 13.5 Symmetry planes I and II of the benzene molecule.

By dividing the entire equation by β and writing $x = (\alpha - \epsilon)/\beta$, we get the "topological matrix" problem

$$\begin{bmatrix} x & 1 & 0 & 0 & 0 & 1 \\ 1 & x & 1 & 0 & 0 & 0 \\ 0 & 1 & x & 1 & 0 & 0 \\ 0 & 0 & 1 & x & 1 & 0 \\ 0 & 0 & 0 & 1 & x & 1 \\ 1 & 0 & 0 & 0 & 1 & x \end{bmatrix} \begin{bmatrix} c_1 \\ c_2 \\ c_3 \\ c_4 \\ c_5 \\ c_6 \end{bmatrix} = \begin{bmatrix} 0 \\ 0 \\ 0 \\ 0 \\ 0 \\ 0 \end{bmatrix}. \tag{13.103}$$

We now use limited symmetry to solve this problem and introduce the two symmetry planes, I and II (see Fig. 13.5).

We first look for MOs that are symmetric under reflection in both planes. This puts the following conditions on the MO coefficients

$$S_I S_{II}:\ c_1 = c_4; \qquad c_2 = c_3 = c_5 = c_6. \tag{13.104}$$

Multiplying out the secular equation [Eq. (13.103)] yields the two equations

$$\begin{aligned} xc_1 + c_2 + c_2 &= 0, \\ c_1 + xc_2 + c_2 &= 0. \end{aligned} \tag{13.105}$$

All the others are copies of these. These two equations correspond to a secular problem in matrix form such that

$$\begin{bmatrix} x & 2 \\ 1 & x + 1 \end{bmatrix} \begin{bmatrix} c_1 \\ c_2 \end{bmatrix} = \begin{bmatrix} 0 \\ 0 \end{bmatrix} \tag{13.106}$$

and the solutions

$$x(x + 1) - 2 = 0; \qquad x = 1, -2. \tag{13.107}$$

Putting $x = 1$, that is, $\epsilon_4 = \alpha - \beta$, we get $c_1 = -2c_2$, and

$$u_4 = \frac{2\chi_1 - \chi_2 - \chi_3 + 2\chi_4 - \chi_5 - \chi_6}{\sqrt{12}} \tag{13.108}$$

and the $x = -2$, that is, $\epsilon_1 = \alpha + 2\beta$, we get $c_2 = c_1$ and

$$u_1 = \frac{\chi_1 + \chi_2 + \chi_3 + \chi_4 + \chi_5 + \chi_6}{\sqrt{6}}. \tag{13.109}$$

The MOs that are symmetric in plane I and antisymmetric in plane II are obtained from

$$S_I A_{II}\colon\ c_1 = -c_4,\ c_2 = -c_3 = -c_5 = c_6, \tag{13.110}$$

and

$$\begin{aligned} xc_1 + c_2 + c_2 &= 0, \\ c_1 + xc_2 - c_2 &= 0, \end{aligned} \tag{13.111}$$

or

$$\begin{bmatrix} x & 2 \\ 1 & x-1 \end{bmatrix} \begin{bmatrix} c_1 \\ c_2 \end{bmatrix} = \begin{bmatrix} 0 \\ 0 \end{bmatrix}, \tag{13.112}$$

which yields $x = -1, 2$. For $x = -1 i.e.$ $\epsilon_2 = \alpha + \beta$ and $c_1 = 2c_2$, we get

$$u_2 = \frac{2\chi_1 + \chi_2 - \chi_3 - 2\chi_4 - \chi_5 + \chi_6}{\sqrt{12}}. \tag{13.113}$$

And when $x = 2$ or $\epsilon_6 = \alpha - 2\beta$ $c_1 = -c_2$, the result is

$$u_6 = \frac{\chi_1 - \chi_2 + \chi_3 - \chi_4 + \chi_5 - \chi_6}{\sqrt{12}}. \tag{13.114}$$

The MO that is antisymmetric in plane I and symmetric in plane II is the one for which

$$c_1 = c_4 = 0; \qquad c_2 = c_3 = -c_5 = -c_6, \tag{13.115}$$

and

$$xc_2 + c_2 = 0, \tag{13.116}$$

that is, $x = -1$, $\epsilon_3 = \alpha + \beta$, and

$$u_3 = \frac{\chi_2 + \chi_3 - \chi_5 - \chi_6}{2}. \tag{13.117}$$

The remaining MO, which is antisymmetric under reflection in both planes, has

$$c_1 = c_4 = 0; \qquad c_2 = -c_3 = c_5 = -c_6, \tag{13.118}$$

and

$$xc_2 - c_2 = 0, \tag{13.119}$$

or $x = 1$, $\epsilon_5 = \alpha - \beta$, yielding

$$u_5 = \frac{\chi_2 - \chi_3 + \chi_5 - \chi_6}{2}. \tag{13.120}$$

These benzene π molecular orbitals can be readily labeled by the irreps of the full D_{6h} point group. By referring to the character table, (Ch. 8) it is easy to conclude that u_1 is an a_{2u} orbital. Since u_2 and u_3 are degenerate, they span an e irrep; and by considering the actions of, say, the inversion and the σ_h operation, one can determine that the irrep is e_{1g}. Similarly, the degenerate u_4 and u_5 orbitals are of e_{2u} symmetry. The u_6 orbital is obviously a b_{2g} molecular orbital.

The ground-state configuration of benzene for the six electrons in the least bound π MOs is then $(a_{2u})^2(e_{1g})^4$, and low-lying excited states can be obtained from the excited electron configuration $(a_{2u})^2(e_{1g})^3(e_{2u})^1$. Using the character table, we can compute the characters for the reducible representation $E_{1g} \times E_{2u}$ and resolve it into its irreps as $B_{1u} + B_{2u} + E_{1u}$, which means that we should expect to find some excited states of symmetries

$${}^{1,3}B_{1u},\, {}^{1,3}B_{2u},\, {}^{1,3}E_{1u} \tag{13.121}$$

Four of these six states have been identified in UV absorption (${}^3B_{1u}, {}^1B_{2u}, {}^1B_{1u}, {}^1E_{1u}$), in that order.

13.8.3 Cyclic Systems

In general, for a system with cyclic symmetry, the Hückel secular problem can be expressed as

$$a_k + a_{k+1}x + a_{k+2} = 0, \text{ modulo } N, \tag{13.122}$$

where the a_k are coefficients such that the molecular orbitals are

$$\psi_{\text{MO}} = \sum_{k=0}^{N-1} \phi_k a_k \tag{13.123}$$

in terms of a set of atomic orbitals $\{\phi_k\}$ centered on N atoms labeled by $k = 0, 1, \ldots, N-1$. A cyclic system of N identical atoms has at least the symmetry of the Abelian point group C_N. There may, of course, be additional symmetry elements but we ignore them for now.

The cyclic group C_N with elements $\{C_N^n\}, n = 0, 1, \ldots, N-1$, has only one-dimensional irreps with the characters

$$\chi^{(m)}(C_N^n) = e^{2\pi imn/N}, \qquad m = 0, 1, \ldots, N-1. \tag{13.124}$$

The irreps are thus labeled by $m = 0, 1, \ldots, N - 1$, where $m = 0$ is obviously the identity irrep.

We use the superscript m to label a molecular orbital spanning a particular irrep, such that

$$\psi^{(m)} = \sum_{k=0}^{N-1} a_k^{(m)} \phi_k . \tag{13.125}$$

Now, since by definition $C_N^n \phi_k = \phi_{k+n}$, modulo N, and

$$C_N^n \psi^{(m)} = e^{2\pi imn/N} \psi^{(m)}, \tag{13.126}$$

we find that

$$\begin{aligned} C_N^n \psi^{(m)} &= \sum_{k=0}^{N-1} a_k^{(m)} \phi_{k+n} \\ &= e^{2\pi imn/N} \sum_{k=0}^{N-1} a_{k+n}^{(m)} \phi_{k+n}, \text{ modulo } N. \end{aligned} \tag{13.127}$$

This means that $a_{k+n}^{(m)} e^{2\pi imn/N} = a_k^{(m)}$, which, when inserted into Eq. (13.122), yields

$$a_k^{(m)} + a_k^{(m)} e^{-2\pi im/N} x_m + a_k^{(m)} e^{-4\pi im/N} = 0 \tag{13.128}$$

or

$$x_m = -2 \cos \frac{2\pi m}{N} . \tag{13.129}$$

Since the topological Hückel matrix is formed by defining $x = (\alpha - \epsilon)/\beta$, we find the MO energies to be

$$\epsilon_m = \alpha + 2\beta \cos \frac{2\pi m}{N} . \tag{13.130}$$

This general result should be checked for $N = 6$, $N = 7$, and so on. One readily finds that since

$$\cos \frac{2\pi m}{N} = \cos \left(2\pi \frac{N - m}{N} \right), \tag{13.131}$$

there must be $N - 1$ pairwise degenerate energies for N odd and $N - 2$ for N even. This indicates that there are additional symmetry elements and that C_N is only a subgroup of the full symmetry group of the Hückel Hamiltonian for these systems.

In the character tables of the C_N groups the one-dimensional irreps corresponding to degenerate energy levels are commonly paired to form two-dimensional (E) representations that then are reducible. It is also notable how the preceding discussion is relevant for general systems with translational symmetry and periodic boundary conditions.

13.8.4 The CNDO Method

The complete neglect of differential overlap (CNDO), the intermediate neglect of differential overlap (INDO), and related schemes are similar to the PPP model, but all valence electrons are treated, not only those in π molecular orbitals. These models apply to arbitrary molecules, but parametrizations have to be supplied for atoms of all elements in the compounds of interest.

The assumptions of the CNDO method are described below:

- Only valence orbitals are treated explicitly.
- The ZDO approximation is applied.
- Electron interaction integrals $\langle\mu\kappa|\mu\kappa\rangle$ and penetration integrals $V_{\kappa\mu}$ are calculated using spherical atomic charge distributions (i.e., atomic Slater-type s orbitals). This is necessary to ensure invariance of the results to rotations of local coordinate systems.
- Other integrals are calculated over nonorthogonal STOs. One could argue that orthogonalized atomic orbitals would correspond more closely to the ZDO assumption. Some integrals are simply parametrized. The Fock matrix elements are

$$\begin{aligned} F_{\mu\mu} &= H_{\mu\mu} - \tfrac{1}{2}P_{\mu\mu}\langle\mu\mu|\mu\mu\rangle + \sum_{\kappa} P_{\kappa\kappa}\langle\mu\kappa|\mu\kappa\rangle, \\ F_{\mu\nu} &= H_{\mu\nu} - \tfrac{1}{2}P_{\mu\nu}\langle\mu\nu|\mu\nu\rangle. \end{aligned} \tag{13.132}$$

- The specific CNDO choices include the symbol $\gamma_{AB} = \langle\mu\nu|\mu\nu\rangle$ for χ_μ on atom A and χ_ν on atom B or vice versa. These integrals then depend only on the kind of atoms A and B and the distance R_{AB} between them.
- Furthermore

$$\begin{aligned} H_{\mu\mu} &= \left(\chi_\mu \,\middle|\, \left[-\frac{1}{2}\vec{\nabla}^2 - \frac{Z_A}{r_A}\right]\chi_\mu\right) + G_{\mu\mu} \\ &\quad - \sum_{B\neq A}\left(\chi_\mu \,\middle|\, \left[\frac{Z_B}{r_B}\right]\chi_\mu\right) = U_{\mu\mu} - \sum_{B\neq A} V_{AB}, \end{aligned} \tag{13.133}$$

  that is, the notation $V_{AB} = V_{\mu B}$ for every orbital χ_μ on atom A.
- $H_{\mu\nu} = 0$ when $\chi_\mu \neq \chi_\nu$ and both on atom A, while $H_{\mu\nu} = \beta_{AB}S_{\mu\nu}$, one is on A and the other on B. The parameters β_{AB} depend only on the kind of atoms and actual values are usually determined from ground state calculations on small molecules of atoms A and B.
- The Fock matrix elements then can be expressed as

$$\begin{aligned} F_{\mu\mu} &= U_{\mu\mu} + (P_{AA} - \tfrac{1}{2}P_{\mu\mu})\gamma_{AA} + \sum_{B\neq A}(P_{BB}\gamma_{AB} - V_{AB}), \\ F_{\mu\nu} &= \beta_{AB}S_{\mu\nu} - \tfrac{1}{2}P_{\mu\nu})\gamma_{AB}, \end{aligned} \tag{13.134}$$

where it is understood that the off-diagonal elements for both χ_μ and χ_ν on atom A are given by putting $S_{\mu\nu} = 0$ and replacing γ_{AB} by γ_{AA}. The parameters $U_{\mu\mu}$ are determined from ionization potentials and electron affinities for valence states, such that

$$\begin{aligned} -I_\mu &= U_{\mu\mu} + (Z_A - 1)\gamma_{AA}, \\ -A_\mu &= U_{\mu\mu} + Z_A\gamma_{AA}, \end{aligned} \tag{13.135}$$

with Z_A the nuclear charge screened by the core. This yields

$$U_{\mu\mu} = -\tfrac{1}{2}(I_\mu + A_\mu) - (Z_A - \tfrac{1}{2})\gamma_{AA}. \tag{13.136}$$

The integrals γ_{AB} and γ_{AA} are calculated with simple Slater-type s orbitals, and the penetration integrals are usually given by $V_{AB} = Z_A\gamma_{AB}$.

The INDO scheme, in addition to the one-center integrals γ_{AA}, uses all one-center two-electron integrals $\langle\mu\kappa|\nu\lambda\rangle$, which means that exchange integrals are included. This is important for spin densities, and reflects the fact that INDO was first introduced for applications to radicals. INDO can also be used for excited-state calculations.

A generic form of the total energy of molecules with stationary nuclei can be given as

$$E = \frac{1}{2}\sum_{\mu,\nu} P_{\mu\nu}[H_{\mu\nu} + F_{\mu\nu}] + \frac{1}{2}\sum_{A\neq B}\frac{Z_A Z_B}{R_{AB}}. \tag{13.137}$$

The repulsion energy between cores is calculated as between point charges. This result holds for the CNDO, INDO, and PPP models when the appropriate expression for each quantity is used.

13.8.5 The Extended Hückel Model

A very simple model for all valence electrons of a molecular system is obtained when the choice of Fock matrix elements is

$$\begin{aligned} F_{\mu\mu} &= -I_\mu, \\ F_{\mu\nu} &= \tfrac{1}{2}k(F_{\mu\mu} + F_{\nu\nu})S_{\mu\nu} \end{aligned} \tag{13.138}$$

The overlap integrals are calculated between atomic Slater-type orbitals, and the parameter k is chosen by fitting to some experimental spectra. A generic choice is $k = 1.75$.

BIBLIOGRAPHY

Numerous texts available in the literature give details about molecular orbital theory, and the variety of styles is sufficient to suit any particular taste. This author has enjoyed reading some older texts and papers such as the following:

Hoffman, R., *J. Chem. Phys.* **39**, 1397 (1963).

Parr, R. G., *The Quantum Theory of Molecular Electronic Structure*, Benjamin, New York, 1963.

Pople, J. A., and G. A. Segal, *J. Chem. Phys.* **43**, S136 (1965).

Pople, J. A., and G. A. Segal, *J. Chem. Phys.* **44**, 3289 (1966).

Slater, J. C., *Quantum Theory of Matter*, McGraw-Hill, New York, 1968.

14

ELECTRON CORRELATION

A many-electron wavefunction consisting of a single determinant often provides a reasonable qualitative representation of the molecular ground state. The total energy of such a wavefunction determined from the variational principle, namely, the Hartree–Fock energy, when compared with the "exact" energy, is too high by about 1%. This is a considerable amount of energy for most molecular systems, in particular when compared to typical energy differences, such as bond energies, potential-energy barriers, and the like. So, while a single determinant wavefunction in many cases provides a reasonable description of molecular electronic states and their properties, the errors associated with the Hartree–Fock approximation, by definition called *correlation errors*, are important and must be addressed in order to obtain a quantitative description of many-electron systems.

14.1 CONFIGURATION INTERACTION

Consider the ground-state electron configuration of the oxygen molecule

$$(1\sigma_g)^2(1\sigma_u)^2(2\sigma_g)^2(2\sigma_u)^2(3\sigma_g)^2(1\pi_u)^4(1\pi_g)^2, \tag{14.1}$$

where all the orbitals except the last one are completely filled. Suppressing the closed-shell part of the wavefunction, we can consider all the various choices of determinantal wavefunctions for the last two electrons. For instance, both electrons can be put in the $1\pi_{+g}$ orbital or both in the degenerate $1\pi_{-g}$ orbital [we are using the complex form $\pi_{\pm g} = p_{\pm A} - p_{\pm B}$, here with

$p_\pm = R(r)r\sin\theta e^{\pm i\varphi}$]. This leads to the two (degenerate) determinantal wavefunctions

$$\begin{aligned}\det\{1\pi_{+g}\alpha 1\pi_{+g}\beta\} &= 1\pi_{+g}(1)1\pi_{+g}(2)\{\alpha\beta - \beta\alpha\},\\ \det\{1\pi_{g-}\alpha 1\pi_{g-}\beta\} &= 1\pi_{g-}(1)1\pi_{g-}(2)\{\alpha\beta - \beta\alpha\},\end{aligned} \tag{14.2}$$

which obviously make up the $^1\Delta_g$ term. We recognize the characteristic spin singlet function $(\alpha\beta - \beta\alpha)$ and the symbol Δ associated with the absolute value of the sum over magnetic quantum numbers $(|\sum_i m_i| = 2)$[1]. Other choices are

$$\det\{1\pi_{+g}\alpha 1\pi_{g-}\alpha\} = \{1\pi_{+g}(1)1\pi_{g-}(2) - 1\pi_{g-}(1)1\pi_{+g}(2)\}\alpha\alpha \tag{14.3}$$

and

$$\det\{1\pi_{+g}\beta 1\pi_{g-}\beta\} = \{1\pi_{+g}(1)1\pi_{g-}(2) - 1\pi_{g-}(1)1\pi_{+g}(2)\}\beta\beta, \tag{14.4}$$

which belong to the $^3\Sigma_g$ term. The third function for this triply degenerate energy is

$$\begin{aligned}&\det\{1\pi_{+g}\alpha 1\pi_{g-}\beta\} + \det\{1\pi_{+g}\beta 1\pi_{g-}\alpha\}\\ &= \{1\pi_{+g}(1)1\pi_{g-}(2) - 1\pi_{g-}(1)1\pi_{+g}(2)\}\{\alpha\beta + \beta\alpha\},\end{aligned} \tag{14.5}$$

which is not possible to write as a single determinant, since it must have the same orbital part as its partners, and since it must be orthogonal to the wavefunction

$$\begin{aligned}&\det\{1\pi_{+g}\alpha 1\pi_{g-}\beta\} - \det\{1\pi_{+g}\beta 1\pi_{g-}\alpha\}\\ &= \{1\pi_{+g}(1)1\pi_{g-}(2) + 1\pi_{g-}(1)1\pi_{+g}(2)\}\{\alpha\beta - \beta\alpha\},\end{aligned} \tag{14.6}$$

of the $^1\Sigma_g$ term. Wavefunctions such as these, which must be expressed as fixed linear combinations of two or more determinants in order to obtain the proper symmetry (in this case the correct total spin), are referred to as *symmetry-adapted configurations*. Obviously, a single determinant, as, say

$$\det\{1\pi_{+g}\alpha 1\pi_{g-}\beta\}, \tag{14.7}$$

is a mixture of a singlet state and a triplet state.

We consider the water molecule (see Fig. 14.1) with the ground-state electron configuration

$$(1a_1)^2(2a_1)^2(3a_1)^2(1b_1)^2(1b_2)^2 \tag{14.8}$$

[1] $|\sum_i m_i| = 0$ denotes Σ terms and $|\sum_i m_i| = 1$ Π terms.

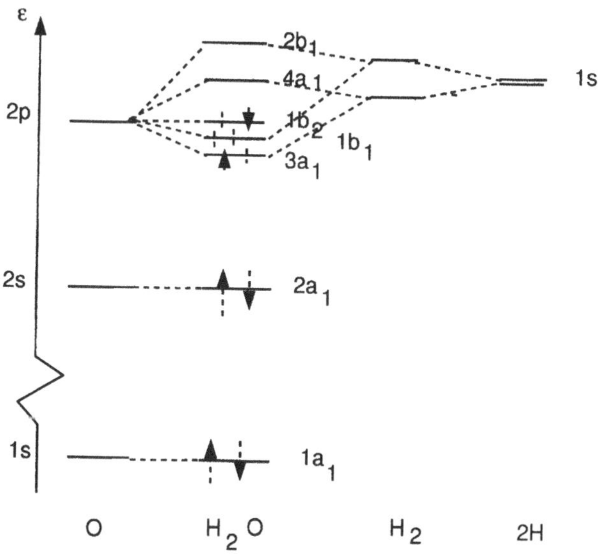

Figure 14.1 MO energy levels of the water molecule.

at equilibrium.

The corresponding closed-shell determinant

$$\det\{\cdots 1b_2\alpha 1b_2\beta\} \tag{14.9}$$

of the ground-state ${}^1A_1$ term can be used as a reference from which to create other determinantal wavefunctions. A *single excitation* means a promotion of an electron from a spin orbital *occupied* in the ground-state determinant to an *unoccupied or virtual* orbital. This creates a singly excited determinant. A single determinant, as we have seen, is not necessarily an eigenfunction of total spin angular momentum. Using the appropriate projection operators, it is always possible to construct from such a determinant a fixed combination of determinants that have the desired eigenfunction properties. In the case of the water molecule we can construct the spin singlet

$$\det\{\cdots 1b_2\alpha 4a_1\beta\} - \det\{\cdots 1b_2\beta 4a_1\alpha\}, \tag{14.10}$$

which has the symmetry ${}^1B_2$, or we can form the singly excited determinant

$$\det\{\cdots 1b_2\alpha 2b_1\alpha\}, \tag{14.11}$$

which is a ${}^3A_2$ state. In this manner we can obtain approximate wavefunctions for excited states. SCF procedures can be used to determine the "best" single determinant wavefunction for the lowest state of a given symmetry. We can also consider *double excitations, triple excitations*, and so on out of the given ground-state determinant. Examples of double excitations would be

$$\det\{\cdots 3a_1\alpha 3a_1\beta 4a_1\alpha 4a_1\beta\} \quad (14.12)$$

of $^1A_1$ symmetry, and

$$\det\{\cdots 3a_1\alpha 3a_1\beta 2b_1\alpha 4a_1\beta\} - \det\{\cdots 3a_1\alpha 3a_1\beta 2b_1\beta 4a_1\alpha\} \quad (14.13)$$

of $^1B_1$ symmetry.

The error in the total energy for a single determinantal wavefunction, namely, in the Hartree–Fock energy, is called the *correlation energy*, since the single determinant produces a mean field theory, which neglects the detailed correlations between electrons in their intricate dynamics by insisting on a one-to-one correspondence between electrons and spin orbitals.

One way to produce a more flexible variational wavefunction that would give lower total energies when used with the variational principle, and, thus would be a better approximation to the solution of the electronic Schrödinger equation, is to form linear combinations of determinants with adjustable mixing coefficients. This approach is referred to as *configuration interaction* (CI) or as the method of *superposition of determinants*.

The CI approximation can take many forms. One common way is to start from a reference determinant $|D_0\rangle$, which is usually a ground-state determinantal wavefunction with orbitals determined by an SCF procedure. When the MO-LCAO approximation is carried out in terms of an orbital basis of size m, it corresponds to a basis of $M = 2m$ spin orbitals. If we have, say, a closed-shell system with $N = 2n$ electrons, there are n occupied orbitals in the reference determinant, which could be an RHF determinant. There are then $M - N$ unoccupied spin orbitals available to which electrons can be promoted to form singly, doubly, and further excited determinants or symmetry-adapted configurations. By constructing all possible excited determinants to all orders including those where all N spin orbitals of $|D_0\rangle$ are replaced in all different ways, we get what is referred to as the *full CI*. This is the best wavefunction that can be constructed within a finite orbital basis and consists of

$$\binom{M}{N} \quad (14.14)$$

determinants. This can become a prohibitively large number for even a modest-sized orbital basis, and truncations are used.

We label the spin orbitals occupied in $|D_0\rangle$ by $a, b, c, \ldots$ and the unoccupied spin orbitals by $p, q, r, \ldots$. A singly excited determinant, created from the reference determinant by replacing spin orbital a by spin orbital p, is denoted D_a^p. Similarly a doubly excited determinant is D_{ab}^{pq}, and so on. The CI expansion can then be written as

$$|\Psi\rangle = |D_0\rangle + \sum_{a,p} c_a^p |D_a^p\rangle + \sum_{\substack{a<b \\ p<q}} c_{ab}^{pq} |D_{ab}^{pq}\rangle + \sum_{\substack{a<b<c \\ p<q<r}} c_{abc}^{pqr} |D_{abc}^{pqr}\rangle + \cdots, \quad (14.15)$$

where the restrictions on the summation indices assure that a given determinant is included only once. CI expansions of this type can be quite long. Even for such an inadequate basis as $M = 20$ and $N = 10$, the full CI contains 184,756 determinants. For extended basis sets one easily gets expansions in the billions of determinants. There are obvious reductions in these expansions due to symmetry, since the secular problem is symmetry blocked. Full CI calculations for good basis sets are important benchmarks, but can only be attempted for small systems.

Normally one has to consider severe truncations, and one talks about special cases such as *singles only* or *singles and doubles*. There are inherent weaknesses in truncated CI expansions. For instance, the lack of *size extensivity* is bothersome. This means that the desired property of the total energy of a system consisting of, say, n identical noninteracting subsystems is in general lacking for a truncated CI, and not equal to n times that of a single subsystem, as it should.

Another desirable property is that of correct behavior of a system dissociating into two subsystems. For instance, the breaking of a bond should give two subsystems in the correct states. Even truncated CI wavefunctions can be constructed to behave correctly in this regard. The simplest possible system for which this can be illustrated is H_2. The ground state of the molecule should separate into two ^2S ground-state hydrogen atoms. We let a and b denote the $1s$ atomic orbital on each of the two hydrogen atoms, and form the two possible molecular orbitals $1\sigma_g = a + b$ and $1\sigma_u = a - b$. The spin singlet ground state determinant is

$$\begin{aligned} D_0 &= |1\sigma_g(1)\alpha(1)1\sigma_g(1)\beta(1)| \\ &\quad [a(1)a(2) + b(1)b(2) + a(1)b(2) + b(1)a(2)](\alpha\beta - \beta\alpha), \end{aligned} \tag{14.16}$$

which, when expanded into products of atomic orbitals, clearly exhibits the equal weights for the system breaking up into an ionic state ($H^+ + H^-$) with wavefunction

$$[a(1)a(2) + b(1)b(2)](\alpha\beta - \beta\alpha), \tag{14.17}$$

and into two separated hydrogen atoms with the wavefunction

$$[a(1)b(2) + b(1)a(2)](\alpha\beta - \beta\alpha). \tag{14.18}$$

This 50% nonphysical ionic character of the closed-shell single determinant can be correced by using also the determinant

$$\begin{aligned} D_1 &= |1\sigma_u(1)\alpha(1)1\sigma_u(1)\beta(1)| \\ &\quad [a(1)a(2) + b(1)b(2) - a(1)b(2) - b(1)a(2)](\alpha\beta - \beta\alpha), \end{aligned} \tag{14.19}$$

and forming the fixed combination

$$D_0 - D_1 \propto [a(1)b(2) + b(1)a(2)](\alpha\beta - \beta\alpha), \tag{14.20}$$

which has the correct behavior on bond breaking. So, if both D_0 and D_1 are made part of a CI expansion, the correct behavior should be guaranteed. For general molecular systems with double and triple bonds, it is not as easy to construct wavefunctions with the correct behavior. For instance, the separation of the nitrogen molecule into two ground-state ^4S nitrogen atoms requires a bit more care.

The many-electron Hamiltonian can be expressed as

$$H = \sum_{i=1}^{N} h(i) + \frac{1}{2} \sum_{i \neq j}^{N} \frac{1}{r_{ij}}, \tag{14.21}$$

with the one-electron operators

$$h(i) = -\frac{1}{2}\vec{\nabla}_i^2 - \sum_{k=1}^{Q} \frac{Z_k}{|\vec{r}_i - \vec{R}_k|}. \tag{14.22}$$

The average with respect to the reference determinant has already been discussed, and we now write it as

$$\begin{aligned} (D_0|HD_0) &= \sum_{a=1}^{N} (a|h|a) + \frac{1}{2} \sum_{a,b=1}^{N} \langle ab||ab\rangle \\ &= \sum_{a=1}^{N} \epsilon_a - \frac{1}{2} \sum_{a,b} \langle ab||ab\rangle \end{aligned} \tag{14.23}$$

with one-electron integrals and the two-electron (Coulomb minus exchange) integrals, and in the second expression the spin orbital energies

$$\epsilon_a = (a|h|a) + \sum_{b=1}^{N} \langle ab||ab\rangle = (a|Fa), \tag{14.24}$$

where [in Eq. (14.24)] we recognize that the molecular spin orbitals are eigenfunctions of the Fock operator F.

Similar reasoning can be used to obtain the matrix element of the many-electron Hamiltonian between the reference determinant and a singly excited determinant:

$$\begin{aligned} &(D_0|HD_a^p) \\ &= \sum_P (-1)^P (\phi_1(1) \cdots \phi_a(a) \cdots \phi_N(N)|HP\phi_1(1) \cdots \phi_p(a) \cdots \phi_N(N)) \\ &= (a|h|a) + \sum_{b=1}^{N} \langle ab||pb\rangle = (a|Fp) = 0, \end{aligned} \tag{14.25}$$

where, in the last step, it is recognized that this matrix element is the same as a matrix element of the Fock operator between the two SCF (eigen)functions

ϕ_a and ϕ_p, which is zero. This lack of interaction between the reference determinant and the singly excited ones when the determinants are expressed in terms of SCF spin orbitals is called the *Brillouin theorem.*

Similarly, we can deduce that the matrix elements between the reference determinant and the doubly excited ones are

$$(D_0|HD_{ab}^{pq}) = \langle ab||pq\rangle, \tag{14.26}$$

while triply excited determinants and higher do not interact across the Hamiltonian with the reference determinant, since the Hamiltonian contains at most two-electron operators. It is also straightforward to show that

$$(D_a^p|HD_a^p) = (D_0|HD_0) + \epsilon_p - \epsilon_a - \langle ap||ap\rangle \tag{14.27}$$

and that

$$(D_a^p|HD_b^q) = \langle pb||aq\rangle. \tag{14.28}$$

Writing the CI expansion symbolically as

$$|\Psi\rangle = |0\rangle + |S\rangle + |D\rangle + |T\rangle + |Q\rangle + \cdots \tag{14.29}$$

with $|0\rangle$ for the reference determinant and $|S\rangle$ for the sum of all the singly excited determinants and so on, then the Hamiltonian matrix takes the following block form (we show only the upper triangular part):

	$\lvert 0\rangle$	$\lvert S\rangle$	$\lvert D\rangle$	$\lvert T\rangle$	$\lvert Q\rangle$	$\cdots$
$\langle 0\rvert$	$\langle 0\lvert H0\rangle$	0	$\langle 0\lvert HD\rangle$	0	0	$\cdots$
$\langle S\rvert$		$\langle S\lvert HS\rangle$	$\langle S\lvert HD\rangle$	$\langle S\lvert HT\rangle$	0	$\cdots$
$\langle D\rvert$			$\langle D\lvert HD\rangle$	$\langle D\lvert HT\rangle$	$\langle D\lvert HQ\rangle$	$\cdots$
$\langle T\rvert$				$\langle T\lvert HT\rangle$	$\langle T\lvert HQ\rangle$	$\cdots$
$\langle Q\rvert$					$\langle Q\lvert HQ\rangle$	$\cdots$
$\vdots$						$\ddots$

The pattern of the secular problem suggests that if the CI expansion is truncated to only $|D_0\rangle = |0\rangle$ and the single excitations $|S\rangle = \sum_{a,p} c_a^p|D_a^p\rangle$, then the ground state is simply approximated by $|0\rangle$ and its energy is unaffected by the single excitations. One could conclude that even when the CI expansion is not truncated in such a severe fashion, the effect of the single excitations (in terms of SCF orbitals) on the ground state is not as great as that of, say, double excitations, while excited states are greatly influenced by the singles.

Vertical excitation energies (those for which the molecular geometry does not change during the transition) as well as the corresponding transition moments are often quite successfully modeled with a singles only CI:

$$|\Psi\rangle = |0\rangle + \sum_{a,p} c_a^p|D_a^p\rangle. \tag{14.30}$$

This approach is less successful for the shapes of potential-energy surfaces and for excited-state properties but often yields quite reliable predictions for electronic spectra involving low-lying states. The transition moment for a transition in absorption from the ground state to an excited state represented as

$$|X\rangle = \sum_{a,p} c_a^p |D_a^p\rangle \tag{14.31}$$

is calculated using the electric dipole moment operator

$$\vec{d} = \sum_{i=1}^{N} \vec{r}_i \,, \tag{14.32}$$

giving the expression

$$\langle 0|\vec{d}X\rangle = \sum_{a,p} (a|\vec{r}|p) c_a^p \,. \tag{14.33}$$

Nonzero contributions occur for those orbitals that belong to irreps (of the appropriate point group) such that $\Gamma^J \in \Gamma^a \times \Gamma^{(x,y,z)} \times \Gamma^p$.

The CI wavefunction does not have to be expressed in terms of an SCF reference determinant or configuration, but can equally well be developed in terms of a multideterminantal reference wavefunction. It can also be expressed in terms of orbitals different from the SCF molecular orbitals. For instance, orthogonalized atomic orbitals can be used. A particularly useful set of orbitals, which can be obtained for any CI wavefunction, is the set of *natural orbitals.* Natural orbitals are obtained from the diagonalization of the one-electron reduced density matrix

$$\gamma(x,x') = N \int \Psi(x,x_2,\ldots,x_N)\Psi^*(x',x_2,\ldots,x_N)\, dx_2\, dx_3 \cdots dx_N \,, \tag{14.34}$$

with the normalization condition

$$1 = \langle\Psi|\Psi\rangle = \int \Psi^*(x_1,x_2,\ldots,x_N)\Psi^*(x_1,x_2,\ldots,x_N)\, dx_1\, dx_2 \cdots dx_N \,. \tag{14.35}$$

When the many-electron wavefunction is written as a CI expansion

$$|\Psi\rangle = \sum_{K=1}^{M} C_K |D_K\rangle \,, \tag{14.36}$$

that is, as a linear combination orthonormal determinants

$$D_K = (N!)^{1/2} Q^{(1^N)} \prod_{i=1}^{N} \phi_{k_i} \,, \tag{14.37}$$

where the expansion coefficients C_K are determined from the secular problem, the density matrix is expressed as

$$\gamma(x, x') = \sum_{k,l} \phi_k(x)\phi_l^*(x)\gamma_{kl}\,, \tag{14.38}$$

and thc normalization is

$$\sum_{K=1}^{M} |C_K|^2 = 1\,. \tag{14.39}$$

The matrix

$$\gamma - \{\gamma_{kl}\} \tag{14.40}$$

is self-adjoint and can be diagonalized by a unitary transformation yielding

$$\gamma(x, x') = \sum_{k} \chi_k(x)\chi_k^*(x)n_k\,, \tag{14.41}$$

where $\{\chi_k\}$ are the natural spin orbitals for this particular CI wavefunction, and where the density matrix eigenvalues n_k are called *occupation numbers* and satisfy

$$0 \leq n_k \leq 1\,. \tag{14.42}$$

A CI expansion expressed in terms of natural spin orbitals contains fewer terms for a given accuracy than in terms of any other basis. Comparison of wavefunctions computed in a variety of different orbital basis sets for a given system is facilitated by first expressing the wavefunctions in terms of natural spin orbitals [see P. O. Löwdin, *Phys. Rev.* **97**, 1490 (1955)]. Sometimes using a basis of atomic natural orbitals for molecular calculations has been successful.

14.2 DETAILED EXAMPLE OF VARIOUS APPROXIMATE METHODS APPLIED TO ELECTRONIC SPECTRA

The ethene molecule has a dipole-allowed $\pi \rightarrow \pi^*$ transition (absorption at 161.5 nm = 7.68 eV). The oscillator strength[2] is $f = 0.3$. These experimental facts can be compared to simple theoretical calculations at different levels of sophistication. The strategy is to use semiempirical theory.

[2] The oscillator strength is a dimensionless quantity that has historical connotations and is defined as

$$f = \frac{2}{3}\frac{m_e h\nu}{e^2\hbar^2}|(\psi_f|\vec{d}\psi_i)|^2$$

We consider the secular problem in the $2p_z$ (π) atomic orbital basis on carbon. We denote these atomic orbitals χ_A on carbon atom A and χ_B on carbon atom B. The overlap integral is $S = (\chi_A|\chi_B)$, and the Fock matrix elements are $F_{AA} = (\chi_A|F\chi_A)$, $F_{BB} = (\chi_B|F\chi_B)$, and $F_{AB} = (\chi_A|F\chi_B)$. The secular problem is

$$\begin{bmatrix} F_{AA} - \epsilon & F_{AB} - \epsilon S \\ F_{BA} - \epsilon S & F_{BB} - \epsilon \end{bmatrix} \begin{bmatrix} c_A \\ c_B \end{bmatrix} = \begin{bmatrix} 0 \\ 0 \end{bmatrix}. \tag{14.43}$$

The symmetry of the problem with two equivalent carbon atoms yields

$$F_{AA} = F_{BB}, \qquad F_{AB} = F_{BA}, \qquad c_A = \pm c_B\,, \tag{14.44}$$

and the normalized MOs are

$$\begin{aligned} \phi_1 &= \frac{\chi_A + \chi_B}{(2 + 2S)^{1/2}} \\ \phi_2 &= \frac{\chi_A - \chi_B}{(2 - 2S)^{1/2}} \end{aligned} \tag{14.45}$$

with the MO energies

$$\epsilon_1 = \frac{F_{AA} + F_{AB}}{1 + S}, \qquad \epsilon_2 = \frac{F_{AA} - F_{AB}}{1 - S}. \tag{14.46}$$

14.2.1 Extended Hückel Method

The parametrization of the Fock matrix elements within the extended Hückel scheme is such that

$$\begin{aligned} F_{AA} &= F_{BB} = -I_C = -11.26\ \text{eV}\,, \\ F_{AB} &= F_{BA} = -kI_C S = -1.7 \times 11.26 \times 0.27 = -5.16\ \text{eV}, \end{aligned} \tag{14.47}$$

where k is an empirical constant and S is calculated with STOs at $R = R_{CC} = 0.1366$ nm. The excitation energy is then simply the differnce of the two orbital energies

$$\Delta E = \epsilon_2 - \epsilon_1 = \frac{2(SF_{AA} - F_{AB})}{1 - S^2} = 4.57\ \text{eV} \tag{14.48}$$

The MOs can be labeled according to the irreps of the full point group of ethene in its ground-state equilibrium geometry (see coordinate system in Fig. 14.2). This is the D_{2h} group with the following character table:

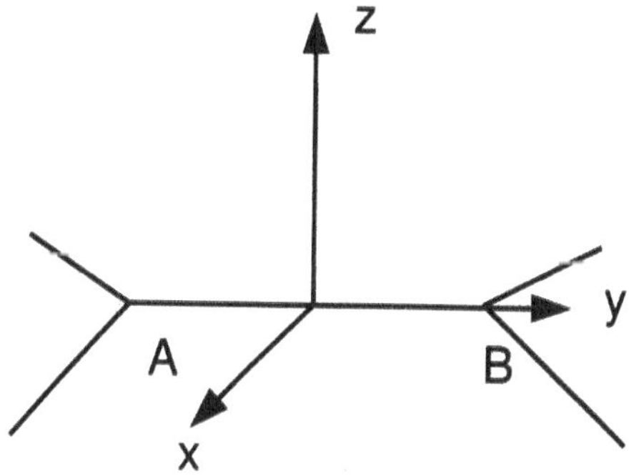

Figure 14.2 Coordinate system for the ethene molecule.

D_{2h}	E	$C_2(z)$	$C_2(y)$	$C_2(x)$	i	$\sigma(xy)$	$\sigma(xz)$	$\sigma(yz)$
A_g	1	1	1	1	1	1	1	1
B_{1g}	1	1	−1	−1	1	1	−1	−1
B_{2g}	1	−1	1	−1	1	−1	1	−1
B_{3g}	1	−1	−1	1	1	−1	−1	1
A_u	1	1	1	1	−1	−1	−1	−1
B_{1u}	1	1	−1	−1	−1	−1	1	1
B_{2u}	1	−1	1	−1	−1	−1	1	1
B_{3u}	1	−1	−1	1	−1	1	1	−1

It follows that $\phi_1 = b_{1u}$ and $\phi_2 = b_{3g}$. We use the notations $\phi = \phi\alpha$, $\bar{\phi} = \phi\beta$, and $\det\{\cdots\} = |\cdots|$.

The various states possible to describe with these two MOs for the two least bound electrons of the ethene molecule are

$$
\begin{aligned}
{}^1A_g &= |N\rangle = \frac{1}{\sqrt{2}}|\phi_1\bar{\phi}_1| = b_{1u}b_{1u}\frac{\alpha\beta - \beta\alpha}{\sqrt{2}}, \\
{}^1B_{2u} &= |V\rangle = \frac{1}{2}[|\phi_1\bar{\phi}_2| - |\bar{\phi}_1\phi_2|] \\
&= \frac{1}{\sqrt{2}}(b_{1u}b_{3g} + b_{3g}b_{1u})\frac{\alpha\beta - \beta\alpha}{\sqrt{2}}, \qquad (14.49) \\
{}^1A_g &= |Z\rangle = \frac{1}{\sqrt{2}}|\phi_2\bar{\phi}_2| = b_{3g}b_{3g}\frac{\alpha\beta - \beta\alpha}{\sqrt{2}}, \\
{}^3B_{2u} &= |T\rangle = \frac{1}{\sqrt{2}}|\phi_1\phi_2| = \frac{1}{\sqrt{2}}(b_{1u}b_{3g} - b_{3g}b_{1u})\alpha\alpha .
\end{aligned}
$$

There are, of course, two more possible ${}^3B_{2u}$ wavefunctions with the two partner spin functions to $\alpha\alpha$.

The transition moment is $\vec{M} = (M_x \quad M_y \quad M_z) = (0 \quad M_y \quad 0)$, where the x and z components are zero because of symmetry, and where

$$
\begin{aligned}
M_y &= \langle V|y_1 + y_2|N\rangle \\
&= \frac{1}{\sqrt{2}}\langle(b_{1u}b_{3g} + b_{3g}b_{1u})|y_1 + y_2|b_{1u}b_{1u}\rangle \\
&= \sqrt{2}\,(b_{3g}|yb_{1u}) \qquad (14.50)\\
&= \frac{1}{(2-2S^2)^{1/2}}[(\chi_A|y\chi_A) - (\chi_B|y\chi_B)] \\
&= -R(2-2S^2)^{-1/2}.
\end{aligned}
$$

Putting $R = 0.1366$ nm $= 2.582$ au, and $S = 0.27$ yields $M_y = -1.896$ au, and the oscillator strength is

$$
f = \frac{2}{3}M^2\Delta E = \frac{2}{3}(1.896)^2\frac{4.57}{27.2} = 0.4. \qquad (14.51)
$$

(Note that 1.0 au = 27.2 eV.)

14.2.2 Pariser–Parr–Pople Method

This treatment employs the zero differential overlap (ZDO) approximation and thus $S = 0$, and

$$
\begin{aligned}
\phi_1 &= \frac{\chi_A + \chi_B}{\sqrt{2}}, \qquad \epsilon_1 = F_{AA} + F_{AB} \\
\phi_2 &= \frac{\chi_A - \chi_B}{\sqrt{2}}, \qquad \epsilon_2 = F_{AA} - F_{AB}.
\end{aligned} \qquad (14.52)
$$

Parametrization of the Fock matrix elements is such that

$$
\begin{aligned}
F_{\mu\mu} &= -I_\mu + \sum_{\sigma\neq\mu}(P_{\sigma\sigma} - Z_\sigma^\pi)\gamma_{\mu\nu} + \tfrac{1}{2}P_{\mu\mu}\gamma_{\mu\mu}, \\
F_{\mu\nu} &= \beta_{\mu\nu} - \tfrac{1}{2}P_{\mu\nu}\gamma_{\mu\nu},
\end{aligned} \qquad (14.53)
$$

where the effective nuclear charge $Z_C^\pi = 1$ for a carbon atom and where the charge–bond-order matrix has the elements $P_{\mu\nu} = 2\sum_j c_{\mu j}c_{\nu j}^*$. The electron repulsion integrals over atomic orbitals are $\gamma_{\mu\nu} = \langle\chi_\mu\chi_\nu|r_{12}^{-1}|\chi_\mu\chi_\nu\rangle$. The molecular orbital energies then are

$$
\begin{aligned}
\epsilon_1 &= -I_C + \tfrac{1}{2}\gamma_{AA} + \beta - \tfrac{1}{2}\gamma_{AB}, \\
\epsilon_2 &= -I_C + \tfrac{1}{2}\gamma_{AA} - \beta + \tfrac{1}{2}\gamma_{AB}.
\end{aligned} \qquad (14.54)
$$

We now let $|D\rangle = |N\rangle$ be the reference determinant, and consider singly excited $|D_a^p\rangle$ and doubly excited $|D_{ab}^{pq}\rangle$ determinants relative to the reference in order to calculate matrix elements. For instance, the result that

$$\langle D_a^p|HD_a^p\rangle = \langle D|HD\rangle + \epsilon_p - \epsilon_a - \langle ap||ap\rangle \tag{14.55}$$

can be used to show that

$$\begin{aligned}\langle T|HT\rangle &= \langle N|HN\rangle + \epsilon_2 - \epsilon_1 - \langle \bar{1}2||\bar{1}2\rangle \\ &= E_N + \epsilon_2 - \epsilon_1 - \langle 12|12\rangle ,\end{aligned} \tag{14.56}$$

where, in the last step, we have put $\langle N|HN\rangle = E_N$ and have summed over spin so that the exchange integral $\langle \bar{1}2|2\bar{1}\rangle$ vanishes.

The result in Eq. (14.55) can again be used together with

$$\langle D|HD_{ab}^{pq}\rangle = \langle ab||pq\rangle \tag{14.57}$$

to calculate the excited-state energy

$$\begin{aligned}E_V &= \tfrac{1}{2}[\tfrac{1}{2}\langle|\phi_1\bar{\phi}_2||H|\phi_1\bar{\phi}_2|\rangle + \tfrac{1}{2}\langle|\bar{\phi}_1\phi_2||H|\bar{\phi}_1\phi_2|\rangle \\ &\quad - \tfrac{1}{2}\langle|\phi_1\bar{\phi}_2||H|\bar{\phi}_1\phi_2|\rangle - \tfrac{1}{2}\langle|\bar{\phi}_1\phi_2||H|\phi_1\bar{\phi}_2|\rangle] \\ &= \tfrac{1}{2}[E_N + \epsilon_2 - \epsilon_1 - \langle\bar{1}\bar{2}||\bar{1}\bar{2}\rangle + E_N + \epsilon_2 - \epsilon_1 \\ &\quad - \langle 12||12\rangle - \langle 1\bar{2}||\bar{1}2\rangle - \langle\bar{1}2||1\bar{2}\rangle] \\ &= E_N + \epsilon_2 - \epsilon_1 - \langle 12|12\rangle + 2\langle 12|21\rangle ,\end{aligned} \tag{14.58}$$

where, in the last step, we have again summed over spin. The excitation energy for the allowed $\pi \rightarrow \pi^*$ transition is then calculated to be

$$\begin{aligned}\Delta E = E_V - E_N &= \epsilon_2 - \epsilon_1 - \langle 12|12\rangle + 2\langle 12|21\rangle \\ &= -2\beta + \gamma_{AB} - \tfrac{1}{2}(\gamma_{AA} + \gamma_{AB}) + \gamma_{AA} - \gamma_{AB} \\ &= -2\beta + \tfrac{1}{2}(\gamma_{AA} - \gamma_{AB}) \\ &= 2(2.93) + \tfrac{1}{2}(10.53 - 7.38) = 7.44 \text{ eV} ,\end{aligned} \tag{14.59}$$

where, in the last step, the recommended parameter values have been used. The transition moment is now $M_y R/\sqrt{2}$, and the oscillator strength

$$f = \tfrac{2}{3}\left(\frac{1.366}{0.529\sqrt{2}}\right)^2 \frac{7.44}{27.2} = 0.61 . \tag{14.60}$$

The *singlet–triplet separation* is zero in the extended Hückel treatment, but here we can calculate

$$E(^1B_{2u}) - E(^3B_{2u}) = 2\langle 12|12\rangle = \gamma_{AA} - \gamma_{AB} = 3.15 \text{ eV} . \tag{14.61}$$

14.2.3 Pariser–Parr–Pople Configuration Interaction

Configuration interaction within the PPP formalism is possible. The two configurations of ${}^1A_g$ symmetry will mix to yield a ground-state wavefunction

$$|N,CI\rangle = |N\rangle C_N + |Z\rangle C_Z\,, \tag{14.62}$$

that can be determined from a secular problem with the matrix elements of the many-electron Hamiltonian H

$$\begin{aligned}\langle N|HN\rangle &= E_N = 2h_{11} + \langle 11|11\rangle\,,\\ \langle Z|HZ\rangle &= 2h_{22} + \langle 22|22\rangle\,,\\ \langle N|HZ\rangle &= \langle Z|HN\rangle = \langle 1\bar{1}||2\bar{2}\rangle = \langle 11|22\rangle\,,\end{aligned} \tag{14.63}$$

where we have, in the final expressions, summed over spin. We have again used the general results

$$\langle D|HD\rangle = \sum_a h_{aa} + \tfrac{1}{2}\sum_{a,b}\langle ab||ab\rangle\,, \tag{14.64}$$

and

$$\langle D|HD_{ab}^{pq}\rangle = \langle ab||pq\rangle\,. \tag{14.65}$$

Another general result is

$$\begin{aligned}\epsilon_a &= h_{aa} + \sum_b\langle ab||ab\rangle\,,\\ \epsilon_p &= h_{pp} + \sum_b\langle pb||pb\rangle\,,\end{aligned} \tag{14.66}$$

which can be used to write

$$\begin{aligned}\epsilon_1 &= h_{11} + \langle 11|11\rangle\,,\\ \epsilon_2 &= h_{22} + 2\langle 12|12\rangle - \langle 12|21\rangle\,.\end{aligned} \tag{14.67}$$

The CI matrix elements can then be expressed in terms of the PPP parameters such that

$$\begin{aligned}\langle Z|HZ\rangle = E_N &+ 2(\epsilon_2 - \epsilon_1) - 4\langle 12|12\rangle + 2\langle 12|21\rangle\\ &+ \langle 11|11\rangle + \langle 22|22\rangle = E_N - 4\beta\,,\end{aligned} \tag{14.68}$$

where the end result is obtained by expressing the molecular orbitals explicitly as the sum and difference of the atomic orbitals as in Eq. (14.52) and using the ZDO approximation. Similarly

$$\langle N|HZ\rangle = \langle 1\bar{1}|2\bar{2}\rangle = \langle 11|22\rangle = \tfrac{1}{2}(\gamma_{AA} - \gamma_{AB})\,. \tag{14.69}$$

The CI secular problem then becomes

$$\begin{bmatrix} E_N - E & \frac{1}{2}(\gamma_{AA} - \gamma_{AB}) \\ \frac{1}{2}(\gamma_{AA} - \gamma_{AB}) & E_N - 4\beta - E \end{bmatrix} \begin{bmatrix} C_N \\ C_Z \end{bmatrix} = \begin{bmatrix} 0 \\ 0 \end{bmatrix}, \tag{14.70}$$

which yields the energies

$$\begin{aligned} E_{\mp} - E_N &= -2\beta \mp 2\beta\left[1 + \frac{1}{16}\left(\frac{\gamma_{AA} - \gamma_{AB}}{\beta}\right)^2\right]^{1/2} \\ &= \begin{cases} 11.928 \ \text{eV} \\ -0.208 \ \text{eV} \end{cases} \end{aligned} \tag{14.71}$$

and the coefficients $C_N = 0.9914$ and $C_Z = 0.1309$. The transition moment is

$$\begin{aligned} M_y &= \langle N, \text{CI} | y_1 + y_2 | V \rangle \\ &= (0.9914 - 0.1309)\frac{R}{\sqrt{2}} = 1.54 \text{ au}, \end{aligned} \tag{14.72}$$

while the excitation energy is

$$\Delta E = E_V - E_N - (E_+ - E_N) = 7.44 + 0.208 = 7.65 \text{ eV}, \tag{14.73}$$

and the oscillator strength

$$f = \frac{2}{3}(M_y)^2 \Delta E = 1.5(1.54)^2 \frac{7.65}{27.2} = 0.46. \tag{14.74}$$

The correlation energy of the ethene ground state within the PPP model for this simple two-determinant CI is -0.208 eV.

15

MOLECULAR VIBRATIONS

The vibrational motion of atomic nuclei in molecules are, under normal conditions, small displacements about the nuclear equilibrium positions, for which the molecule has the symmetry described by a particular point group. The displacements can be considered to form a basis of the representations of the group, and thus be labeled by the irreps. One may proceed by using the displacements along the Cartesian x, y, and z directions of each of the nuclei as a basis. Then one must realize that such displacements include the three degrees of freedom that are translations and the three that are rotations of the molecule as well as all possible vibrational modes.

15.1 SYMMETRY MODES OF THE WATER MOLECULE

We use the C_{2v} point group for the water molecule. The character table has all the information we need:

C_{2v}	E	C_2	$\sigma(xz)$	$\sigma(yz)$		
A_1	1	1	1	1	z	x^2, y^2, z^2
A_2	1	1	-1	-1	R_z	xy
B_1	1	-1	1	-1	x, R_y	xz
B_2	1	-1	-1	1	y, R_x	yz
Γ^{xyz}	9	-1	3	1		

In the last row we list the character values of the reducible representation that is spanned by the nine Cartesian displacement coordinates (three for each atom). Using the expression for the frequency of an irrep in a general representation, we can deduce which irreps are spanned by the various combinations of displacement coordinates. These combinations include those that are pure translations and rotations of the entire water molecule. The symmetries of these modes are listed in the character table, since the three translational degrees of freedom transform as x, y, and z, while the rotational degrees of freedom transform as the rotations about the Cartesian coordinate axes R_x, R_y, R_z. Each rotation can be depicted as a curved (right-handed) arrow about the rotation axis, and the transformation of these curved arrows under the group operations determines the corresponding irrep.

	$3A_1 + A_2 + 3B_1 + 2B_2$
Translations	$A_1 + B_1 + B_2$
Rotations	$A_2 + B_1 + B_2$
Vibrations	$2A_1 + B_1$

We recognize that vibrations are made up of stretches and bends and first concentrate on the OH stretches (see Fig. 15.1).

We can see that these stretches transform among themselves under the operations of the group and span a two-dimensional reducible representation with the character

C_{2v}	E	C_2	$\sigma(xz)$	$\sigma(yz)$	
$\Gamma^{\Delta r}$	2	0	2	0	$A_1 + B_1$

Using the character projectors, we can readily show that

$$Q^{A_1}\Delta r_1 \propto (\Delta r_1 + \Delta r_2) \tag{15.1}$$

and

$$Q^{B_1}\Delta r_1 \propto (\Delta r_1 - \Delta r_2). \tag{15.2}$$

Figure 15.1 The OH bond stretches.

These two modes can be given the graphical representations shown in Figure 15.2 and are called the *symmetric stretch* and *asymmetric stretch*, respectively. The bond contraction $-\Delta r$ is represented as two opposing arrows. It should be obvious that these two figures treated as *objects subjected to the operations of the group* exhibit the symmetries of the irreps that label them. The basis for the in-plane bending modes can be graphically represented as arrows such as shown in Figure 15.3 and the projector of the totally symmetric irrep yields $Q^{A_1}\Delta\alpha_1 \propto \Delta\alpha_1 - \Delta\alpha_2$, and the corresponding graphical representation of the symmetry mode is shown in Figure 15.4.

The two bending "arrows" as basis elements span the reducible representation with the character

C_{2v}	E	C_2	$\sigma(xz)$	$\sigma(yz)$	
$\Gamma^{\Delta\alpha}$	2	0	2	0	$A_1 + B_1$

However, it is clear that $Q^{B_1}\Delta\alpha_1 \propto \Delta\alpha_1 + \Delta\alpha_2$, or graphically as in Figure 15.5, which clearly is a rotation R_y about the y axis. We have added a small arrow on the oxygen atom as well, since the rotation occurs about the

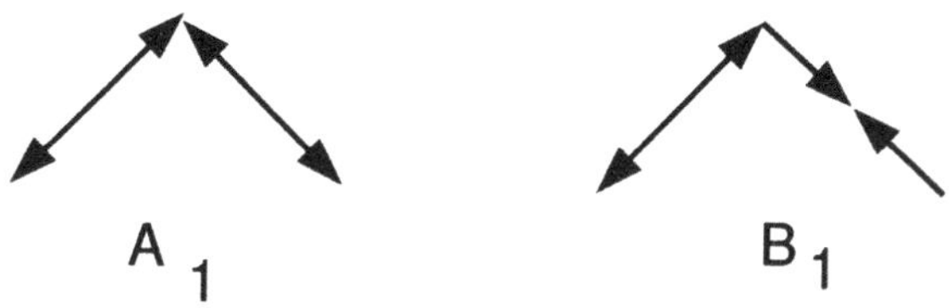

Figure 15.2 The stretch symmetry modes of water.

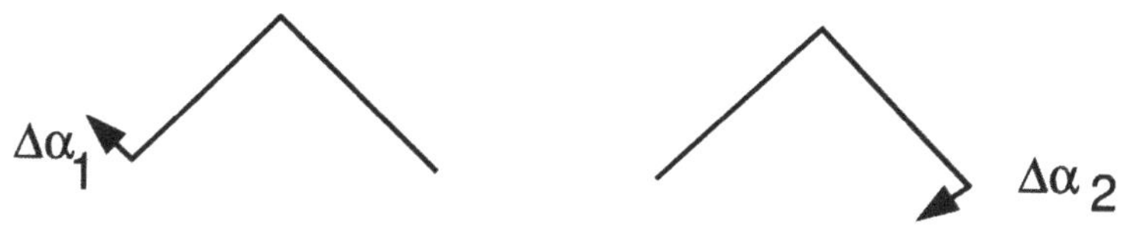

Figure 15.3 The in-plane bending coordinates.

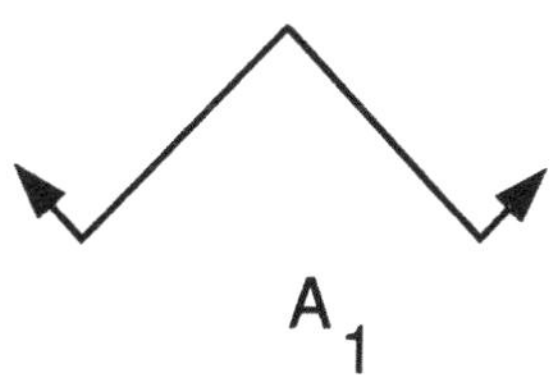

Figure 15.4 The bending symmetry mode of water.

Figure 15.5 The combinations of bending coordinates leading to rotation.

center of mass. This arrow also clearly has B_1 symmetry. It is, of course, possible to express the bond stretches and bends in terms of the Cartesian displacements of each nucleus involved. In analogy to the case of symmetry-adapted orbitals, these symmetry modes will combine to form what are called "normal" modes. Details of that problem are addressed in the next section. Only symmetry modes of the same irrep combine, and only symmetry modes of similar energies mix to any great extent.

At room temperature by far the majority of the molecules in a sample are in the vibrational ground state, which is totally symmetric. In absorption spectroscopy in the infrared regime the allowed transitions are identified by nonzero transition moments involving the electric dipole operator, which transforms as the Cartesian coordinates (x, y, z). In Raman spectroscopy, they involve the induced electric dipole moment that is proportional to the polarizability, which, in turn, transforms as the components of

$$\begin{bmatrix} 3x^2 - r^2 & xy & xz \\ yx & 3y^2 - r^2 & yz \\ zx & zy & 3z^2 - r^2 \end{bmatrix}. \tag{15.3}$$

The final state of a transition is expressed in terms of the vibrational symmetry modes of a particular kind.

15.2 SYMMETRY MODES OF SQUARE PLANAR MOLECULES

We consider the hypothetical square planar molecule C_4H_4 with the point group D_{4h}. The Cartesian displacements of the eight nuclei span a reducible representation with the character

D_{4h}	E	$2C_4$	C_2	$2C_2'$	$2C_2''$	i	$2S_4$	σ_h	$2\sigma_v$	$2\sigma_d$
Γ^{xyz}	24	0	0	−4	0	0	0	8	4	0
Total		$2A_{1g} + 2A_{2g} + 2B_{1g} + 2B_{2g} + 2E_g + 2A_{2u} + 2B_{2u} + 4E_u$								
Rotation			A_{2g}			E_g				
Translation							A_{2u}		E_u	
Vibration		$2A_{1g} + A_{2g} + 2B_{1g} + 2B_{2g} + E_g + A_{2u} + 2B_{2u} + 3E_u$								

The vibrational symmetry modes can again be separated into C—C stretches, C—H stretches, in-plane bends, and out-of-plane bends. The four C—C stretches shown in Figure 15.6 span the reducible representation with the character

D_{4h}	E	$2C_4$	C_2	$2C_2'$	$2C_2''$	i	$2S_4$	σ_h	$2\sigma_v$	$2\sigma_d$
$\Gamma^{\Delta rCC}$	4	0	0	0	2	0	0	4	0	2

$$A_{1g} + B_{1g} + E_u$$

The graphical representations of these symmetry modes are given in Figure 15.7, where the two degenerate E_u modes can be obtained as

$$\begin{aligned} &Q^{E_u}\Delta r_1, \\ &Q^{E_u}\Delta r_2. \end{aligned} \tag{15.4}$$

The four C—H stretches (see Fig. 15.8) span a reducible representation with the character

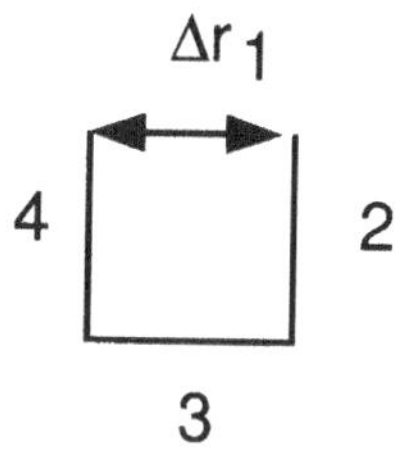

Figure 15.6 The C—C stretch coordinate.

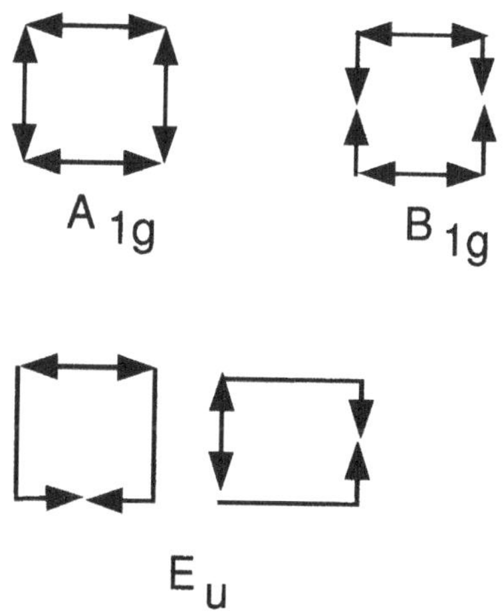

Figure 15.7 The C—C stretch symmetry modes.

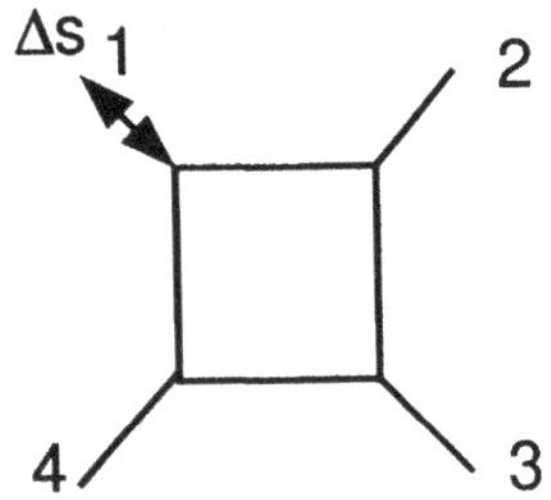

Figure 15.8 The C—H stretch coordinate.

D_{4h}	E	$2C_4$	C_2	$2C_2'$	$2C_2''$	i	$2S_4$	σ_h	$2\sigma_v$	$2\sigma_d$
$\Gamma^{\Delta s_{CH}}$	4	0	0	2	0	0	0	4	2	0
					$A_{1g} + B_{1g} + E_u$					

Graphical representations of these symmetry modes are shown in Figures 15.9 and 15.10 and with the two degenerate E_u modes as in Figure 15.11. We note how these particular symmetry modes "look like" the (x,y) coordinate pair (see character table).

The in-plane bending modes can be expressed in terms of the Cartesian x and y coordinates of the carbons and hydrogens and represented as arrows perpendicular to the C—H bonds (see Fig. 15.2).

The four distortions labeled $\Delta\alpha$ transform among each other under the group as do the four $\Delta\beta$. Either set spans a reducible representation with the character

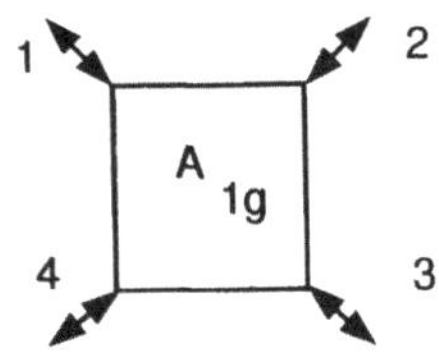

Figure 15.9 The A_{1g} C—H stretch symmertry mode.

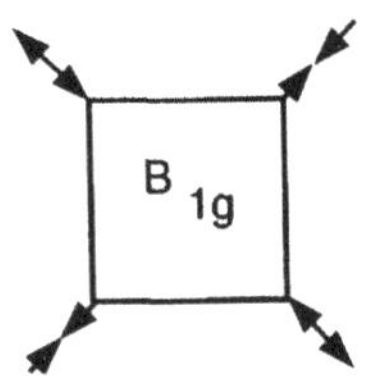

Figure 15.10 The B_{1g} C—H stretch symmetry mode.

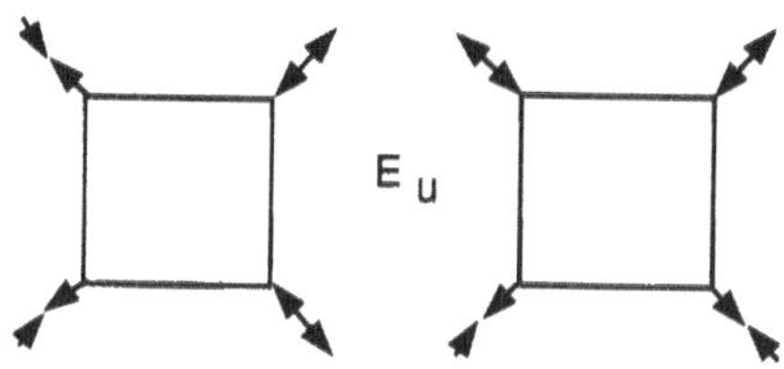

Figure 15.11 The E_u C—H stretch symmetry mode.

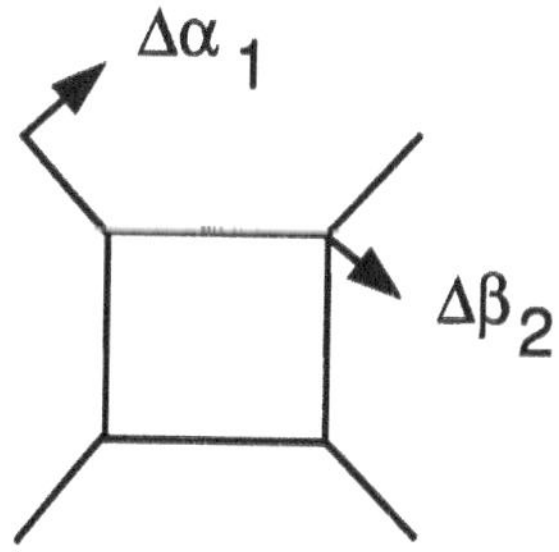

Figure 15.12 The in-plane bend coordinates.

D_{4h}	E	$2C_4$	C_2	$2C_2'$	$2C_2''$	i	$2S_4$	σ_h	$2\sigma_v$	$2\sigma_d$
$\Gamma^{\Delta\alpha}$	4	0	0	−2	0	0	0	4	−2	0
					$A_{2g} + B_{2g} + E_u$					

The corresponding graphical representation of the A_{2g} symmetry mode, shown in Figure 15.13, when combined with the one for the hydrogens, clearly constitutes a rotation (about the z axis) for the plus combination, while the difference yields the in-plane "rocking mode" shown in Figure 15.14.

Similarly, we obtain the B_{2g} rocking modes as the sum and difference of the hydrogen and carbon displacements, shown in Figure 15.15.

The in-plane rocking modes of E_u symmetry are shown in Figure 15.16.

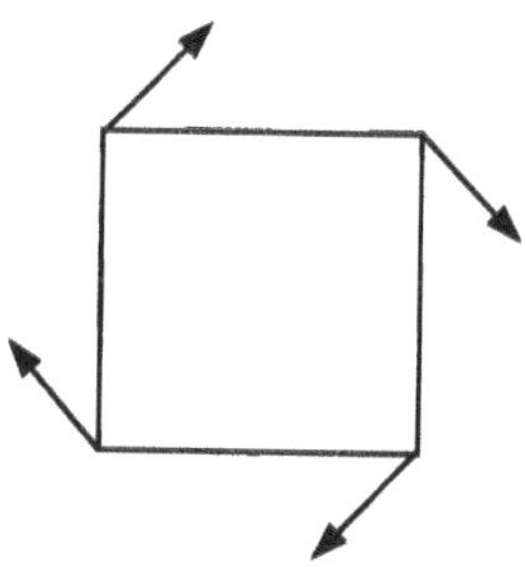

Figure 15.13 The A_{2g} symmetry mode (rotation) of the carbons.

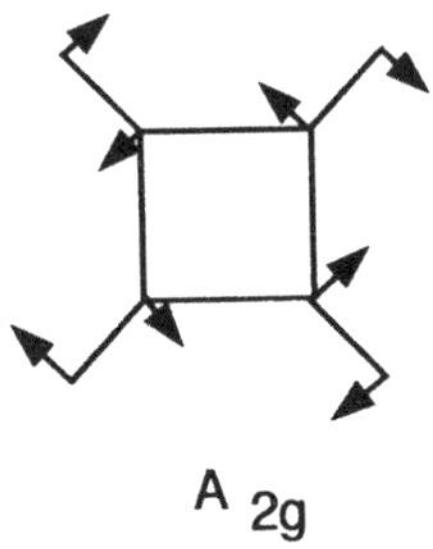

Figure 15.14 The in-plane A_{2g} symmetry bending (rocking) mode.

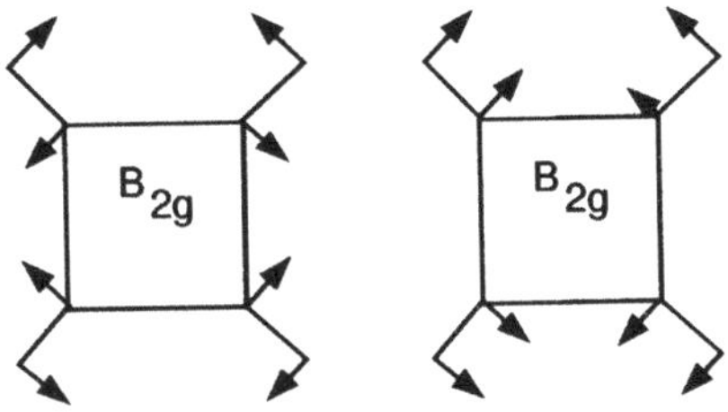

Figure 15.15 Two different in-plane B_{2g} symmetry bending modes.

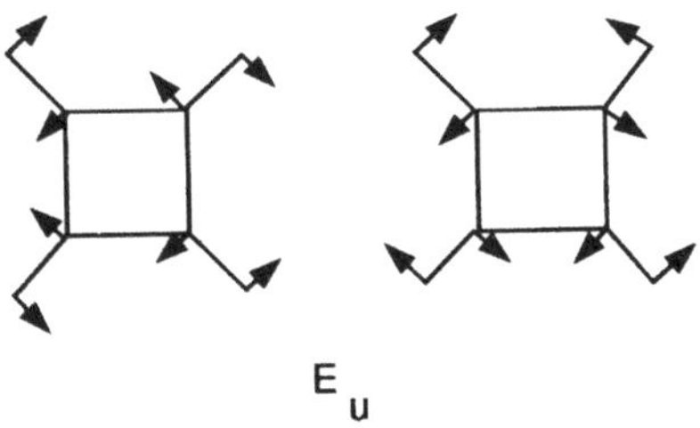

Figure 15.16 The in-plane E_u symmetry bending mode.

This molecule exhibits both CCH and CCC bond angle distortions. The latter support the A_{1g}, B_{1g}, and E_u irreps, but both the A_{1g} and E_u modes are only different representations of the corresponding C—C stretches, which one can see also may involve CCC bond angle changes. The B_{1g} mode can be represented as shown in Figure 15.17.

The out-of-plane modes are described by the nuclear displacements along the z direction, and these displacements of the hydrogens span the same reducible four-dimensional representation as do the displacements of the carbons. The character of this representation is

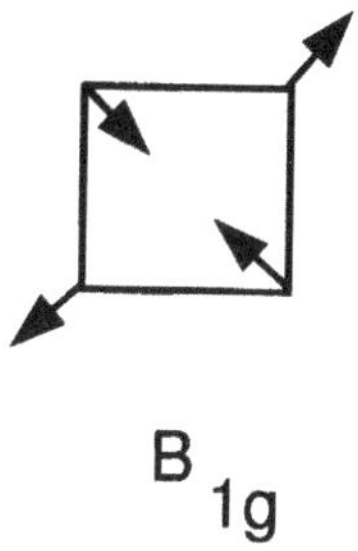

Figure 15.17 The CCC bond angle bending symmetry mode.

D_{4h}	E	$2C_4$	C_2	$2C_2'$	$2C_2''$	i	$2S_4$	σ_h	$2\sigma_v$	$2\sigma_d$
$\Gamma^{\Delta\alpha}$	4	0	0	−2	0	0	0	−4	2	0
					$A_{2u} + B_{2u} + E_g$					

The hydrogen and the carbon A_{2u} modes can be combined to yield the structure shown in Figure 15.18, where the filled and open circles have opposite phases. The B_{2u} mode of the hydrogens can be combined with the one for the carbons in two ways (sum and difference) to yield two vibrational symmetry modes as shown in Figure 15.19.

It is straightforward to construct the combination of A_{2u} carbon and hydrogen displacements, which is simply a translation, and also the E_g displacements,

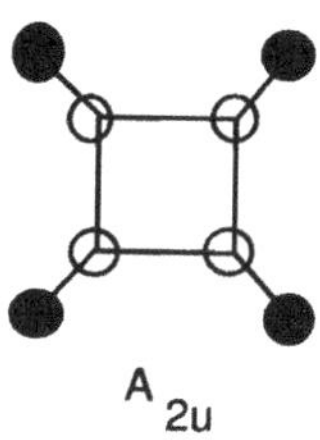

Figure 15.18 Out-of-plane bending A_{2u} symmetry mode.

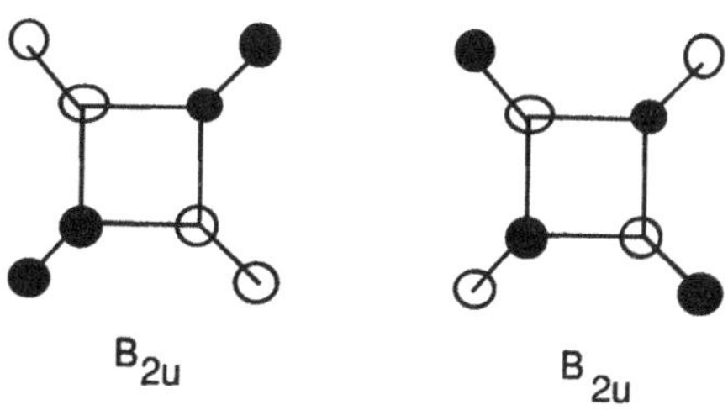

Figure 15.19 Two different out-of-plane bending B_{2u} symmetry modes.

which are rotations. The two degenerate E_g vibrational out-of-plane bending symmetry modes are shown in Figure 15.20.

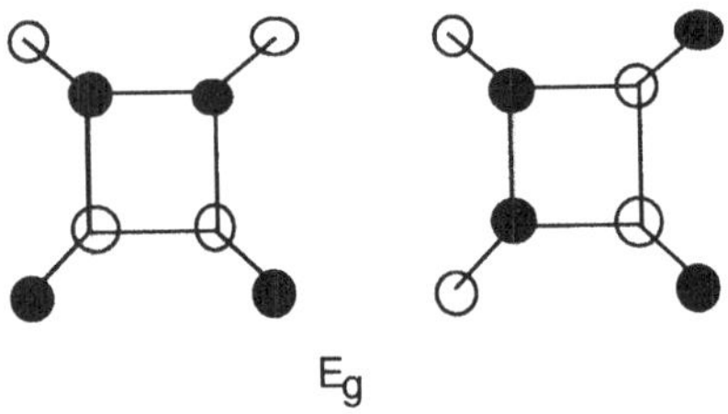

Figure 15.20 The out-of-plane bending E_g symmetry mode.

In general, the symmetry modes of the same irrep can combine to form so-called normal modes, which are true representations of small vibrational distortions of a molecular system. Whenever symmetry modes of the same kind are widely separate in energy, such as is the case of, say, CH stretches and some of the bending modes, the mixing is weak.

15.3 NORMAL MODES

The vibrational modes discussed in the previous section are so-called symmetry modes. For instance, consider the arrangement shown in Figure 15.21. The C—C stretch of the ethene molecule can be expressed as

$$Q_1^{A_g} = \bar{x}_{\mathrm{II}} - \bar{x}_{\mathrm{I}} \tag{15.5}$$

and spans the A_g irrep of the point group D_{2h}. The bar over the coordinate symbol means mass weighting

$$\bar{x}_i = \sqrt{m_i} x_i \tag{15.6}$$

and the same for the y and z coordinates. The in-plane bend shown in Figure 15.22

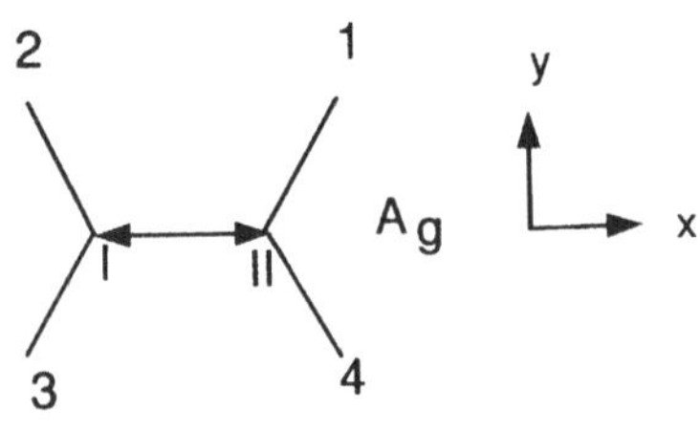

Figure 15.21 The ethene C—C stretch symmetry mode.

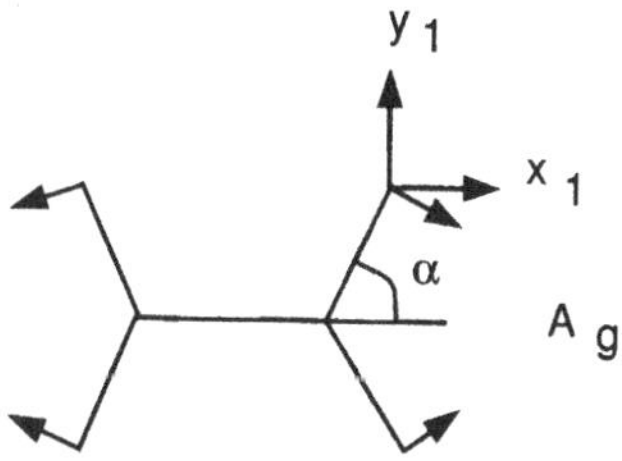

Figure 15.22 The coordinates for the ethene in-plane bends.

has the same symmetry and can be expressed as

$$Q_2^{A_g} = \Delta\alpha_1 - \Delta\alpha_2 + \Delta\alpha_3 - \Delta\alpha_4 \tag{15.7}$$

with, say

$$\Delta\alpha_1 = \bar{x}_1 \sin\alpha - \bar{y}_1 \cos\alpha\,. \tag{15.8}$$

Another ethene A_g symmetry mode is the C—H stretch shown in Figure 15.23. This mode can be expressed as

$$Q_3^{A_g} = \Delta u_1 + \Delta u_1 + \Delta u_3 + \Delta u_4\,, \tag{15.9}$$

with, say

$$\Delta u_1 = (\bar{x}_1 - \bar{x}_{\mathrm{I}})\cos\alpha + (\bar{y}_1 - \bar{y}_{\mathrm{I}})\sin\alpha\,. \tag{15.10}$$

Similarly, for this molecule there are other symmetry modes: three B_{3u} modes; two B_{1g}; two B_{2u}; and one each of B_{2g}, B_{1u}, and A_u symmetry. In this manner there are a total of 12 symmetry modes, which is the same as the number of internal nuclear degrees of freedom for a six-atom system as it should. The B_{1u} mode is the out-of-plane bending mode. In Figure 15.24 the open circle and the full circle indicate that the atom is moving in the positive and negative z directions, respectively. The rest of the symmetry modes are readily obtained. If we let q be a common notation for the coordinate combinations $\Delta\alpha, \Delta u, \bar{z}$, and so on, we can write for a particular symmetry mode

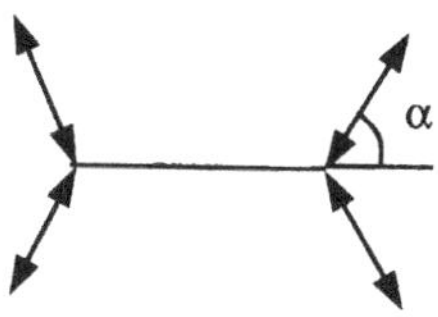

Figure 15.23 The C—H stretch A_g symmetry mode.

$$Q_i^\alpha = \mathrm{M}_g f^\alpha \chi^\alpha(g) g^{-1} q_i \tag{15.11}$$

using the character projection operators[1].

A product of two symmetry modes is

$$Q_i^\alpha Q_j^\beta = \delta^{\alpha\beta} \mathrm{M}_g f^\alpha \chi^\alpha(g) g^{-1} q_i q_j , \tag{15.12}$$

a result that will prove useful. A potential-energy surface of the ethene molecule is a function V of internal coordinates in 12 dimensions. These internal coordinates can be chosen to be the 12 symmetry modes Q_i^α. The simplest treatment possible of the molecular vibrations is the so-called harmonic approximation. It is then assumed that the potential-energy surface (locally) can be accurately represented in terms of a second order power series expansion about the nuclear equilibrium positions

$$V = V(0) + \sum_{i,\alpha} \frac{\partial V}{\partial Q_i^\alpha}\Big|_{Q=0} Q_i^\alpha + \frac{1}{2} \sum_{i,\alpha} \sum_{j,\beta} \frac{\partial^2 V}{\partial Q_i^\alpha \partial Q_j^\beta}\Big|_{Q=0} Q_i^\alpha Q_j^\beta , \tag{15.13}$$

which is often a reasonable approximation for small nuclear displacements. Higher-order terms may be included to account for anharmonicity effects. We do not discuss this here.

At equilibrium, which we denote by 0, the first derivatives vanish and $V(0)$ is just an additive constant that may be set to zero. Because of the result of Eq. (15.12), we can consider each symmetry block of the Hermitian force constant matrix with elements

$$F_{ij}^\alpha = \frac{\partial^2 V}{\partial Q_i^\alpha \partial Q_j^\alpha}\Big|_{Q=0} . \tag{15.14}$$

Finding the unitary matrix $\mathbf{U}$ that diagonalizes a particular symmetry block

$$\mathbf{U}^\dagger \mathbf{F}^\alpha \mathbf{U} = \boldsymbol{\lambda}^\alpha , \tag{15.15}$$

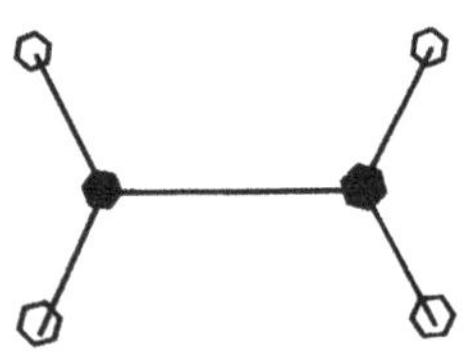

Figure 15.24 The out-of-plane bending B_{1u} symmetry mode.

[1] The D_{2h} symmetry has only one-dimensional irreps, and for a more general situation we can use the projectors P_{11}^α, and the appropriate shift operators.

yields the so-called normal modes N

$$\mathrm{N}_k^\alpha = \sum_{l=1}^{n^\alpha} Q_l^\alpha U_{lk} \tag{15.16}$$

and the corresponding eigenvalues λ_k^α.

In terms of these normal modes the potential-energy function (in the harmonic approximation) can be expressed as

$$V = \tfrac{1}{2}\sum_{\alpha,k} \lambda_k^\alpha (\mathrm{N}_k^\alpha)^2. \tag{15.17}$$

It is also quite straightforward to show that the kinetic-energy operator for the internal dynamics of the nuclei is diagonal in terms of the normal modes, having the form

$$T = \frac{1}{2}\sum_{\alpha,k} \frac{\partial^2}{\partial \mathrm{N}_k^{\alpha 2}}. \tag{15.18}$$

This means that the normal-mode vibrations are simply a set of decoupled harmonic oscillators whose wavefunctions satisfy the Schrödinger equation

$$\left[-\frac{1}{2}\frac{\partial^2}{\partial \mathrm{N}_k^{\alpha 2}} + \frac{1}{2}\lambda_k^\alpha \mathrm{N}_k^{\alpha 2}\right]\psi_n(\mathrm{N}_k^\alpha) = E_n^\alpha(k)\psi_n(\mathrm{N}_k^\alpha). \tag{15.19}$$

It follows directly that

$$\psi_n(\mathrm{N}_k^\alpha) = N \exp\left[-\frac{\omega_k^\alpha \mathrm{N}_k^{\alpha 2}}{2}\right] H_n(\sqrt{\omega_k^\alpha}\,\mathrm{N}_k^\alpha). \tag{15.20}$$

Here

$$\sqrt{\lambda_k^\alpha} = 2\pi\nu_k^\alpha = \omega_k^\alpha \tag{15.21}$$

refers to the frequency of the normal-mode N_k^α ($\hbar = 1$). The first few Hermite polynomials are

$$\begin{aligned} &H_0(x) = 1;\ H_1(x) = 2x;\ H_2(x) = 4x^2 - 2; \\ &H_3(x) = 8x^3 - 12x \end{aligned} \tag{15.22}$$

and the energies for each normal mode are

$$E_n^\alpha(k) = \omega_k^\alpha(n + \tfrac{1}{2}), \quad n = 0, 1, 2, \ldots. \tag{15.23}$$

To zeroth order the molecular wavefunction apart from translational and rotational degrees of freedom is a product of an electronic part and a vibrational part. If we assume only vibrational excitations to take place, we need to consider the vibrational wavefunction of the total molecule to be a product of such harmonic oscillator states, one for each normal mode. We write

$$\Psi_{\text{vib}}(n) = \psi_1(n_1)\psi_2(n_2)\cdots\psi_{\bar{n}}(n_{\bar{n}}) \tag{15.24}$$

with $\bar{n} = 3N - 6$, where N is the number of atoms in the molecule. In this manner we have just numbered all the normal modes in some convenient order.

The total electric dipole moment operator of the molecule is

$$\vec{D}_{\text{op}} = -e\sum_i \vec{r}_i + \sum_k eZ_k\vec{R}_k\,, \tag{15.25}$$

and after averaging over the electronic wavefunction, we see that the electric dipole function depends on the normal-mode coordinates

$$\vec{D} = (\psi_{\text{el}}|\vec{D}_{\text{op}}\psi_{\text{el}}) = \vec{D}(0) + \sum_k \frac{\partial\vec{D}}{\partial\text{N}_k}\bigg|_{N=0}\text{N}_k + \cdots, \tag{15.26}$$

where we have suppressed the symmetry labels. Through first order we can then write

$$\vec{D}_{nn'} = \sum_k \frac{\partial\vec{D}}{\partial\text{N}_k}\bigg|_{N=0}(\psi_k(n_k)|\text{N}_k\psi_k(n_k')) \tag{15.27}$$

for the transition moment associated with a vibrational excitation. Then for the case in which only a single normal mode is excited, say, the kth one, $\vec{D}_{nn'} \neq 0$ only if $n_i = n_i'$, $i = 1, 2, \ldots (i \neq k)$ and

$$(\psi_k(n_k)|\text{N}_k\psi_k(n_k')) \neq 0\,. \tag{15.28}$$

We can calculate this last integral since it is proportional to ($n_k = n$, $n_k' = n'$)

$$\int_{-\infty}^{\infty} H_n(\xi)H_{n'}(\xi)\xi e^{-\xi^2}d\xi\,, \tag{15.29}$$

where we have used the solutions from Eq. (15.20). The Hemite polynomials satisfy the relations

$$\begin{gathered} H_{n+1} - 2\xi H_n + 2nH_{n-1} = 0, \\ \int_{-\infty}^{\infty} H_n(\xi)H_{n'}(\xi)e^{-\xi^2}d\xi \propto \delta_{nn'}\,. \end{gathered} \tag{15.30}$$

We can thus write

$$\xi H_{n'} \propto (H_{n'+1} + 2n'H_{n'-1}) \tag{15.31}$$

and conclude that

$$(\psi_k(n_k)|\mathrm{N}_k\psi_k(n_k')) \neq 0. \tag{15.32}$$

only if

$$n_k' = n_k + 1, \quad \text{or} \quad n_k' - n_k - 1 \tag{15.33}$$

leading to the selection rule $\Delta n = \pm 1$ for vibrational absorption or emission spectroscopy (IR). It is also notable that the derivative

$$\left.\frac{\partial \vec{D}}{\partial \mathrm{N}_k}\right|_{N=0} \tag{15.34}$$

must be different from zero in order for the transition moment to be nonzero. This means that even a molecule with vanishing permanent electric dipole moment can absorb radiation in the infrared region and become vibrationally excited as long as the normal-mode distortion changes the electric dipole moment.

From Eq. (15.25) we can see that the electric dipole function can also be expressed in terms of Cartesian coordinates such that $D_x, D_y,$ and D_z components of the electric dipole function are linear functions of the coordinates $x, y,$ and z, respectively. When the molecule in its vibrational ground state is excited, we consider transition moment integrals such as

$$(\psi_k(0)|(x \quad y \quad z)\psi_k(1)) \tag{15.35}$$

with

$$\psi_k(0) \propto \exp\left[-\frac{\omega \mathrm{N}_k^2}{2}\right] \tag{15.36}$$

and with

$$\psi_k(1) \propto \mathrm{N}_k \exp\left[-\frac{\omega \mathrm{N}_k^2}{2}\right]. \tag{15.37}$$

This means that $\psi_k(0)$ tranforms according to the totally symmetric irrep Γ^J of the relevant point group, while $\psi_k(1)$ transforms as the irrep Γ^{N_k} of the normal-mode coordinate N_k, and for the transition moment integral to be nonzero, the integrand must transform according to a representation that contains the totally symmetric irrep

$$\Gamma^J \otimes \Gamma^s \otimes \Gamma^{\mathrm{N}_k} \ni \Gamma^J, \tag{15.38}$$

where $s = x, y,$ or z. Obviously this is the same as

$$\Gamma^s \otimes \Gamma^{\mathrm{N}_k} \ni \Gamma^J, \tag{15.39}$$

which, in terms of characters, means that

$$M_g \chi^s(g)\chi^{N_k}(g)\chi^J(g^{-1}) = M_g \chi^s(g)\chi^{N_k*}(g^{-1}) \neq 0 . \qquad (15.40)$$

For real characters, this leads to $s = N_k$, or the excited normal mode must span the same irrep as the Cartesian coordinates $s = x, y$, or z.

For Raman spectra one deals with the polarizability tensor rather than the electric dipole moment. Otherwise the same symmetry arguments apply with $s = x^2$, y^2, z^2, xy, xz, or yz.

It can very well happen that more than one normal mode is excited in an absorption process. The transition moment Eq. (15.27) will then contain integrals such as

$$(\psi_k(0)\psi_l(0)|s\psi_k(1)\psi_l(1)) , \qquad (15.41)$$

where s refers to the appropriate Cartesian combinations for IR or Raman spectra. For instance, for ethene the two normal modes B_{1u}, shown in Fig. 15.24 with a frequency of 943 cm^{-1} and the "internal rotation" inactive mode A_u (see Fig. 15.25) at 1007 cm^{-1}, may be excited in a Raman process since

$$xy \propto N^{A_u} N^{B_{1u}} \in \Gamma^{B_{1g}} \qquad (15.42)$$

at approximately

$$1007 + 943 = 1950 \text{ cm}^{-1} . \qquad (15.43)$$

Higher-order (harmonic) excitations are also possible.

Obviously, pure vibrational spectra occur only approximately, since an exact separation of rotational and vibrational degrees of freedom is not possible. Nevertheless, the simple theory and associated rules discussed here are useful and give a basis on which to start to interpret spectra.

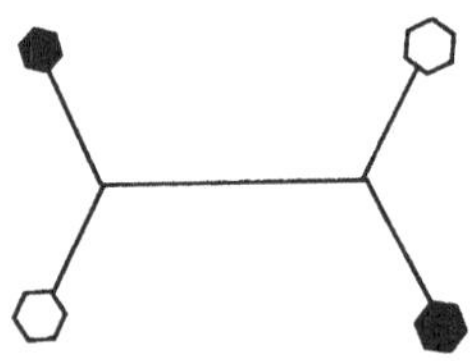

Figure 15.25 The internal rotation A_u symmetry mode.

16

SPACE GROUPS

16.1 INTRODUCTION

Crystal structure determination is an important part of chemistry, yet the symmetry of crystals is a much neglected topic in chemistry texts. The brief treatment of space groups given in this chapter is intended as an appetizer for the interested reader. The student who wants to know the details will have to go to a special text on symmetry in the solid state.

One normally approaches this subject by first discussing translational symmetry in general and then combining the translational symmetry with point group symmetry to obtain the so-called space groups. The author follows that approach here and introduces the common terminology as well as some very simple examples of symmetry functions and energy bands.

16.2 TRANSLATIONAL SYMMETRY

In a homogeneous (three-dimensional) space all points are equivalent and a general translation by a vector

$$\vec{a} = \alpha_1\vec{a}_1 + \alpha_2\vec{a}_2 + \alpha_3\vec{a}_3 \tag{16.1}$$

(where $\vec{a}_1, \vec{a}_2$, and $\vec{a}_3$ are three linearly independent vectors) for arbitrary (real) numbers α_1, α_2, α_3 is a symmetry operation. The collection of all such symmetry operations labeled by three numbers $(\alpha_1, \alpha_2, \alpha_3)$ form an infinite group called the *translation group* $T(3)$ in 3-space with elements

$$\{e|\vec{a}\} \tag{16.2}$$

acting on a three-dimensional vector $\vec{r}$ such that

$$\{e|\vec{a}\}\vec{r} = \vec{r} + \vec{a}. \tag{16.3}$$

A lattice group $L(3)$ is an infinite, proper, discrete subgroup of $T(3)$ with elements

$$\{e|\vec{m}\}, \tag{16.4}$$

where

$$\vec{m} = \mu_1\vec{a}_1 + \mu_2\vec{a}_2 + \mu_3\vec{a}_3, \tag{16.5}$$

with μ_1, μ_2, μ_3 as integers. Obviously generalizations to an arbitrary dimensional space with k linearly independent vectors $\{\vec{a}_i\}$ exist. We are interested in $k = 1$, 2, or 3.

Choosing $\vec{a}_1, \vec{a}_2$, and $\vec{a}_3$ such that

$$\vec{a}_1 \cdot (\vec{a}_2 \times \vec{a}_3) = \vec{a}_3 \cdot (\vec{a}_1 \times \vec{a}_2) = \vec{a}_2 \cdot (\vec{a}_3 \times \vec{a}_1) = V \tag{16.6}$$

is the smallest possible (positive) volume permits the identification of eight points forming the corners of a basic parallelepiped, or *primitive cell* of the lattice by the (seven) vectors $\vec{a}_1, \vec{a}_2, \vec{a}_3, \vec{a}_1 + \vec{a}_2, \vec{a}_1 + \vec{a}_3, \vec{a}_2 + \vec{a}_3, \vec{a}_1 + \vec{a}_2 + \vec{a}_3$. Applying all the elements of $L(3)$ to a nonzero vector $\vec{r}$ generates a lattice of points in 3-space.

16.3 TRANSLATIONAL AND POINT GROUP SYMMETRY

Such lattices of points are of fundamental importance in the study of extended crystalline substances. A general (stereoregular) crystal is invariant under translation operations on the lattice that normally would be defined by the equilibrium positions of atomic nuclei. However, a general crystal lattice is also invariant under certain point group operations, such as reflections and rotations. The combination of point group operations and translations form a *space group* G with elements

$$\{g|\vec{m}\} \tag{16.7}$$

where $g \in G_0$ is a point group operation and $\{e|\vec{m}\} \in L(3)$ and e is the identity element of the point group G_0. Acting with a space group element on a nonzero vector $\vec{r}$ yields

$$\{g|\vec{m}\}\vec{r} = g\vec{r} + \vec{m}. \tag{16.8}$$

This means that the binary law of multiplication of a space group is

$$\begin{aligned} \{g|\vec{m}\}\{h|\vec{n}\}\vec{r} &= \{g|\vec{m}\}\vec{r}' = g\vec{r}' + \vec{m} \\ &= g(h\vec{r} + \vec{n}) + \vec{m} = gh\vec{r} + g\vec{n} + \vec{m}. \end{aligned} \tag{16.9}$$

Thus

$$\{g|\vec{m}\}\{h|\vec{n}\} = \{gh|g\vec{n} + \vec{m}\}. \tag{16.10}$$

Comparison with the direct product of two groups reveals a difference with the idea of a space group, which is a semidirect product of a point group G_0 and a lattice group $L(3)$. This technical aspect of the group structure is not further exploited here.

It follows that the inverse of an element is

$$\{g|\vec{m}\}^{-1} = \{g^{-1}| - g^{-1}\vec{m}\} \tag{16.11}$$

while the identity element is

$$\{e|\vec{0}\} \tag{16.12}$$

and

$$\begin{aligned} \{g|\vec{m}\}\{g|\vec{m}\}^{-1} &= \{g|\vec{m}\}\{g^{-1}| - g^{-1}\vec{m}\} \\ &= \{e| - \vec{m} + \vec{m}\} = \{e|\vec{0}\}. \end{aligned} \tag{16.13}$$

The fact that an element of the space group combination of a point group and a lattice group acting on a vector $\vec{r}$ must map a three-dimensional lattice on itself implies certain restrictions on the possible point groups and also on possible forms of lattices. The point group must be one of the 32 crystallographic point groups, which, in turn, leads to 14 lattice types, the so-called Bravais lattices. A point group that fixes a point $\vec{r}$ and maps a three-dimensioanl lattice containing $\vec{r}$ into itself is called a *crystallographic point group.* The maximal crystallographic point group at $\vec{r}$ is called the *holohedry* of the lattice at $\vec{r}$.

Theorem 16.1 If $g \in G_0$ is a rotation (not identity), then g is of order 2, 3, 4, or 6. If $g' = ig \in G_0$ is a rotation-inversion then g is of order 1, 2, 3, 4, or 6.

Proof Any vector $\vec{r}$ can be expressed as a linear combination of the basis vectors $\{\vec{a}_i\}$ that define the lattice. Thus it is sufficient to consider an operation on a basis vector:

$$g\vec{a}_i = \sum_{j=1}^{3} \vec{a}_j \Gamma_{ji}(g), \tag{16.14}$$

where $\Gamma(g)$ is the transformation g in the basis $\{\vec{a}_i\}$, namely, a representation of G_0. Since the character (the trace of the representation matrices) is invariant to basis transformations, we can choose an orthogonal basis $\{\vec{a}_i\}$ and place one of the vectors, say, $\vec{a}_3$, along the rotation axis. Then it follows that

$$\Gamma(g) = \pm \begin{bmatrix} \cos\varphi & -\sin\varphi & 0 \\ \sin\varphi & \cos\varphi & 0 \\ 0 & 0 & 1 \end{bmatrix} \tag{16.15}$$

for a rotation by an angle φ. (The + or - sign in front refers to the inversion operation.) The character is

$$\mathrm{Tr}\{\mathbf{\Gamma}(g)\} = \pm(1 + 2\cos\varphi) = \text{integer}. \tag{16.16}$$

This restricts the rotation angle to be

$$\varphi = \frac{\pi}{2}, \quad \frac{3\pi}{2}, \quad \text{or} \quad \frac{n\pi}{3} \quad (n = 0,1,2,3,4,5). \tag{16.17}$$

One can then conclude that the following rotations are possible:

$$\begin{aligned} &\frac{\pi}{2}: C_4; \quad \frac{3\pi}{2}: C_4^3; \quad n = 0: C_1; \quad n = 1: C_6; \\ &n = 2: C_3; \quad n = 3: C_2; \quad n = 4: C_3^2; \quad n = 5: C_6^5; \end{aligned} \tag{16.18}$$

and a point group with rotations of order 5 or greater[1] than 6 cannot be a crystallographic point group. This restricts the crystallographic point groups (CPGs) to the following:

$$C_1, C_2, C_3, C_4, C_6, D_2, D_3, D_4, D_6, T, O, \tag{16.19}$$

which do not contain the inversion operation and are designated as CPGs of the first kind, and

$$\begin{aligned} C_i = S_2, S_4, S_6, C_{1h}, C_{2h}, C_{3h}, C_{4h}, C_{6h}, C_{2v}, C_{3v}, C_{4v}, C_{6v} \\ D_{2h}, D_{3h}, D_{4h}, D_{6h}, D_{2d}, D_{3d}, T_h, T_d, O_h, \end{aligned} \tag{16.20}$$

designated as CPGs of the second kind.

The translation part of the space group must be associated with a Bravais lattice. The seven point groups

$$S_2, C_{2h}, D_{2h}, D_{3d}, D_{4h}, D_{6h}, O_h \tag{16.21}$$

are holohedrics (maximal point groups) for each of seven lattice types, leading to seven crystal systems subdivided into 14 Bravais lattices.

The choice of the basis vectors $\{\vec{a}_i\}$ can be worked out for each holohedry leading to the various Bravais lattices (see, e.g., W. Miller, *Symmetry Groups*

---

[1] In 1984 D. Shechtman, I. Blech, D. Gratias, and J. Cahn [*Phys. Rev. Lett.* **17** 1951 produced an aluminum–manganese alloy that appeared to have fivefold symmetry based on its diffraction pattern. This resulted in the recognition that materials can exhibit local fivefold symmetry that makes them approximately periodic structures. Such materials are called *quasi-crystals*.

and Their Applications, Academic Press, New York, 1972). For instance, the simple cubic, body-centered cubic, and face-centered cubic lattices are all Bravais lattices of the cubic holohedry O_h.

Bravais Lattice (Holohedry)	**Crystal Classes**	**Possible Basis Vectors**
Triclinic (S_2)	S_2, C_1	$\vec{a}_1\ \vec{a}_2\ \vec{a}_3$ arbitary
Monoclinic (C_{2h})	C_{2h}, C_2, C_{1h}	
simple		$\vec{a}_2 \perp \vec{a}_3 \perp \vec{a}_1$,
Base-centered		$\vec{a}_1 \perp \vec{a}_2 \perp (2\vec{a}_3 - \vec{a}_2) \perp \vec{a}_1$
Orthorhombic (D_{2h})	D_{2h}, D_2, C_{2v}	
simple		$\vec{a}_1 \perp \vec{a}_2 \perp \vec{a}_3 \perp \vec{a}_1$,
Base-centered		$\vec{a}_3 \perp (\vec{a}_1 + \vec{a}_2) \perp (\vec{a}_1 - \vec{a}_3) \perp \vec{a}_3$,
Body-centered		$\vec{a}_1 \perp \vec{a}_2 \perp (2\vec{a}_3 - \vec{a}_1 - \vec{a}_2) \perp \vec{a}_1$, $\lvert\vec{a}_1\rvert = \lvert\vec{a}_2\rvert$,
Face-centered		$(\vec{a}_1 + \vec{a}_2) \perp (\vec{a}_1 - \vec{a}_2)\vec{a}_3, \perp (2\vec{a}_3 - \vec{a}_1 - \vec{a}_2) \perp (\vec{a}_1 + \vec{a}_2)$
Tetragonal (D_{4h})	D_{4h}, D_4, D_{2d},	
simple	C_{4v}, C_{4h}, C_4,	$\vec{a}_1 \perp \vec{a}_2 \perp \vec{a}_3 \perp \vec{a}_1, \lvert\vec{a}_1\rvert = \lvert\vec{a}_2\rvert$,
Body-centered	S_4	$\vec{a}_1 \perp (2\vec{a}_3 - \vec{a}_1 - \vec{a}_2) \perp \vec{a}_2 \perp \vec{a}_1$, $\lvert\vec{a}_1\rvert = \lvert\vec{a}_2\rvert$
Rhombohedral (D_{3d})	D_{3d}, D_3, C_{3v},	
simple	S_6, C_3	$\vec{a}_1 \perp (2\vec{a}_3 - \vec{a}_1 + \vec{a}_2) \perp \vec{a}_2$, $\lvert\vec{a}_1\rvert = \lvert\vec{a}_2\rvert, < \vec{a}_1\vec{a}_2 = 2\pi/3$
Hexagonal (D_{6h})	D_{6h}), C_{6h}, C_{6v}	
simple	D_{3h}), C_{3h}, C_6, D_6	$\vec{a}_1 \perp \vec{a}_3 \perp \vec{a}_2, \lvert\vec{a}_1\rvert = \lvert\vec{a}_2\rvert$, $< \vec{a}_1\vec{a}_2 = \pi/3$
Cubic (O_h)	O_h, O, T_d, T_h,	
simple	T	$\vec{a}_1 \perp \vec{a}_2 \perp \vec{a}_3 \perp \vec{a}_1$, $\lvert\vec{a}_1\rvert = \lvert\vec{a}_2\rvert = \lvert\vec{a}_3\rvert$,
Body-centered		$\vec{a}_1 \perp \vec{a}_2 \perp \vec{a}_4 \perp \vec{a}_1$, $\lvert\vec{a}_1\rvert = \lvert\vec{a}_2\rvert = \lvert\vec{a}_4\rvert$, $\vec{a}_4 = 2\vec{a}_3 - \vec{a}_1 - \vec{a}_2$,
Face-centered		$\vec{a}_1 \perp \vec{a}_2 \perp \vec{a}_4 \perp \vec{a}_1$, $\lvert\vec{a}_1\rvert = \lvert\vec{a}_2\rvert = \lvert\vec{a}_4\rvert / \sqrt{2}$, $\vec{a}_4 = 2\vec{a}_3 - \vec{a}_1$

The lattice group $L(3)$ is a normal subgroup of the space group G, since

$$\begin{aligned} \{g|\vec{m}\}\{e|\vec{n}\}\{g|\vec{m}\}^{-1} &= \{g|g\vec{n} + \vec{m}\}\{g^{-1}| - g^{-1}\vec{m}\} \\ &= \{e|g\vec{n}\} \end{aligned} \tag{16.22}$$

is a lattice translation in $L(3)$. The space group can therefore be expressed as a union of left cosets of L

$$G = \sum_{i=1}^{h} \{g_i|\vec{0}\}L, \tag{16.23}$$

where the coset representatives are elements of the point group G_0. It is important to note that L contains all the primitive translations associated with the lattice and as such is maximal; that is, all lattice points can be reached from any other by the elements of L.

Space groups containing elements of the type

$$\{g|\vec{m}\}, \tag{16.24}$$

where $\vec{m}$ is always a primitive lattice vector (or a linear combination of primitive lattice vectors) are called *symmorphic*. There are 73 symmorphic space groups in 3-space.

There also exist space groups with elements

$$\{g|\vec{v}\}, \tag{16.25}$$

where $\vec{v}$ is not a primitive lattice vector (or a linear combination of primitive lattice vectors) but as a pure translation arrives at a point that is not a lattice point. Together with a point group operation, such nonprimitive translations are symmetry operations, for instance, when screw axes and glide planes exist. There are 157 *nonsymmorphic* space groups in 3-space. Here we deal only with symmorphic space groups.

The construction of symmetry-adapted functions that span irreps of a space group is considerably more complicated than it is for a point group. However, the rather simple structure of symmorphic space groups makes it rather straightforward at least for that case.

The translation group [lattice group) $L(3)$] is Abelian, and all its irreps are therefore one-dimensional. The elements $\{e|\vec{m}\} \in L(3)$ are labeled by three integers, the coefficients of the basis vectors

$$\vec{m} = \mu_1\vec{a}_1 + \mu_2\vec{a}_2 + \mu_3\vec{a}_3, \tag{16.26}$$

which thus characterize the primitive lattice vectors $\vec{m}$. Using periodic boundary conditions permits a finite lattice to represent an infinite system. According to the so called Born–Kàrmàn periodic boundary conditions, the values of the coefficients are limited to

$$-\frac{N_i}{2} \leq \mu_i \leq \frac{N_i}{2}, \quad i = 1, 2, 3. \tag{16.27}$$

We set

$$N_1 = N_2 = N_3 = N \tag{16.28}$$

and the total number of $\vec{m}$ vectors in this Born–Kàrmàn (BK) region is N^3, which then is the order of the translation group.

16.4 RECIPROCAL SPACE AND BLOCH FUNCTIONS

One says that the *direct lattice* is spanned by the basis vectors $\{\vec{a}_i\}$, and a unit cell (primitive cell) in this lattice has the volume

$$\vec{a}_1 \cdot (\vec{a}_2 \times \vec{a}_3) = \vec{a}_3 \cdot (\vec{a}_1 \times \vec{a}_2) = \vec{a}_2 \cdot (\vec{a}_3 \times \vec{a}_1) = V\,. \tag{16.29}$$

A reciprocal lattice can then be introduced and spanned by the basis vectors

$$\vec{b}_1 = 2\pi\vec{a}_2 \times \frac{\vec{a}_3}{V}; \qquad \vec{b}_2 = 2\pi\vec{a}_3 \times \frac{\vec{a}_1}{V}; \qquad \vec{b}_3 = 2\pi\vec{a}_1 \times \frac{\vec{a}_2}{V} \tag{16.30}$$

such that the two sets of basis vectors satisfy

$$\vec{a}_i \cdot \vec{b}_j = 2\pi\delta_{ij}\,. \tag{16.31}$$

A general vector in the *reciprocal lattice* is of the form

$$\vec{K} = (\nu_1\vec{b}_1 + \nu_2\vec{b}_2 + \nu_3\vec{b}_3), \quad \nu_i = \text{integer}\,, \tag{16.32}$$

that is, the scalar product

$$\vec{K} \cdot \vec{m} = 2\pi(\nu_1 + \nu_2 + \nu_3) \tag{16.33}$$

is an integer times 2π. A general vector in *reciprocal space* is a linear combination of the basis vectors $\{\vec{b}_i\}$ with arbitrary coefficients. The quasi-continuous set of N^3 vectors

$$\vec{k} = \frac{1}{N}(\kappa_1\vec{b}_1 + \kappa_2\vec{b}_2 + \kappa_3\vec{b}_3), \quad \kappa_i = \text{integer}\,, \tag{16.34}$$

with $-N/2 \le \kappa_i < N/2$, lie in the first Brillouin zone (BZ).

For instance, for a body-centered cubic lattice, with cube sides of lengths a, the three vectors

$$\begin{aligned} \vec{a}_1 &= \frac{a}{2}(1 \;\; -1 \;\; 1)\,, \\ \vec{a}_2 &= \frac{a}{2}(1 \;\; 1 \;\; -1)\,, \\ \vec{a}_3 &= \frac{a}{2}(-1 \;\; 1 \;\; 1)\,, \end{aligned} \tag{16.35}$$

are possible basis vectors in direct space with the basic volume $V = a^3/2$. The corresponding reciprocal basis vectors are

$$\begin{aligned} \vec{b}_1 &= \frac{2\pi}{a}(1 \;\; 0 \;\; 1)\,, \\ \vec{b}_2 &= \frac{2\pi}{a}(1 \;\; 1 \;\; 0)\,, \\ \vec{b}_3 &= \frac{2\pi}{a}(0 \;\; 1 \;\; 1)\,, \end{aligned} \tag{16.36}$$

with the basic volume $(2\pi)^3/V$. Lattice vectors in direct space expressed in terms of the coordinates $[\mu_1 \quad \mu_2 \quad \mu_3]$ are (the notation with double signs means that the upper signs go together and the lower signs go together, i.e., *not* all possible combinations)

$$\begin{aligned}
&[0 \quad 0 \quad 0], \text{ of length } 0;\\
&[\pm 1 \quad 0 \quad 0], [0 \quad \pm 1 \quad 0], [0 \quad 0 \quad \pm 1],\\
&[1 \quad -1 \quad 1], [-1 \quad 1 \quad -1], \text{ of length } \frac{a\sqrt{3}}{2};\\
&[\pm 1 \quad 0 \quad \pm 1], [1 \quad -1 \quad 0],\\
&[-1 \quad 1 \quad 0], [0 \quad 1 \quad -1], [0 \quad 1 \quad -1], \text{ of length } a,
\end{aligned} \tag{16.37}$$

while the lattice vectors in reciprocal space are in terms of the coordinates $\{\nu_1 \quad \nu_2 \quad \nu_3\}$

$$\begin{aligned}
&\{0 \quad 0 \quad 0\}, \text{ of length } 0;\\
&\{\pm 1 \quad 0 \quad 0\}, \{0 \quad \pm 1 \quad 0\}, \{0 \quad 0 \quad \pm 1\}, \{\pm 1 \quad \pm 1 \quad 0\},\\
&\{\pm 1 \quad 0 \quad \pm 1\}, \{1 \quad 0 \quad -1\}, \{-1 \quad 0 \quad 1\}, \text{ of length } 2\pi\frac{\sqrt{2}}{a};\\
&\text{and}\\
&\{\pm 1 \quad \pm 1 \quad \pm 1\}, \{1 \quad 1 \quad -1\}, \{-1 \quad -1 \quad 1\},\\
&\{1 \quad -1 \quad -1\}, \{-1 \quad 1 \quad 1\}, \text{ of length } 2\pi\frac{2}{a}.
\end{aligned} \tag{16.38}$$

Translations in one dimension, say, along the $\vec{a}_1$ direction in direct space by elements $\{e|\mu_1\vec{a}_1\} = \{e|\vec{m}_1\}$ due to the periodic boundary conditions, satisfy

$$\{e|\vec{m}_1\}^N = 1 = (\text{identity}) \tag{16.39}$$

and thus the (one-dimensional) representation matrices or the character becomes

$$\chi^{\vec{k}_1}(\vec{m}_1) = e^{-i\vec{k}_1\cdot\vec{m}_1} \tag{16.40}$$

with $\vec{k}_1 = \kappa_1\vec{b}_1/N$. Translations in three dimensions are obtained as the triple direct product of such translations in the three linearly independent directions, and the character of the direct product representation is the product of the characters of the representations referring to translations in each of the three directions. Since the representations are one-dimensional and the groups commutative, everything is trivial. The irreducible representations $\Gamma^{\vec{k}}$ of this direct product are one-dimensional and labeled by the vectors

$$\vec{k} = \frac{\kappa_1 \vec{b}_1 + \kappa_2 \vec{b}_2 + \kappa_3 \vec{b}_3}{N}, \qquad \kappa_i = \text{integer}, \tag{16.41}$$

such that the character at the element $\{e|\vec{m}\}$ is

$$\chi^{\vec{k}}(\vec{m}) = e^{-i\vec{k}\cdot\vec{m}}. \tag{16.42}$$

Thus, the *wavevector* $\vec{k}$ labels the irreps and $\vec{m}$ the classes of the pure translation group.

Functions $\psi(\vec{k};\vec{r})$ that transform according to the irreps of $L(3)$, that satisfy

$$\{e|\vec{m}\}\psi(\vec{k};\vec{r}) = \psi(\vec{k};\{e|\vec{m}\}^{-1}\vec{r}) = \psi(\vec{k};\vec{r} - \vec{m}) = e^{-i\vec{k}\cdot\vec{m}}\psi(\vec{k};\vec{r}), \tag{16.43}$$

are called *Bloch functions*. Under a general space group operation a Bloch function transforms as

$$\{g|\vec{m}\}\psi(\vec{k};\vec{r}) = \psi(\vec{k};g^{-1}\vec{r} - g^{-1}\vec{m}) = \psi(\vec{k};g^{-1}(\vec{r} - \vec{m})), \tag{16.44}$$

and it is possible to show that this is a Bloch function characterized by the wavevector $g\vec{k}$. This is so since a scalar product $\vec{k}\cdot\vec{r}$ obviously satisfies

$$\vec{k}\cdot(g^{-1}\vec{r}) = (g\vec{k})\cdot\vec{r}. \tag{16.45}$$

Bloch functions can be constructed by the (character) projection operators

$$Q^{\vec{k}} = \frac{1}{N^3}\sum_{\vec{m}}^{\text{BK}} e^{i\vec{k}\cdot\vec{m}}\{e|\vec{m}\}. \tag{16.46}$$

A great variety of Bloch functions are used in solid-state theory. Two typical kinds are plane waves (PWs),

$$\eta(\vec{k};\vec{r}) = \frac{1}{\sqrt{N^3 V}}e^{i\vec{k}\cdot\vec{r}} \tag{16.47}$$

and the LCAO (linear combination of atomic orbitals) type

$$\phi(\vec{k};\vec{r}) = N_c \sum_{\vec{m}}^{\text{BK}} \phi(\vec{m},\vec{r})e^{i\vec{k}\cdot\vec{m}}, \tag{16.48}$$

where

$$\phi(\vec{m},\vec{r}) = \phi(\vec{r} - \vec{m}) \tag{16.49}$$

is an atomic orbital (AO) or any other localized function associated with or centered on lattice site $\vec{m}$. The constant N_c is the normalization, which depends on the overlaps of these localized functions.

For PWs it then holds that

$$\{g|\vec{m}\}\eta(\vec{k};\vec{r}) = \frac{1}{\sqrt{N^3V}} e^{i(g\vec{k})\cdot\vec{r}} e^{-i\vec{k}\cdot(g^{-1}\vec{m})}, \tag{16.50}$$

that is, a Bloch function with wavevector $g\vec{k}$. Since it is assumed that $\{g|\vec{m}\}$ is an element of a symmorphic space group, $g^{-1}\vec{m}$ is a lattice vector. The special case $g = e$ yields the Bloch condition

$$\{e|\vec{m}\}\eta(\vec{k};\vec{r}) = e^{-i\vec{k}\cdot\vec{m}}\eta(\vec{k};\vec{r}). \tag{16.51}$$

For a Bloch function of LCAO type one gets

$$\begin{aligned}
\{g|\vec{m}\}\phi(\vec{k};\vec{r}) &= N_c \sum_{\vec{n}}^{\mathrm{BK}} \{g|\vec{m}\}\{e|\vec{n}\}\phi(\vec{r})e^{i\vec{k}\cdot\vec{n}} \\
&= N_c \sum_{\vec{n}}^{\mathrm{BK}} \{g|g\vec{n} + \vec{m}\}\phi(\vec{r})e^{i\vec{k}\cdot\vec{n}} \\
&= N_c \sum_{\vec{n}}^{\mathrm{BK}} \{e|\vec{n}'\}\{g|g\vec{n} + \vec{m} - \vec{n}'\}\phi(\vec{r})e^{i\vec{k}\cdot\vec{n}} \\
&= N_c \sum_{\vec{n}'}^{\mathrm{BK}} \{e|\vec{n}'\}\{g|\vec{0}\}\phi(\vec{r})e^{i\vec{k}\cdot g^{-1}(\vec{n}'-\vec{m})} \\
&= N_c \sum_{\vec{n}}^{\mathrm{BK}} \{e|\vec{n}\}\phi(g^{-1}\vec{r})e^{i(g\vec{k})\cdot(\vec{n}-\vec{m})},
\end{aligned} \tag{16.52}$$

or a Bloch function with wavevector $g\vec{k}$. One should note that the AO part is also affected by the point group part of the space group element. When

$$g\vec{k} = \vec{k} + \vec{K}, \tag{16.53}$$

the transformed LCAO function reduces to

$$\{g|\vec{m}\}\phi(\vec{k};\vec{r}) = N_c \sum_{\vec{n}}^{\mathrm{BK}} \{e|\vec{n}\}\phi(g^{-1}\vec{r})e^{i\vec{k}\cdot(\vec{n}-\vec{m})}. \tag{16.54}$$

16.5 CONSTRUCTION OF SYMMETRY FUNCTIONS

Those elements $g \in G_0$ that satisfy (16.53) form the group of the wavevector $\vec{k}$. This is a subgroup, $G_0(\vec{k})$, of G_0, proper or improper. When $G_0(\vec{k}) = G_0$, as is the case for $\vec{k} = \vec{0}$, or when $G_0(\vec{k})$ is a proper subgroup of G_0, one calls $\vec{k}$ (or rather its endpoint) a *special point* in BZ. Those points $\vec{k}$ for which $G_0(\vec{k})$ contains only the identity element are called *general points* in BZ. The subgroup $G_0(\vec{k})$ of G_0 is called the little cogroup of $\vec{k}$. The union of the distinct (left) cosets of $L(3)$ by elements $\{h_i|\vec{0}\} \in G_0(\vec{k})$ defines the *little group* of $\vec{k}$:

$$G(\vec{k}) = \sum_{i=1}^{b} \{h_i|\vec{0}\}L. \tag{16.55}$$

The point group G_0 can be expressed in terms of the (left) cosets of the little cogroup as

$$G_0 = \sum_{i=1}^{q} \{g_i|\vec{0}\}G_0(\vec{k}) \equiv \sum_{i=1}^{q} g_i G_0(\vec{k}), \tag{16.56}$$

and obviously qb is the order of G_0. We let $g_1 = \{g_1|\vec{0}\} = e$, which is the identity of G_0. The other coset representatives must satisfy

$$g_i\vec{k} \neq \vec{k} + \vec{K}, \quad i = 2, 3, 4, ..q. \tag{16.57}$$

The q coset representatives $g_i = \{g_i|\vec{0}\}$ generate a set of $g_i\vec{k}$ that together form the *star* of $\vec{k}$. These particular operations can also be used advantageously as coset representatives in the decomposition of the full space group in terms of cosets with respect to the little group of $\vec{k}$:

$$G = \sum_{i=1}^{q} \{g_i|\vec{0}\}G(\vec{k}). \tag{16.58}$$

All the necessary ingredients for the construction of symmetry functions of a space group should now be available. A Bloch function that transforms according to an irrep of $G_0(\vec{k})$ can be created by a projector

$$P^{\alpha}_{ss} = f^{\alpha}\mathbf{M}_{g\in G_0(\vec{k})}\Gamma^{\alpha}_{ss}(g)g^{-1}, \tag{16.59}$$

which commutes with the projector $Q^{\vec{k}}$ [Eq. (16.46)], which creates Bloch functions. This can be seen from

$$\begin{aligned} P^{\alpha}_{ss}Q^{\vec{k}} &= \frac{f^{\alpha}}{N^3}\mathbf{M}_{g\in G_0(\vec{k})}\sum_{\vec{m}}^{\mathrm{BK}}\Gamma^{\alpha}_{ss}(g)e^{i\vec{k}\cdot\vec{m}}\{g^{-1}|\vec{0}\}\{e|\vec{m}\} \\ &= \frac{f^{\alpha}}{N^3}\mathbf{M}_{g\in G_0(\vec{k})}\sum_{\vec{m}}^{\mathrm{BK}}\Gamma^{\alpha}_{ss}(g)e^{i\vec{k}\cdot\vec{m}}\{e|g^{-1}\vec{m}\}\{g^{-1}|\vec{0}\}. \end{aligned} \tag{16.60}$$

With the use of $\vec{n} = g^{-1}\vec{m}$ as summation index instead of $\vec{m}$, the exponential becomes

$$e^{i\vec{k}\cdot\vec{m}} = e^{i\vec{k}\cdot(g\vec{n})} = e^{i(g^{-1}\vec{k})\cdot\vec{n}} = e^{i\vec{k}\cdot\vec{n}}, \tag{16.61}$$

when g belongs to the little cogroup of $\vec{k}$. Therefore

$$P^{\alpha}_{ss}Q^{\vec{k}} = Q^{\vec{k}}P^{\alpha}_{ss} \tag{16.62}$$

when P^{α}_{ss} belongs to the group algebra of $G_0(\vec{k})$. One can prove that the functions

$$\Phi^{\alpha}_{ss,i}(\vec{k};\vec{r}) = \{g_i|\vec{0}\}P^{\alpha}_{ss}\Phi(\vec{k};\vec{r}) \tag{16.63}$$

span an irrep of the full space group of dimension $f^{\alpha}q$. This procedure applies to all irreps of the space group G.

Given a space group, one can construct symmetry adapted functions for it using the following procedure:

1. Choose the basis set.
2. Choose a wavevector $\vec{k}$.
3. Construct a Bloch function using $Q^{\vec{k}}$.
4. Find the little cogroup $G_0(\vec{k})$ of that $\vec{k}$.
5. Find the star of that $\vec{k}$.
6. Choose an irrep Γ^{α} of $G_0(\vec{k})$ and form P^{α}_{ss}.
7. Use Eq. (16.63).

Example 16.1 Consider a body-centered cubic lattice, typical for an alkali metal or for iron. The space group is called O_h^9. It is symmorphic, and its point group is the cubic group O_h. The BZ of the body-centered cubic lattice is shown in Figure 16.1.

Each special point indicated in Figure 16.1 is the endpoint of a particular wavevector $\vec{k}$ and has a group $G_0(\vec{k})$ and a star of order $q(\vec{k})$. It is important to remember that the point group elements that transform a wavevector $\vec{k}$ in BZ into itself plus a reciprocal lattice vector $\vec{K}$ make up the group of the wavevector. Table 16.1 lists the group of the wavevector, its order, and the order of the star of the wavevector for the special points of the BZ.

Constructing symmetry adapted Bloch functions for selected points in BZ can now be done as follows. Starting from an atomic orbital $\phi_{lm}(\vec{r}) = R_{nl}(r)Y_{lm}(\theta, \varphi)$, $m = -l, -l+1, \ldots, l$ $(l = 0, 1, 2, \ldots)$ a Bloch function can be produced as

$$\psi_{lm}(\vec{k}, \vec{r}) = Q^{\vec{k}}\phi_{lm}(\vec{r}) = \frac{1}{N^3}\sum_{\vec{n}}^{\mathrm{BK}} \phi_{lm}(\vec{n}, \vec{r})e^{i\vec{k}\cdot\vec{n}}. \tag{16.64}$$

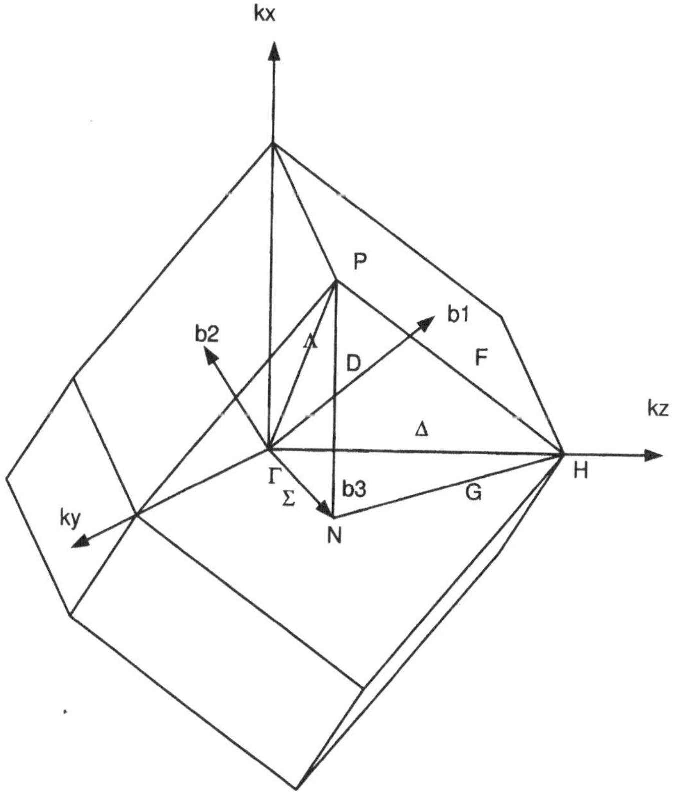

Figure 16.1 The first Brillouin zone (BZ) of the body-centered cubic lattice. It has the shape of a cube with a pyramid on each face. The special points in the BZ are of various kinds such as high-symmetry points Γ and *H*, and the low-symmetry point Λ, which is any point on the line between Γ and *P*. The vectors *b*1, *b*2, and *b*3 are the vectors $\frac{1}{2}\vec{b}_1$, $\frac{1}{2}\vec{b}_2$, and $\frac{1}{2}\vec{b}_3$ respectively.

The $(2l + 1)$ functions for a fixed l in general form a reducible representation of the group of the wavevector $\vec{k}$.

We consider first the point Γ ($\vec{k} = \vec{0}$). Then $G_0(\vec{k})$ is the full group O_h. The characters of the representations $\gamma^l \downarrow O_h$ with γ^l the irrep of the three-dimensional rotation group $SO(3)$ spanned by the $(2l + 1)$ functions $\{\phi_{lm}(\vec{r})\}$ are given here for the first few values of l:

O_h	E	$8C_3$	$6C_2$	$6C_4$	$3C_4^2$	i	$6S_4$	$8S_6$	$3\sigma_h$	$6\sigma_d$
γ^0	1	1	1	1	1	1	1	1	1	1
γ^1	3	0	−1	1	−1	−3	0	1	−1	1
γ^2	5	−1	1	−1	1	5	−1	1	−1	1

TABLE 16.1 Special Points in Brillouin Zone with Their Coordinates, Group of the Wavevector, Its Order, and Order of Star of Wavevector.

Point	Coordinate[a]	$G_0(\vec{k})$	$\lvert G_0(\vec{k})\rvert$	$q(\vec{k})$
Γ	$(0\ \ 0\ \ 0)$	O_h	48	1
H	$\pi(0\ \ 0\ \ 2)/a$	O_h	48	1
N	$\pi(0\ \ 1\ \ 1)/a$	D_{2h}	8	6
P	$\pi(1\ \ 1\ \ 1)/a$	T_d	24	2
Δ	$\pi(0\ \ 0\ \ 2\kappa)/a$	C_{4v}	8	6
Λ	$\pi(\kappa\ \ \kappa\ \ \kappa)/a$	C_{3v}	6	8
Σ	$\pi(0\ \ \kappa\ \ \kappa)/a$	C_{2v}	4	12
D	$\pi(\kappa\ \ 1\ \ 1)/a$	C_{2v}	4	12
F	$\pi(1-\kappa\ \ 1-\kappa\ \ 1+\kappa)/a$	C_{3v}	6	8
G	$\pi(0\ \ 1-\kappa\ \ 1+\kappa)/a$	C_{2v}	4	12

[a]Here a is the side of the cube in direct space.

These representations reduce to

$$\begin{aligned} \gamma^0 &= A_{1g} \\ \gamma^1 &= T_{1u} \\ \gamma^2 &= E_g + T_{2g} \end{aligned} \tag{16.65}$$

This means that for $\vec{k} = \vec{0}$ an s function on each atom in a body-centered cubic (bcc) lattice locally generates only a_{1g} functions, while p functions generate only t_{1u} symmetry functions, and d functions yield both e_g and t_{2g} symmetry functions. Since for $\vec{k} = \vec{0}$, $q(\vec{k}) = 1$ it follows that for the space group O_h^9, correspondingly the irreps are of dimensions 1, 2, or 3. A notation system due to Bouckaert–Smoulchowski–Wigner (BSW) for the irreps of the space group is listed below and compared to the corresponding point group notation for the irreps of the group of the wavevector.

BSW	Molecular Notation
Γ_1	A_{1g}
Γ_2	A_{2g}
Γ_{12}	E_g
$\Gamma_{15'}$	T_{1g}
$\Gamma_{25'}$	T_{2g}
$\Gamma_{1'}$	A_{1u}
$\Gamma_{2'}$	A_{2u}
$\Gamma_{12'}$	E_u
Γ_{15}	T_{1u}
Γ_{25}	T_{2u}

Thus, the function $\phi_{00}(\vec{r}) = \phi(\vec{r})$ spans the irrep A_{1g} of O_h, while the Bloch function

$$\psi(\vec{0},\vec{r}) = Q^{\vec{k}}\phi(\vec{r}) = \frac{1}{N^3}\sum_{\vec{n}}^{\text{BK}}\phi(\vec{n},\vec{r}) = \frac{1}{N^3}\sum_{\vec{n}}^{\text{BK}}\phi(|\vec{r}-\vec{n}|) \tag{16.66}$$

transforms according to the irrep Γ_1 of O_h^9. Note that the BSW notation (i.e., using the symbol Γ) also denotes the wavevector considered.

Explicitly, a space group operation acting on $\phi_{00}(\vec{r}) = \phi(\vec{r})$ yields

$$\begin{aligned}\{g|\vec{m}\}\phi(\vec{r}) &= \phi(\{g|\vec{m}\}^{-1}\vec{r}) = \phi(g^{-1}\vec{r} - g^{-1}\vec{m}) \\ &= \phi(g^{-1}|\vec{r}-\vec{m}|) = \phi(|\vec{r}-\vec{m}|)\end{aligned} \tag{16.67}$$

where we have used that it is an s function. Similarly

$$\begin{aligned}\{g|\vec{m}\}\psi(\vec{0};\vec{r}) &= \frac{1}{N^3}\sum_{\vec{n}}^{\text{BK}}\{g|\vec{m}\}\{e|\vec{n}\}\phi(\vec{r}) \\ &= \frac{1}{N^3}\sum_{\vec{n}}^{\text{BK}}\phi(\{g^{-1}|-\vec{n}-g^{-1}\vec{m}\}\vec{r}) = \frac{1}{N^3}\sum_{\vec{n}'}^{\text{BK}}\phi(g^{-1}(\vec{r}-\vec{n}')).\end{aligned} \tag{16.68}$$

Using AOs of d type raises the need for the projectors

$$P_{11}^{E_g}, P_{11}^{T_{2g}} \tag{16.69}$$

and the corresponding shift operators or the character projectors. The point group character tables usually gives a hint as to which combinations of AOs have the desired transformation properties. For instance, the e_g symmetry Bloch functions are

$$\begin{aligned}\psi_{20;(2z^2-x^2-y^2)}(\vec{0};\vec{r}) &= O^{\vec{0}}\phi_{20}(\vec{r}) = \frac{1}{N^3}\sum_{\vec{n}}^{\text{BK}}\phi_{20}(\vec{n},\vec{r}) \\ \psi_{22;(x^2-y^2)}(\vec{0};\vec{r}) &= O^{\vec{0}}\frac{\phi_{22}(\vec{r})+\phi_{2-2}(\vec{r})}{\sqrt{2}} \\ &= \frac{1}{N^3}\sum_{\vec{n}}^{\text{BK}}\frac{\phi_{22}(\vec{n},\vec{r})+\phi_{2-2}(\vec{n},\vec{r})}{\sqrt{2}}.\end{aligned} \tag{16.70}$$

In band calculations of LCAO type the final solution for a crystal orbital is in general a linear combination of Bloch functions obtained from many different AO basis functions. Our analysis shows that if we include AOs in our basis, say, through f functions [at the center point ($\vec{k} = \vec{0}$) in BZ], only AOs of s type contribute for a state of Γ_1 symmetry, only p AOs for a state of Γ_{15} symmetry, and so forth.

Considering another point, say, Δ, in the BZ, one has $G_0(\Delta) = C_{4v}$. This group contains rotations about only one of the coordinate axes, and two pairs of reflections: one in the planes normal to the other two coordinate axes, and in two diagonal planes. The BSW notation for the space group irreps and their corresponding point group irreps are given in the character table below.

C_{4v}	E	C_2	$2C_4$	$2\sigma_v$	$2\sigma_d$
$\Delta_1(A_1)$	1	1	1	1	1
$\Delta_2(B_1)$	1	1	−1	1	−1
$\Delta_{1'}(A_2)$	1	1	1	−1	−1
$\Delta_{2'}(B_2)$	1	1	−1	−1	1
$\Delta_5(E)$	2	−2	0	0	0

The characters of the representations $\gamma^l \downarrow C_{4v}$ are listed below.

C_{4v}	E	C_2	$2C_4$	$2\sigma_v$	$2\sigma_d$
γ^0	1	1	1	1	1
γ^1	3	−1	1	1	1
γ^2	5	1	−1	1	1
γ^3	7	−1	−1	1	1

The reduction of these representations into irreps is as follows:

$$\begin{aligned}\gamma^0 &= A_1\\ \gamma^1 &= A_1 + E\\ \gamma^2 &= A_1 + B_1 + B_2 + E\\ \gamma^3 &= A_1 + B_1 + B_2 + 2E\,,\end{aligned} \tag{16.71}$$

showing that the Bloch functions of Δ_1 symmetry may contain s, p, d, f AOs, while Bloch functions of Δ_2 and $\Delta_{2'}$ symmetry can contain only d and f AOs, and so forth.

Band calculations are typically carried out for discrete wavevectors, and one chooses various symmetry points. In connecting these discrete points into continuous band energies for intermediate wavevectors, it is useful to know something about the symmetry. Starting, say, at a high-symmetry point as Γ and moving toward a lower-symmetry point as Δ, one finds the various representations subduced by the irreps of O_h into C_{4v}. One finds

C_{4v}	E	C_2	$2C_4$	$2\sigma_v$	$2\sigma_d$
Γ_1	1	1	1	1	1
Γ_2	1	1	-1	1	-1
Γ_{12}	2	2	0	2	0
$\Gamma_{15'}$	3	-1	1	-1	-1
$\Gamma_{25'}$	3	-1	-1	-1	1

for the even (gerade) irreps, and

C_{4v}	E	C_2	$2C_4$	$2\sigma_v$	$2\sigma_d$
$\Gamma_{1'}$	1	1	1	-1	-1
$\Gamma_{2'}$	1	1	-1	-1	1
$\Gamma_{12'}$	2	2	0	-2	0
Γ_{15}	3	-1	1	1	1
Γ_{25}	3	-1	-1	1	-1

for the odd (ungerade).

The reduction into irreps of C_{4v} is as follows:

$$\begin{aligned}\Gamma_1 &= \Delta_1\\ \Gamma_2 &= \Delta_2\\ \Gamma_{12} &= \Delta_1 + \Delta_2\\ \Gamma_{15'} &= \Delta_{1'} + \Delta_5\\ \Gamma_{25'} &= \Delta_{2'} + \Delta_5\end{aligned} \tag{16.72}$$

and

$$\begin{aligned}\Gamma_{1'} &= \Delta_{1'}\\ \Gamma_{2'} &= \Delta_{2'}\\ \Gamma_{12'} &= \Delta_{1'} + \Delta_{2'}\\ \Gamma_{15} &= \Delta_1 + \Delta_5\\ \Gamma_{25} &= \Delta_2 + \Delta_5\end{aligned} \tag{16.73}$$

There is then a consistency in the notation. This means that an energy band that has symmetry Γ_1 at Γ must continue in a Δ_1 band in the Δ direction, while a band of Γ_{12} symmetry must split up into two bands in the same direction, one of Δ_1 symmetry and one of Δ_2 symmetry. In Figure 16.2, note that point H in the Brillouin zone BZ (a corner where four sides meet) has the same group of the wavevector as Γ.

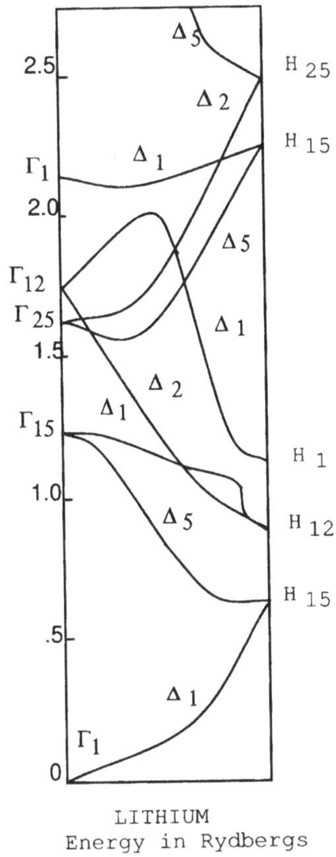

Figure 16.2 Sketched energy bands for lithium metal.

REFERENCES

The fundamentals of space groups and their use is covered in many texts.

Burns, G., and A. M. Glazer, *Space Groups for Solid State Scientists*, Academic Press, Boston, 1990 (an accessible source).

Kettle, S. F. A., *Symmetry and Structure*, 2nd ed., Wiley, New York, 1995 (contains a pictorial approach to space groups that might appeal to some).

Miller, W., *Symmetry Groups and Their Application*, Academic Press, New York, 1972 (a rather mathematical treatment of space groups; this text contains a wealth of information about group theory in general).

17

CONTINUOUS GROUPS

17.1 DEFINITIONS

While the group elements of discrete groups (finite or infinite) are labeled by integers, the elements of continuous groups are labeled by a set of continuous parameters defined on some domain.

Example 17.1 The two-parameter group of collinear transformations on the straight line of real numbers is considered in detail below.

Group Operations

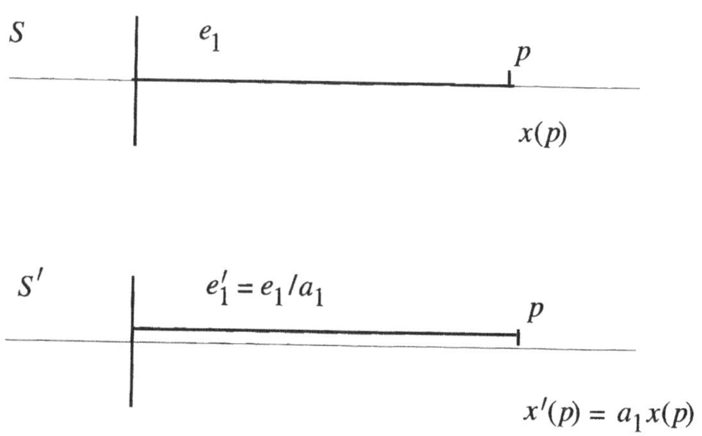

The length of the basis vector can be changed as follows.

1. Under scaling of the basis vector, $\vec{e}_1 \rightarrow \vec{e}_1/a_1$ for $a_1 > 0$. The coordinate of point p is multiplied by a_1, $x'(p) = a_1 x(p)$.
2. Shift of origin. Under a shift of the origin, $0 \rightarrow -a_2$, the coordinate of p shifts: $x'(p) = x(p) + a_2$.
3. Reversal of the orientation of the basis vector (one operation). Under reversal of the direction of $\vec{e}_1$ the coordinate of p changes: $x'(p) = -x(p)$.

The basis vector $\vec{e}_1$ and the coordinate $x(p)$ of some fixed point p transform in the contravariant fashion. We define the group operations on $x(p)$:

$$(a_1, a_2)x = a_1 x + a_2, \qquad a_1 \neq 0. \tag{17.1}$$

The underlying *topological space* for this group is the real plane $\{a_1, a_2\}$ excluding the line $a_1 = 0$. The *geometric space* on which the group acts is the real line x. We write the *composition function* $f(x; a_1, a_2)$ for this group of linear transformations:

$$x' = f(x; a_1, a_2) \equiv a_1 x + a_2 \tag{17.2}$$

The group multiplication becomes

$$\begin{aligned} x'' &= f(x'; b_1, b_2) = b_1 x' + b_2 \\ &= b_1(a_1 x + a_2) + b_2 = f(x; b_1 a_1, b_1 a_2 + b_2) \end{aligned} \tag{17.3}$$

Group Postulates We can check that the four group postulates are satisfied for this type of construction:

I. Closure: $b_1 \neq 0;\ a_1 \neq 0 \rightarrow b_1 a_1 \neq 0$

II. Associativity:

$$\begin{aligned} (c_1, c_2)[(b_1, b_2)(a_1, a_2)] &= [(c_1, c_2)(b_1, b_2)](a_1, a_2) \\ (c_1, c_2)(b_1 a_1, b_1 a_2 + b_2) &= (c_1 b_1, c_1 b_2 + c_2)(a_1, a_2) \\ (c_1 b_1 a_1, c_1 b_1 a_2 + c_1 b_2 + c_2) &= (c_1 b_1 a_1, c_1 b_1 a_2 + c_1 b_2 + c_2) \end{aligned} \tag{17.4}$$

III. Identity $(a_1, a_2)(1, 0) = (1, 0)(a_1, a_2) = (a_1, a_2)$

IV. Inverse $(a_1, a_2)(1/a_1, -a_2/a_1) = (1/a_1, -a_2/a_1)(a_1, a_2) = (1, 0)$

Realizations Linear transformations are most commonly realized by mappings on matrices:

$$(a_1, a_2) \rightarrow \begin{pmatrix} a_1 & a_2 \\ 0 & 1 \end{pmatrix}. \tag{17.5}$$

In terms of these matrix representatives, the transformations are

$$\begin{pmatrix} x' \\ 1 \end{pmatrix} = \begin{pmatrix} a_1 & a_2 \\ 0 & 1 \end{pmatrix} \begin{pmatrix} x \\ 1 \end{pmatrix} = \begin{pmatrix} xa_1 + a_2 \\ 1 \end{pmatrix}. \tag{17.6}$$

Since for a positive integer N

$$(x')^N = (a_1 x + a_2)^N = \sum_{r=0}^{N} \binom{N}{r} (a_1)^r (a_2)^{N-r} x^r, \tag{17.7}$$

the $N + 1$ powers (homogeneous polynomials) $(x^N, x^{N-1}, \ldots, x, 1)$ can be used as a basis (carrier space) for a matrix representation of dimension $N + 1$ of this group. In this way a number of representations can be generated. For example, consider $N = 3$:

$$(a_1, a_2) \rightarrow \begin{pmatrix} a_1^3 & 3a_1^2 a_2 & 3a_1 a_2^2 & a_2^3 \\ 0 & a_1^2 & 2a_1 a_2 & a_2^2 \\ 0 & 0 & a_1 & a_2 \\ 0 & 0 & 0 & 1 \end{pmatrix} \tag{17.8}$$

Topological Concepts The properties of the parameters are commonly referred to as the *topology* of the parameter space.

For instance, a space is said to be *connected* if any two points in the space can be joined by a line and all points of the line lie in the space

By this definition the topological space, T_2, of $f(x;\, a_1,\, a_2)$ (T_2 = the real plane R_2 except the line $a_1 = 0$) is not connected since any line connecting, say, (1,0) and (−1,0) must contain point $(a_1 = 0, a_2)$, which is not in T_2.

A parameter space of a continuous group can be divided into various parts called *components*. A connected component of a continuous group cannot itself be a group unless it contains the identity. It is then often called the *connected component*. Thus the component of $f(x;\, a_1,\, a_2)$ with the topological space $a_1 > 0$ (i.e., the right half-plane of T_2) is a subgroup and is called the *connected component*, while the component with the topological space $a_1 < 0$ is not a group.

17.2 SUBGROUPS

Let H be a subgroup of $G = \{(a_1, a_2)\}$. Then for any $g \varepsilon G$ the set of group operations

$$gHg^{-1} = \{ghg^{-1},\ h\varepsilon H\} \tag{17.9}$$

also forms a group, *conjugate* to H.

As an example, consider $\{(a_1 > 0, a_2 = 0)\}$. This is a one-dimensional Abelian subgroup of G, and

$$\begin{aligned}(b_1, b_2)(a_1, 0)(b_1, b_2)^{-1} &= (b_1, b_2)(a_1, 0)\left(\frac{1}{b_1}, -\frac{b_2}{b_1}\right) \\ &= (b_1, b_2)\left(\frac{a_1}{b_1}, -\frac{a_1 b_2}{b_1}\right) = (a_1, b_2(1 - a_1))\end{aligned} \tag{17.10}$$

is a series of conjugate subgroups.

It is not difficult to show that $\{(a_1, b_2(1 - a_1))\}$ satisfies the group postulates under the combination law of G.

As another example, consider $\{(1, a_2)\}$. This is a one-parameter abelian subgroup of G. Now

$$\begin{aligned}&(b_1, b_2)\,(1, a_2)\left(\frac{1}{b_1}, -\frac{b_2}{b_1}\right) = \\ &(b_1, b_2)\left(\frac{1}{b_1}, -\frac{b_2}{b_1 + a_2}\right) = (1, b_1 a_2),\end{aligned} \tag{17.11}$$

which, of course, is the same group as $\{(1, a_2)\}$. Thus, in this case $gHg^{-1} = H$. As we have learned, such subgroups are called *self-conjugate*, *invariant*, or *normal*.

We state without proof the following theorem: *The connected component G_0 (containing the identity) of a continuous group G is an invariant subgroup of G.*

Consider a subgroup H of G and the right cosets

$$Hc_0 + Hc_1 + \cdots = G; \tag{17.12}$$

where the elements $c_0, c_1, \ldots, \in G$ are the coset representatives. Every element of G can be written $g = hc_i$ with $h \in H$. Left cosets can be treated analogously.

We also state without proof the following theorem: *If H is invariant, then the coset representatives $c_0, c_1, \ldots$ can be chosen in such a manner that they are closed under multiplication and form a group called the factor group G/H.*

As an example, consider $G = \{(a_1, a_2)\}$ and the subgroups $H_1 = \{(a_1, 0)\}$ and $H_2 = \{(1, a_2)\}$. We can then write

$$\begin{array}{ccccc}(c_1, c_2) & = & (c_1, 0) & \left(1, \frac{c_1}{c_2}\right) \\ G & & H_1 & C_R \\ & & C_L & H_2\end{array} \tag{17.13}$$

Here it is fortuitous that C_R (the right coset representatives) form a group since H_1 is not an invariant subgroup, but by this theorem C_L must have the form of a group.

17.3 LIE GROUPS

A Lie group is defined as the *connected component* of a continuous group in which the composition function (or transformation) is analytic on its domain (actually, the existence of first and second derivatives is sufficient).

Consider groups of linear transformations in an n-dimensional vector space; the transformation is characterized by a finite number r of parameters:

$$x_i' = f_i(x_1, x_2, \ldots, x_n; a_1, a_2, \ldots, a_r) \tag{17.14}$$

The basic theorems and concepts already established are applicable to *compact* Lie groups.

A parameter space T is compact if (1) every infinite sequence of points $t_1, t_2, \ldots (t_i \in T)$ contains a subset of points that converges to a point and (2) this limit point is in T.

This intuitive concept can be illustrated by the following examples:

1. The real line R_1 is not compact since, say, $t_i = i$ (i.e., $1, 2, 3, \ldots$) does not have a subsequence that converges.
2. The circumference of the unit circle is compact.
3. The interior of the unit circle is not compact, since the sequence $t_i = 1 - 1/i$, $i = 1, 2, \ldots$ converges to a point on the circumference, not in the space.

Those theorems and statements discussed for finite groups that depend on the finiteness, of course, cease to have a meaning for continuous groups.

For compact Lie groups the number of nonequivalent irreps is infinite, although they form a discrete set and are of finite order, while for *noncompact* Lie groups this does not necessarily hold.

Summation over group elements such as is done in the formation of group algebra elements must, for continuous groups, involve integration over the domain over which the parameters vary. The volume element of the integration is chosen in such a manner that the integral of an arbitrary function over the group is invariant under any transformation of the group. This is known as *invariant integration*. Compare the invariant mean for finite groups

$$\int f(g)d\tau_g = \int f(gh)d\tau_g = \int f(hg)d\tau_g = \int f(hg)d\tau_{hg} \tag{17.15}$$

for any h of G. The right invariant $d\tau_{gh}$ and the left invariant measure $d\tau_{hg}$ are not necessarily the same for an arbitrary continuous group and the integrals may not even exist.

Example 17.2

$$\int d\tau_g = V_g$$
$$\frac{1}{V_g}\int \Gamma^{\alpha}_{km}(g)\Gamma^{\beta}_{nl}(g^{-1})d\tau_g = \frac{\delta^{\alpha\beta}\delta_{kl}\delta_{nm}}{f^{\alpha}}, \qquad (17.16)$$

which is the master orthogonality relation for irreps and where the volume V_g of the parameter space is well defined for a compact Lie group.

The invariant mean becomes

$$\frac{1}{|G|}\sum_{g\in G} \rightarrow \frac{1}{V_g}\int d\tau_g \qquad (17.17)$$

17.4 MATRIX GROUPS

Consider a vector space of n dimensions and a general nonsingular transformation $\mathbf{A} = \{A_{ik}\}$ on this space:

$$\mathbf{x}' = \mathbf{A}\mathbf{x}, \quad \det \mathbf{A} \neq 0. \qquad (17.18)$$

Obviously, the set of all such transformations form a continuous group under matrix multiplication. The matrix $\mathbf{A} = \{A_{ik}\}$ multiplying a column vector $\mathbf{x}$ defines the composition functions. It becomes important to know how many independent parameters it takes to fully specify an element of such a matrix group.

The most general group of this kind occurs when the matrix elements A_{ik} belong to the field of complex numbers, C. All such matrices form the *general linear group* **GL**(n,C), with each group element labeled by $2n^2$ real parameters. Various subgroups of **GL**(n,C) can be distinguished by putting restrictions on the matrix elements.

For instance, the *unitary group* **U**(n) is obtained by restricting the matrices to those satisfying

$$\mathbf{A}^{\dagger}\mathbf{A} = \mathbf{A}\mathbf{A}^{\dagger} = \mathbf{1} \qquad (17.19)$$

or in more detail

$$\sum_i A^*_{ij}A_{ik} = \sum_i A_{ji}A^*_{ki} = \delta_{jk}\,. \qquad (17.20)$$

The number of relations are n for $j = k$ and $n(n-1)/2$ for $j \neq k$ for the real part and the same for the imaginary part. That makes a total of n^2 relations and thus results in n^2 free real parameters.

The *special unitary group* **SU**(n) has the further restriction that $\det \mathbf{A} = 1$, thus fixing the phase of the determinant of a unitary matrix. This leaves $n^2 - 1$ free parameters.

The *orthogonal group* **O**(n) has n^2 real parameters with the restrictions

$$\mathbf{A}^T\mathbf{A} = \mathbf{A}\mathbf{A}^T = \mathbf{1}, \tag{17.21}$$

with $n(n+1)/2$ relations among them, leaving $n(n-1)/2$ free parameters.

The *special orthogonal group* **SO**(n) has, in addition $\det \mathbf{A} = 1$, which imposes no restriction here, since in **O**(n), $\mathbf{A}^T\mathbf{A} = \mathbf{1}$ means $\det \mathbf{A} = \pm 1$. Thus **O**(n) has two components and the one with $\det \mathbf{A} = 1$ is **SO**(n).

Thus we can form "chains" of subgroups, such as

$$\mathbf{GL}(n) \supset \mathbf{U}(n) \supset \{\mathbf{O}(n), \mathbf{SU}(n)\} \supset \mathbf{SO}(n). \tag{17.22}$$

17.5 INFINITESIMAL TRANSFORMATIONS

Any element of a Lie group can be represented as a succession of infinitesimal transformations. The theory of Lie groups shows that finite transformations are uniquely determined by specifying infinitesimal transformations. In short, a compact Lie group is completely specified by its infinitesimal operations about the identity element. General proofs of this can be found in texts on Lie groups (see, e.g., R. Gilmore, *Lie Groups, Lie Algebras and Some of Their Applications*, Wiley, New York, 1974).

One can always arrange the parametrization of the group such that the identity corresponds to the parameters being zero: $\mathbf{a} = \mathbf{0}$.

Example 17.3

$$\begin{aligned} x' &= (a_1, a_2)x = a_1 x + a_2 \rightarrow (1,0) = \text{identity} \\ x' &= (a_1, a_2)x = \exp(a_1)x + a_2 \rightarrow (0,0) = \text{identity} \end{aligned} \tag{17.23}$$

Consider $x_i' = f_i(\mathbf{x};\mathbf{a}) = f_i(x_1, x_2, \ldots, x_n; a_1, a_2, \ldots, a_r)$, with the identity $x_i' = f_i(\mathbf{x}, \mathbf{0})$.

For an infinitesimal transformation, one can write

$$x_i' + \mathrm{d}x_i' = f_i(\mathbf{x}; \delta\mathbf{a}) = f_i(\mathbf{x};\mathbf{0}) + \sum_{p=1}^{r} \left[\frac{\partial f_i(\mathbf{x};\mathbf{a})}{\partial a_p}\right]_{\mathbf{a}=\mathbf{0}} \delta a_p. \tag{17.24}$$

Introducing the notation

$$\left[\frac{\partial f_i(\mathbf{x};\mathbf{a})}{\partial a_p}\right]_{\mathbf{a}=\mathbf{0}} = u_{ip}(\mathbf{x}), \tag{17.25}$$

one can write

$$dx_i = \sum_{p=1}^{r} u_{ip}(\mathbf{x})\delta a_p , \tag{17.26}$$

showing that the r vectors $\mathbf{u}_p(\mathbf{x}) = \{u_{ip}(\mathbf{x})\}$ in an n-dimensional vector space ($i = 1,2,\ldots,n$) define an infinitesimal change $d\mathbf{x}$ in any point $\mathbf{x}$. A corresponding change in an arbitrary function $F(\mathbf{x})$ can now be determined.

In a coordinate system S the point p has the coordinates $x_1(p)$, $x_2(p),\ldots,x_n(p) = \mathbf{x}$, and in another coordinate system S' the coordinates of point p are $x_1'(p), x_2'(p),\ldots,x_n'(p) = \mathbf{x}'$. Since under the transformation from one coordinate system to the other by some Lie group operation, the form of F must change in such a manner as to preserve its value at p. Such a transformation is

$$x_i'(p) = f_i(\mathbf{x};\mathbf{a}), \tag{17.27}$$

and its inverse is

$$x_i(p) = f_i(\mathbf{x}';\mathbf{a}^{-1}) \tag{17.28}$$

($i = 1,2,\ldots,n$). Thus

$$\begin{aligned} &F^{S'}(x_1'(p), x_2'(p),\ldots,x_n'(p)) = \\ &F^{S}(f_1(\mathbf{x}';\mathbf{a}^{-1}), f_2(\mathbf{x}';\mathbf{a}^{-1}),\ldots,f_n(\mathbf{x}';\mathbf{a}^{-1})) . \end{aligned} \tag{17.29}$$

Since for compact Lie groups we can concentrate on the infinitesimal transformations about the identity, we can write

$$x_i(p) = f_i(\mathbf{x}';(\delta\mathbf{a})^{-1}) = f_i(\mathbf{x}';\mathbf{0}) + \sum_p u_{ip}(\mathbf{x}')(-\delta a_p), \tag{17.30}$$

where we have used the fact that near the identity (i.e., $\mathbf{a} = \mathbf{0}$) the inverse is given by $(\delta a_p)^{-1} = -\delta a_p$.

This can be illustrated by an example for the group

$$(a_1,a_2)x = f(x;a_1,a_2) = \exp{(a_1)}x + a_2 \tag{17.31}$$

by considering

$$\begin{aligned} x' &= (\delta a_1, \delta a_2)x = \exp(\delta a_1)x + \delta a_2 , \\ x'' &= (b_1,b_2)x' = \exp(b_1)\exp(\delta a_1)x + \exp(b_1)\delta a_2 + b_2 , \end{aligned} \tag{17.32}$$

and for $b_1 = -\delta a_1,\ \ b_2 = -\delta a_2$, we obtain

$$x'' = \exp(-\delta a_1)\exp(\delta a_1)x + \exp(-\delta a_1)\delta a_2 - \delta a_2 = x \tag{17.33}$$

when terms of order higher than the first are neglected.

Thus

$$x_i(p) = f_i(\mathbf{x}';\mathbf{0}) - \sum_{p=1}^{r} \delta a_p u_{ip}(\mathbf{x}') = x_i'(p) - \Delta_i, \tag{17.34}$$

and we obtain

$$\begin{aligned} F^{S'}(\mathbf{x}') &= F^S(\mathbf{x}) - F^S(\mathbf{x}' - \boldsymbol{\Delta}) \\ &= F^S(\mathbf{x}') - \sum_{i=1}^{n} \frac{\Delta_i \partial F^S(\mathbf{x}')}{\partial x_i'} . \end{aligned} \tag{17.35}$$

To lowest order we then find

$$\begin{aligned} F^{S'}(\mathbf{x}') - F^S(\mathbf{x}') &= \sum_{p=1}^{r} \delta a_p \Big[-\sum_{i=1}^{n} u_{ip}(\mathbf{x}') \Big] \frac{\partial F^S(\mathbf{x}')}{\partial x_i'} \\ &= \sum_{p=1}^{r} \delta a_p X_p(\mathbf{x}') F^S(\mathbf{x}'), \end{aligned} \tag{17.36}$$

where

$$X_p = -\sum_{i=1}^{n} \left[\frac{\partial f_i(\mathbf{x}';\mathbf{a})}{\partial a_p} \right]_{\mathbf{a}=0} \frac{\partial}{\partial x_i'} \tag{17.37}$$

are the generators of infinitesimal displacements of coordinate axes by $\delta\mathbf{a}$ or simply *generators*.

Example 17.4 $x' = \exp(a_1)x + a_2 = f_1(x; a_1, a_2)$;

$$\begin{aligned} u_{11} &= \left[\frac{\partial f_1}{\partial a_1} \right]_{\mathbf{a}=0} = x , \\ u_{12} &= \left[\frac{\partial f_1}{\partial a_2} \right]_{\mathbf{a}=0} = 1 , \end{aligned} \tag{17.38}$$

and

$$X_1 = -x \frac{\partial}{\partial x}, \qquad X_2 = -\frac{\partial}{\partial x} . \tag{17.39}$$

The commutator of these generators is

$$[X_1, X_2]_- = x \frac{\partial^2}{\partial x^2} - \frac{\partial}{\partial x} - x \frac{\partial^2}{\partial x^2} = X_2 . \tag{17.40}$$

This is an example of a general result. It can, for instance, be shown that the infinitesimal generators satisfy the generic relation

$$[X_p, X_q]_- = \sum_s C^s_{pq} X_s . \tag{17.41}$$

The generators in this way form a linear algebra (a Lie algebra) in which the commutation operation is the algebra multiplication law. The coefficients C^s_{pq} are the so-called structure constants.

17.5.1 Connection between Infinitesimal and Finite Transformations

Examples 17.5 and 17.6 show the connection between infinitesimal and finite transformations using the simple two-parameter group of linear transformations introduced before.

Example 17.5 Consider the group $f(x;a_1,a_2)$ and the function $F^S(x) = (x - d)^2$. An infinitesimal transformation in the subgroup $(0, a_2)$ would result in $x' = x + \delta a_2$. Thus $F^{S'}(x') = (x' - d - \delta a_2)^2$. This trivial result can also be obtained from the generator:

$$\begin{aligned}
F^{S'}(x') &= [I + \delta a_2 X_2](x' - d)^2 \\
&= \left[I - \delta a_2 \frac{\partial}{\partial x'}\right](x' - d)^2 \\
&= (x' - d)^2 - 2\delta a_2(x' - d) \\
&= [x' - d - \delta a_2]^2 - (\delta a_2)^2 \\
&= [x' - d - \delta a_2]^2.
\end{aligned} \tag{17.42}$$

Terms higher than first order in the infinitesimals are neglected.

In the limit of large N, a_2/N may be considered as an infinitesimal quantity. The finite displacement $x' = x + a_2$ can now be accomplished by N consecutive displacements a_2/N:

$$\begin{aligned}
F^{S'}(x') &= \lim_{N\to\infty} \left\{I + \left(\frac{a_2}{N}\right)X_2\right\}^N (x' - d)^2 \\
&= \exp(a_2 X_2)(x' - d)^2 = \exp\left(-a_2 \frac{\partial}{\partial x'}\right)(x' - d)^2 \\
&= (x' - d)^2 - 2a_2(x' - d) + a_2^2 = (x' - d - a_2)^2.
\end{aligned} \tag{17.43}$$

An infinitesimal transformation in the subgroup $(a_1, 0)$ would be

$$x' = \exp(\delta a_1)x = (1 + \delta a_1)x \tag{17.44}$$

and

$$\begin{aligned}
F^{S'}(x') &= \{I + \delta a_1 X_1\}(x' - d)^2 = \left\{I - \delta a_1 x' \frac{\partial}{\partial x'}\right\}(x' - d)^2 \\
&= (x' - d)^2 - 2\delta a_1 x'(x' - d) + (\delta a_1 x')^2 - (\delta a_1 x')^2 \\
&= [x'(1 - \delta a_1) - d]^2 = (\exp(-\delta a_1)x' - d)^2,
\end{aligned} \tag{17.45}$$

where again the infinitesimals only to first order count. For a finite transformation we proceed as before (dropping the prime on x'):

$$\begin{aligned}
\lim_{N\to\infty}&\left\{I - \left(\frac{a_1}{N}x\frac{\partial}{\partial x}\right)\right\}^N (x-d)^2 \\
&= \exp\left(-a_1 x\frac{\partial}{\partial x}\right)(x-d)^2 \\
&= \left[\sum_{n=0}^{\infty}\frac{1}{n!}\left(-a_1 x\frac{\partial}{\partial x}\right)^n\right](x-d)^2 \qquad (17.46)\\
&= \exp(-2a_1)x^2 - 2\exp(-a_1)xd + d^2 \\
&= (\exp(-a_1)x - d)^2.
\end{aligned}$$

This is so since $(x\partial/\partial x)^n(x-d)^2 = 2^n x^2 - 2xd$.

The same procedure can in principle be followed for arbitrary continuous groups. As another example we consider a matrix group.

Example 17.6 We consider the matrix group $SO(3)$ $(r = n(n-1)/2 = 3)$. This compact Lie group of 3×3 orthogonal matrices has a three-dimensional parameter space.

Consider the one-parameter subgroup of rotations about the Cartesian x axis:

$$\begin{aligned}
x' &= f_1(x,y,z;a_1,0,0) = x \\
y' &= f_2(x,y,z;a_1,0,0) = y\cos a_1 + z\sin a_1 \qquad (17.47)\\
z' &= f_3(x,y,z;a_1,0,0) = -y\sin a_1 + z\cos a_1,
\end{aligned}$$

which means that a group element g of this subgroup transforms the coordinates such that

$$g\begin{pmatrix}x\\y\\z\end{pmatrix} = \begin{pmatrix}x'\\y'\\z'\end{pmatrix} = \begin{pmatrix}1 & 0 & 0\\ 0 & \cos a_1 & \sin a_1\\ 0 & -\sin a_1 & \cos a_1\end{pmatrix}\begin{pmatrix}x\\y\\z\end{pmatrix}. \qquad (17.48)$$

Similarly, the one-parameter subgroups of rotations about the y axis and the z axis have the elements

$$\begin{pmatrix}\cos a_2 & 0 & -\sin a_2\\ 0 & 1 & 0\\ \sin a_2 & 0 & \cos a_2\end{pmatrix} \quad\text{and}\quad \begin{pmatrix}\cos a_3 & \sin a_3 & 0\\ -\sin a_3 & \cos a_3 & 0\\ 0 & 0 & 1\end{pmatrix}, \qquad (17.49)$$

respectively. Calculating the $u_{ip} = [\partial f_i/\partial a_p]_{\mathbf{a}=\mathbf{0}}$, we obtain

$$\begin{aligned} &u_{11} = 0, \quad u_{21} = z, \quad u_{31} = -y\,; \\ &u_{12} = -z, \quad u_{22} = 0, \quad u_{32} = x\,; \\ &u_{13} = y, \quad u_{23} = -x, \quad u_{33} = 0\,; \end{aligned} \tag{17.50}$$

which means that the generators are

$$\begin{aligned} X_1 &= -\left[u_{11}\frac{\partial}{\partial x} + u_{21}\frac{\partial}{\partial y} + u_{31}\frac{\partial}{\partial z} \right] \\ &= -z\frac{\partial}{\partial y} + y\frac{\partial}{\partial z}, \end{aligned} \tag{17.51}$$

$$X_2 = -x\frac{\partial}{\partial z} + z\frac{\partial}{\partial x}, \tag{17.52}$$

and

$$X_3 = -y\frac{\partial}{\partial x} + x\frac{\partial}{\partial y}. \tag{17.53}$$

The commutators of these generators can readily be calculated to yield

$$[X_1, X_2]_- = -X_3, \quad [X_2, X_3]_- = -X_1, \quad [X_3, X_1]_- = -X_2. \tag{17.54}$$

We define the concept of *tangent matrix* at the identity, and form

$$L_3 = \frac{\partial}{\partial a_3}\begin{pmatrix} \cos a_3 & \sin a_3 & 0 \\ -\sin a_3 & \cos a_3 & 0 \\ 0 & 0 & 1 \end{pmatrix}_{a_3=0} = \begin{pmatrix} 0 & 1 & 0 \\ -1 & 0 & 0 \\ 0 & 0 & 0 \end{pmatrix} \tag{17.55}$$

and in the same way

$$L_1 = \begin{pmatrix} 0 & 0 & 0 \\ 0 & 0 & 1 \\ 0 & -1 & 0 \end{pmatrix}, \qquad L_2 = \begin{pmatrix} 0 & 0 & -1 \\ 0 & 0 & 0 \\ 1 & 0 & 0 \end{pmatrix}. \tag{17.56}$$

The tangent matrices satisfy the same commutation relations as the generators X_1, X_2, and X_3, and serve as the infinitesimal generators of the matrix group $SO(3)$ and can be used as a basis of the corresponding Lie algebra $so(3)$.

It can be readily shown that

$$\begin{aligned} \exp(a_1 L_1) &= \begin{pmatrix} 1 & 0 & 0 \\ 0 & \cos a_1 & \sin a_1 \\ 0 & -\sin a_1 & \cos a_1 \end{pmatrix}, \\ \exp(a_2 L_2) &= \begin{pmatrix} \cos a_2 & 0 & -\sin a_2 \\ 0 & 1 & 0 \\ \sin a_2 & 0 & \cos a_2 \end{pmatrix}, \\ \exp(a_3 L_3) &= \begin{pmatrix} \cos a_3 & \sin a_3 & 0 \\ -\sin a_3 & \cos a_3 & 0 \\ 0 & 0 & 1 \end{pmatrix}. \end{aligned} \tag{17.57}$$

This is a useful exercise.

In quantum mechanics the change in a state vector due to infinitesimal rotations about an axis is expressed in terms of the orbital angular momentum operator $\vec{J} = (J_1 \quad J_2 \quad J_3)$, the components of which are proportional to the infinitesimal generators X_1, X_2, and X_3:

$$X_p = iJ_p \tag{17.58}$$

(we are putting $\hbar = 1$).

Again, it is perhaps worthwhile to point out that the parameterization of a group can be made in many ways. $SO(3)$ is more commonly parameterized in terms of Euler angles.

A general rotation is then carried out by

1. A rotation by α $(0 \leq \alpha < 2\pi)$ about the z axis $\rightarrow z'$ axis.
2. A rotation by β $(0 \leq \beta \leq \pi)$ about the y' axis $\rightarrow y''$ axis.
3. A rotation by γ $(0 \leq \gamma < 2\pi)$ about the z'' axis.

Thus, the polar angles of the z'' axis in the original frame are (β, α), and a general rotation of $SO(3)$ is

$$D(\alpha, \beta, \gamma) = \exp(\gamma L_3)\exp(\beta L_2)\exp(\alpha L_1). \tag{17.59}$$

This realization should be used when operating on a three-dimensional vector space of coordinates. When operating on functions of x, y, and z, the differential operators are used, and we can write

$$D(\alpha, \beta, \gamma) = \exp(i\gamma J_3)\exp(i\beta J_2)\exp(i\alpha J_1). \tag{17.60}$$

17.6 THE UNITARY GROUP

Unitary matrices of dimension $M \times M$ form a group $U(M)$, *the unitary group*. Any matrix in the group can be written as

$$\mathbf{U} = \exp[i\mathbf{A}], \quad \text{with} \quad \mathbf{A}^\dagger = \mathbf{A}, \tag{17.61}$$

in other words, $\mathbf{A}$ is an $M \times M$ self-adjoint matrix. Since a self-adjoint matrix has $M + 2M(M-1)/2 = M^2$ independent real parameters, the same is true for the unitary group $U(M)$. The matrices $\mathbf{U}(t) = \exp[i\mathbf{H}t]$ with $\mathbf{H}$ a fixed $M \times M$ matrix $(\mathbf{H}^\dagger = \mathbf{H})$, form a one-parameter subgroup of $U(M)$. It is easy to see that $t = 0$ yields the identity, and $-t$ the inverse of t.

We can define the special one-parameter subgroups:

$$\mathbf{U}_{kk}(t) = \exp[it\mathbf{e}_{kk}],$$
$$\mathbf{U}_{kl}(t) = \exp\frac{it(\mathbf{e}_{kl} + \mathbf{e}_{lk})}{2} \tag{17.62}$$

for $k < l$ and

$$\mathbf{U}_{lk}(t) = \exp \frac{it(\mathbf{e}_{kl} - \mathbf{e}_{lk})}{2i} \tag{17.63}$$

for $k < l$.

There are exactly M^2 such one-parameter subgroups, where $\mathbf{e}_{kl}$ is a $M \times M$ matrix unit, specifically, a matrix with all elements equal to 0 except the kl element, which is 1. This can be expressed using Krönecker deltas as $\{\mathbf{e}_{kl}\}_{\mu\nu} = \delta_{\mu k}\delta_{l\nu}$. The matrix units satisfy the commutation relations

$$[\mathbf{e}_{ij}, \mathbf{e}_{kl}] = \delta_{kj}\mathbf{e}_{il} - \delta_{il}\mathbf{e}_{kj}, \tag{17.64}$$

which can be readily shown by using the Krönecker deltas and considering a general matrix element. An arbitrary unitary matrix $\mathbf{U}$ can be written in terms of matrices of the M^2 one-parameter subgroups. For a matrix $\mathbf{U} = \exp[i\mathbf{A}]$ we have

$$\mathbf{A} = \sum\sum A_{ij}\mathbf{e}_{ij} = \begin{pmatrix} A_{11} & A_{12} & \dots & A_{1M} \\ A_{21} & A_{22} & \dots & A_{2M} \\ \dots & \dots & \dots & \dots \\ A_{M1} & A_{M2} & \dots & A_{MM} \end{pmatrix} \tag{17.65}$$

with $A_{ij}^* = A_{ji}$.

One can consider $\mathbf{U}_{kl}(t) = \exp[i(\mathbf{e}_{kl} + \mathbf{e}_{lk})/2]$, $(k \leq l)$ as a representative matrix for the unitary group element $u_{kl} = \exp[it(E_{kl} + E_{lk})/2]$, where the operators $\{E_{ij}\}$ are the infinitesimal generators of the unitary group algebra and must satisfy the same commutation relation as the matrix units $\{\mathbf{e}_{ij}\}$. Compare the results for the generators and the tangent matrices in the discussion of $SO(3)$! A common realization of the generators, which is used in quantum mechanics, is expressed in terms of fermion field operators $\{a_{i\sigma}\}$ and their adjoints for M orbitals:

$$E_{ij} = \sum_{\sigma} a_{i\sigma}^{\dagger} a_{j\sigma} \tag{17.66}$$

where the sum is over the discrete spin labels, for example, α, and β for electrons (or over the four combinations of spin–isospin for dynamically equivalent nucleons). (For an introduction to electron field operators, see, e.g., J. Linderberg and Y. Öhrn, *Propagators in Quantum Chemistry*, Academic Press, London, 1973.)

Each electron field operator $a_{j\sigma}$ and its adjoint $a_{j\sigma}^{\dagger}$ belong to a spin orbital, for instance, a Hartree–Fock molecular orbital with the spin label σ that is either α or β. Such field operators act in a Fock space, namely, a space that is a union of all Hilbert spaces from that pertaining to 0 electrons up through the one relevant for $2M$ electrons, where M is the size of the orbital basis. The 0-electron system has only one state, the so-called vacuum state $|\text{vac.}\rangle$, which is normalized to unity and non-degenerate. A field operator acting on the vacuum state yields either

$$a_{j\sigma}|\text{vac.}\rangle = 0 \qquad (17.67)$$

(by definition) for the annihilators or

$$a^{\dagger}_{j\sigma}|\text{vac.}\rangle = |j\sigma\rangle, \qquad (17.68)$$

which is the creation of a one-electron state (which has a one-to-one correspondence to a spin orbital).

The electron field operators satisfy the anticommutation relations

$$\begin{aligned} [a_{i\sigma}, a_{j\sigma'}]_+ &= a_{i\sigma}a_{j\sigma'} + a_{j\sigma'}a_{i\sigma} = 0, \\ [a^{\dagger}_{i\sigma}, a^{\dagger}_{j\sigma'}]_+ &= a^{\dagger}_{i\sigma}a^{\dagger}_{j\sigma'} + a^{\dagger}_{j\sigma'}a^{\dagger}_{i\sigma} = 0, \\ [a_{i\sigma}, a^{\dagger}_{j\sigma'}]_+ &= a_{i\sigma}a^{\dagger}_{j\sigma'} + a^{\dagger}_{j\sigma'}a_{i\sigma} = \delta_{ij}\delta_{\sigma\sigma'} \end{aligned} \qquad (17.69)$$

when the spin orbital basis is assumed to be orthonormal. The total number of electrons in the system is given by the number operator

$$N_o = \sum_{j,\sigma} a^{\dagger}_{j\sigma}a_{j\sigma}\,. \qquad (17.70)$$

It is straightforward to show, using the preceding anticommutation relations and the result from Eq. (17.67), that

$$N_o a^{\dagger}_{k\sigma}|\text{vac.}\rangle = N_o|k\sigma\rangle = 1|k\sigma\rangle, \qquad (17.71)$$

i.e. $|k\sigma\rangle$ is a one-electron state. Similarly, we can form

$$|k\sigma l\sigma'\rangle = a^{\dagger}_{k\sigma}a^{\dagger}_{l\sigma'}|\text{vac.}\rangle \qquad (17.72)$$

and show that it satisfies

$$N_o a^{\dagger}_{k\sigma}a^{\dagger}_{l\sigma'}|\text{vac.}\rangle = N_o|k\sigma l\sigma'\rangle = 2|k\sigma l\sigma'\rangle; \qquad (17.73)$$

thus it is a two-electron state. In this manner you can form a "ladder" of state vectors for an increasing number of electrons. You can also step down that "ladder" by applying the annihilators, for instance

$$a_{k\sigma}|k\sigma l\sigma'\rangle = |l\sigma'\rangle \qquad (17.74)$$

or

$$a_{l\sigma'}|k\sigma l\sigma'\rangle = -|k\sigma\rangle \qquad (17.75)$$

The anticommutation relations ensure fermion symmetry, where state vectors are antisymmetric in the electron or spin orbital labels. Furthermore, a state vector

$$|k_1\sigma_1 k_2\sigma_2 \cdots k_N\sigma_N\rangle = a^\dagger_{k_1\sigma_1} a^\dagger_{k_2\sigma_2} \cdots a^\dagger_{k_N\sigma_N} |\text{vac.}\rangle \tag{17.76}$$

is an algebraic representative of a single determinant constructed from the N spin orbitals listed.

A completely algebraic approach to electronic structure can thus be introduced, where matrix elements between determinants are obtained by using the rules of the electron field operators. The appropriate many-electron Hamiltonian to use in this connection can be expressed as

$$H = \sum_{k,l,\sigma,\sigma'} (k\sigma|h|l\sigma')a^\dagger_{k\sigma}a_{l\sigma'} + \frac{1}{2} \sum_{i,j,\tau,\tau'} \sum_{k,l,\sigma,\sigma'} (i\tau j\tau'|k\sigma l\sigma')a^\dagger_{i\tau}a^\dagger_{k\sigma}a_{l\sigma'}a_{j\tau'} \tag{17.77}$$

which contains the appropriate one- and two-electron integrals in the spin orbital basis. When there are no spin operators involved in those integrals, one can sum over spin labels and express the Hamiltonian in terms of the unitary group generators

$$E_{ij} = \sum_\sigma a^\dagger_{i\sigma}a_{j\sigma} \,. \tag{17.78}$$

It is a recommended exercise to show this in detail.

The Hamiltonian for a general many-electron system then becomes

$$H = \sum_{k,l=1}^{M} (k|h|l)E_{kl} + \frac{1}{2} \sum_{i,j,k,l=1}^{M} (ij|kl)\left[E_{ij}E_{kl} - \delta_{jk}E_{il}\right], \tag{17.79}$$

where we have used the quantum chemistry notations[1] for the one- and two-electron integrals and where the generator labels refer to orbitals. The operators E_{kl} and u_{kl} operate in a Fock space of dimension 2^{2M}, since for ν electrons there is a Hilbert space of dimension $(2M)!/\nu!(2M - \nu)!$, and the Fock space is the union of all Hilbert spaces from $\nu = 0$ to $\nu = 2M$.

The number operator $a^\dagger_{i\sigma}a_{i\sigma}$ operates in an occupation number space spanned by its two eigenvectors corresponding to the eigenvalues 1 and 0, referring to whether the spin orbital $i\sigma$ is occupied or not occupied, respectively. This means that the possible eigenvalues of E_{ii} in this space are 2, 1, and 0, where the eigenvalue 1 is doubly degenerate. The entire Fock space is

[1] The notation of the two-electron integrals is $(ij|kl) = \int u_i^*(1)u_j(1)(1/r_{12})u_k^*(2)u_l(2)d(1)d(2)$.

a direct product space of all such occupation number spaces for the individual "diagonal" generators $\{E_{ii}\}$, and thus of dimension $4^M = 2^{2M}$.

A general treatment of the unitary group (also the general linear group) is due to Gel'fand, who considers a complete set of commuting operators I_k^M on the Fock space, where

$$I_1^M = \sum_{i=1}^{M} E_{ii}, \ldots, I_k^M = \sum_{i_1=1}^{M} \sum^{M} \cdots \sum_{i_k=1}^{M} E_{i_1 i_2} E_{i_2 i_3} \cdots E_{i_k i_1} \tag{17.80}$$

for $k = 1, 2, \ldots, M$.

Gel'fand showed that $[I_k^M, E_{ij}] = 0$ for all *k, i,* and *j,* and that the irreps of $U(M)$ can be labeled by the eigenvalues of the $\{I_k^M\}$.

To construct the basis of the corresponding invariant subspaces one considers the chain of subgroups: $U(M) \supset U(M-1) \supset \cdots \supset U(1)$ and following Gel'fand introduces the symbol:

$$\begin{matrix} I_1^M & I_2^M & \ldots & I_M^M \\ & I_1^{M-1} & \ldots I_{M-1}^{M-1} & \\ & & \ldots & \\ & & I_1^1 & \end{matrix} \tag{17.81}$$

This triangular pattern is called a Gel'fand symbol, where the *i*th row contains the basic invariants of the generator algebra of *U(i)*. A labeling scheme of the simultaneous eigenvectors of the invariants due to Gel'fand and Tsetlin places integers in this triangular array, to form what is called a Gel'fand state:

$$\begin{matrix} m_{1M} & m_{2M} & \ldots & m_{MM} \\ & m_{1M-1} & \ldots & m_{M-1M-1} \\ & & \ldots & \\ & m_{12} & m_{22} & \\ & & m_{11} & \end{matrix} \tag{17.82}$$

where for electrons the m_{ij} can assume the values 2, 1, and 0.

The first row $\{m_M\} \equiv [m_{1M}, m_{2M}, \ldots, m_{MM}]$ arranged such that $m_{1M} \geq m_{2M} \geq \cdots \geq m_{MM}$ labels an irrep of $U(M)$. The basis of this irrep is then given by all the Gel'fand states obtained by the "in betweenness" condition $m_{ij} \geq m_{ij-1} \geq m_{i+1j}$.

For the application to systems of electrons, one simply labels the irrep by the possible eigenvalues m_{iM} (which, as occupation numbers of orbitals, can only be 2, 1, or 0) of the diagonal generators E_{ii}, and these are used as entries in the first row of the Gel'fand symbols for the basis of the irrep.

As an example, consider the eight basis states of the irrep [2,1,0] for $M = 3$. The

$$
\begin{array}{ccccc} 2 & & 1 & & 0 \\ & 2 & & 1 & \\ & & 2 & & \end{array}
\quad
\begin{array}{ccccc} 2 & & 1 & & 0 \\ & 2 & & 1 & \\ & & 1 & & \end{array}
\quad
\begin{array}{ccccc} 2 & & 1 & & 0 \\ & 2 & & 0 & \\ & & 2 & & \end{array}
\quad
\begin{array}{ccccc} 2 & & 1 & & 0 \\ & 2 & & 0 & \\ & & 1 & & \end{array}
$$
$$
\quad (1) \qquad\qquad\qquad (2) \qquad\qquad\qquad (3) \qquad\qquad\qquad (4)
$$

$$\tag{17.90}$$

$$
\begin{array}{ccccc} 2 & & 1 & & 0 \\ & 2 & & 0 & \\ & & 0 & & \end{array}
\quad
\begin{array}{ccccc} 2 & & 1 & & 0 \\ & 1 & & 1 & \\ & & 1 & & \end{array}
\quad
\begin{array}{ccccc} 2 & & 1 & & 0 \\ & 1 & & 0 & \\ & & 1 & & \end{array}
\quad
\begin{array}{ccccc} 2 & & 1 & & 0 \\ & 1 & & 0 & \\ & & 0 & & \end{array}
$$
$$
\quad (5) \qquad\qquad\qquad (6) \qquad\qquad\qquad (7) \qquad\qquad\qquad (8)
$$

information in these symbols is that (1) the number of orbitals is M (the number or rows), (2) the number of electrons

$$N = \sum_{i=1}^{M} n_i = \sum_{j=1}^{M} m_{jM} \tag{17.91}$$

and (3) the number of electrons in orbital i:

$$n_i = \sum_{j=1}^{i} m_{ji} - \sum_{j=1}^{i-1} m_{ji-1} \tag{17.92}$$

There are several alternative ways of representing the orthogonal basis for the irreps of $U(M)$, particularly for electrons, and recently (at the time of writing) the unitary group approach has been widely applied in electronic structure theory.

17.6.1 Weyl Tableaux, Line Tableaux, *b*-Vector, and Paldus Tableaux

For the irrep $[m_{1M}, m_{2M}, \ldots, m_{MM}]$, one introduces a Young graph (as in the theory of the symmetric group) with m_{1M} boxes in the first row, m_{2M} boxes in the second row, and so on. To obtain a Weyl tableaux notation for a particular Gel'fand state, one then puts

- A 1 in the first row of the graph m_{11} times
- A 2 in the first row of the graph $(m_{12} - m_{11})$ times

  ⋮
- An M in the first row of the graph $(m_{1M} - m_{1M-1})$ times

  ⋮
- A 2 in the second row m_{22} times
- A 3 in the second row $(m_{23} - m_{22})$ times

And so on, until the Young shape is filled.

Since the Weyl tableaux have at most two columns for electrons, an alternative notation is a so-called line tableau: $|x_1, x_2, \ldots, x_q; y_1, y_2, \ldots, y_p\rangle$, where the x symbols represent the entries in the first column of the Weyl tableau and the y are the entries in the second column.

The following example illustrates this

$$\begin{matrix} 2 & & 1 & & 0 \\ & 2 & & 1 & \\ & & 2 & & \end{matrix} = \begin{matrix} 1 & 1 \\ 2 & \end{matrix} = |1,2;1\rangle, \tag{17.93}$$

where the Gel'fand symbol is expressed first as a Weyl symbol and then as a line tableau. It is interesting to note that, say, in the line tableaux the number of x's and the number of y's are related to the total spin quantum number S and to the total number of electrons N as is the case for the Young graphs of the symmetric group:

$$q = \frac{N}{2} + S; \qquad p = \frac{N}{2} - S; \tag{17.94}$$

Still another notation in use is the so-called b vector, which just lists the number of 1s in the first, second, and further row of the Gel'fand symbol. For the preceding example [Eq. (17.93)] the b vector is $|0, 1, 1\rangle$. Finally we have the Paldus tableau that is an array with three columns and M rows, which for the same example reads as follows. The first

$$\begin{matrix} 2 & & 1 & & 0 \\ & 2 & & 1 & \\ & & 2 & & \end{matrix} = \begin{vmatrix} 1 & 1 & 1 \\ 1 & 1 & 0 \\ 1 & 0 & 0 \end{vmatrix} \tag{17.95}$$

column lists the number of 2s, the second the number of 1s, and the third the number of 0s in the corresponding Gel'fand symbol.

As we will see, the calculation of matrix elements of the Hamiltonian can be greatly simplified by the use of the Gel'fand states and the results given in terms of the matrix elements of the generators E_{ij}. Before we discuss this, we shall explore the relationship to the spin adapted orbital products obtained from the symmetric group.

Our example above has $N = 3$, $M = 3$, and $S = \frac{1}{2}$. This corresponds to the irrep [2,1] of the symmetric group S_3:

$$S_3\colon\ e, (12), (13), (23), (123), (321)\,. \tag{17.96}$$

The irrep [2,1] has the following graph:

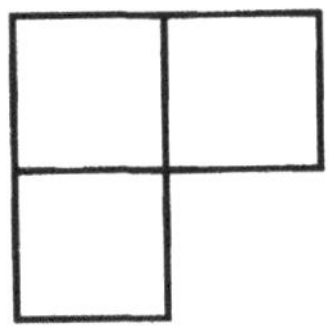

with the standard Young tableaux

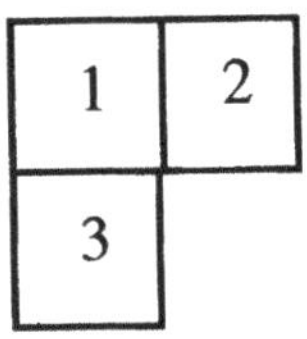

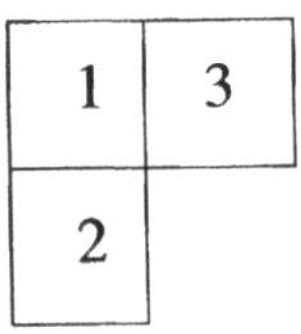

forming the basis of the orthogonal irrep:

$$e = \begin{pmatrix} 1 & 0 \\ 0 & 1 \end{pmatrix}, \quad (12) = \begin{pmatrix} 1 & 0 \\ 0 & -1 \end{pmatrix}, \quad (23) = \begin{pmatrix} -\frac{1}{2} & -\sqrt{\frac{3}{2}} \\ -\sqrt{\frac{3}{2}} & \frac{1}{2} \end{pmatrix},$$

$$(13) = \begin{pmatrix} -\frac{1}{2} & \sqrt{\frac{3}{2}} \\ \sqrt{\frac{3}{2}} & \frac{1}{2} \end{pmatrix}, \quad (123) = \begin{pmatrix} -\frac{1}{2} & -\sqrt{\frac{3}{2}} \\ \sqrt{\frac{3}{2}} & -\frac{1}{2} \end{pmatrix}, \tag{17.97}$$

$$(321) = \begin{pmatrix} -\frac{1}{2} & \sqrt{\frac{3}{2}} \\ -\sqrt{\frac{3}{2}} & -\frac{1}{2} \end{pmatrix},$$

where the last three matrices are formed from (12), and (23) as (13) = (12)(23)(12), (123) = (12)(23), and (321) = (23)(12).

The relationship between a Gel'fand state $|(m)_M\rangle$ and states spanning the corresponding irrep of the symmetric group has been given by A. Matsen [*Int. J. Quantum Chem.* **10**, 525 (1976)]:

$$|(m)_M\rangle = (-1)^{p+q} N_m P_{qp}^{[\lambda]} |R\rangle, \tag{17.98}$$

where $p \geq q, N_m$ is a normalization constant and $|R\rangle$ is the appropriate orbital product. The projection and shift operators are

$$P_{km}^{\alpha} = f^{\alpha} \frac{1}{|G|} \sum_{g \in G} \Gamma_{km}^{\alpha}(s) s^{-1} \tag{17.99}$$

or for the specific case here:

$$P_{11}^{[2,1]} = \frac{e + (12) - \frac{1}{2}\{(13) + (23) + (123) + (321)\}}{3}$$

$$P_{12}^{[2,1]} = \frac{\sqrt{3}}{2} \frac{(13) - (23) - (321) + (123)}{3} \tag{17.100}$$

For the case of three electrons either one orbital is doubly occupied and one is singly occupied or three orbitals are singly occupied:

$$P_{11}^{[2,1]} aab \propto \left[aab + aab - \tfrac{1}{2}\{baa + aba + aba + baa\}\right], \tag{17.101}$$

or in normalized form

$$\frac{2aab - aba - baa}{\sqrt{6}}, \tag{17.102}$$

and

$$P_{12}^{[2,1]}aab = 0;$$
$$P_{11}^{[2,1]}abc \propto \left[abc + bac - \frac{1}{2}\{cba + acb + bca + cab\}\right], \tag{17.103}$$

or normalized

$$\frac{abc + bac - \frac{1}{2}\{cba + acb + bca + cab\}}{\sqrt{3}}, \tag{17.104}$$

and

$$P_{12}^{[2,1]}abc \propto [cba - acb + bca - cab], \tag{17.105}$$

or with the normalization and phase

$$-\tfrac{1}{2}[cba - acb + bca - cab]. \tag{17.106}$$

We can now list the eight Gel'fand states that span the [2,1,0] irrep of $U(3)$ with the corresponding Weyl tableaux, line tableaux, b vectors, Paldus tableaux, and the basis of the irrep [2,1] of S_3, namely, the symmetrized orbital products of a spin doublet state of a three-electron system described in terms of three orbitals [refer to the list in Eq. (17.90) for the labels of Gel'fand states].

Gel'fand	Weyl	line	*b*vector	Paldus	
(1)	1 1 2	|12;1>	|011>	1 1 1 1 1 0 1 0 0	$[2aab - aba - baa]/(6)^{1/2}$
(2)	1 2 2	|12;2>	|111>	1 1 1 1 1 0 0 1 0	$-[2bba - bab - abb]/(6)^{1/2}$
(3)	1 1 3	|13;1>	|001>	1 1 1 1 0 1 1 0 0	$[2aac - aca - caa]/(6)^{1/2}$
(4)	1 2 3	|13;2>	|101>	1 1 1 1 0 1 0 1 0	$[2abc + 2bac - cba -$ $acb - cab - bca]/(12)^{1/2}$
(5)	2 2 3	|23;2>	|001>	1 1 1 1 0 1 0 0 1	$[2bbc - bcb - cbb]/(6)^{1/2}$
(6)	1 3 2	|12;3>	|121>	1 1 1 0 2 0 0 1 0	$[acb - cba - bca + cab]/2$
(7)	1 3 3	|13;3>	|111>	1 1 1 0 1 1 0 1 0	$-[2cca - cac - acc]/(6)^{1/2}$
(8)	2 3 3	|23;3>	|011>	1 1 1 0 1 1 0 0 1	$-[2ccb - cbc - bcc]/(6)^{1/2}$

17.6.2 Matrix Elements of Generators

Orbital a refers to the first (bottom) row of the Gel'fand symbol, orbital b to the second row and so on.

Gel'fand and Tsetlin have given formulas for the matrix elements of $E_{i,i+1}$ between Gel'fand patterns. This is sufficient since results for the other generators can be obtained from

$$E_{ij} = [E_{i,j-1}, E_{j-1,j}] \quad \text{for} \quad i < j \tag{17.107}$$

as a special case of the general commutation relation for the generators.

Define $p_{ij} = m_{ij} + j - i$, and let the Gel'fand states $|(m')_n\rangle$ and $|(m)_n\rangle$ differ precisely in the element $(ij) : m'_{ij} = m_{ij} + 1$ otherwise identical.

The matrix element of $E_{i,i+1}$ between these two Gel'fand states is

$$\langle (m')_n | E_{i,i+1} | (m)_n \rangle =$$

$$\left[-\frac{\prod_{k=1}^{i+1} (p_{k,i+1} - p_{ij} - 1) \prod_{k=1}^{i-1} (p_{k,i-1} - p_{ij})}{\prod_{k=1(k \neq j)}^{i} (p_{ki} - p_{ij} - 1)(p_{ki} - p_{ij})} \right]^{1/2} \tag{17.108}$$

In general, for the raising generators $E_{ij}, i < j$ one has

$$E_{ij} |(m)_M\rangle = \sum_{(m')} |(m')_M\rangle \langle (m')_M | E_{ij} | (m)_M \rangle \tag{17.109}$$

where coefficients (matrix elements) are zero unless the two Gel'fand symbols are identical in all rows except $i.i + 1, i + 2, \ldots, j - 1$. In those rows the allowed final numbers (m') are those that can be obtained from the initial ones (m) by adding 1 in all possible positions:

$$m'_k = m_k \quad \text{for} \quad k < i \quad \text{and} \quad k \geq j$$
$$m'_k = (m_{1k}, m_{2k}, \ldots, m_{rk} + 1, \ldots, m_{kk}) \quad \text{for} \quad i \leq k < j, 1 \leq r \leq k. \tag{17.110}$$

Here m_k denotes the kth row of the Gel'fand symbol. The lowering generators $E_{ji}, i < j$ obviously are the adjoint of the raising ones, and the real orthogonal matrices for them are obtained as the transpose of those for the corresponding E_{ij}.

Example 17.7 Several examples are presented here. First, let

$$|(m_3)\rangle = \begin{array}{ccccc} m_{13} & & m_{23} & & m_{33} \\ & m_{12} & & m_{22} & \\ & & m_{11} & & \end{array} \qquad (17.111)$$

Then

$$E_{12}|(m_3)\rangle = [(m_{12} - m_{11})(m_{11} - m_{22} + 1)]^{1/2}$$
$$\times \begin{array}{ccccc} m_{13} & & m_{23} & & m_{33} \\ & m_{12} & & m_{22} & \\ & & m_{11}+1 & & \end{array} \qquad (17.112)$$

$$E_{23}|(m)_3\rangle$$
$$= \left\{\frac{(m_{13} - m_{12})(m_{23} - m_{12} - 1)(m_{33} - m_{12} - 2)(m_{12} - m_{11} + 1)}{(m_{12} - m_{22} + 2)(m_{12} - m_{22} + 1)}\right\}^{1/2}$$
$$\times \begin{array}{ccccc} m_{13} & & m_{23} & & m_{33} \\ & m_{12}+1 & & m_{22} & \\ & & m_{11} & & \end{array} \qquad (17.113)$$

$$+ \left\{\frac{(m_{13} - m_{22} + 1)(m_{23} - m_{22})(m_{33} - m_{22} - 1)(m_{22} - m_{11})}{(m_{12} - m_{22} + 1)(m_{12} - m_{22})}\right\}^{1/2}$$
$$\times \begin{array}{ccccc} m_{13} & & m_{23} & & m_{33} \\ & m_{12} & & m_{22}+1 & \\ & & m_{11} & & \end{array} \qquad (17.114)$$

From our definition of the occupation numbers we find that

$$E_{11}|(m)_3\rangle = m_{11}|(m)_3\rangle;$$
$$E_{22}|(m)_3\rangle = (m_{12} + m_{22} - m_{11})\,|(m)_3\rangle; \qquad (17.115)$$

and

$$E_{33}|(m)_3\rangle = (m_{13} + m_{23} + m_{33} - m_{12} - m_{22})\,|(m)_3\rangle. \qquad (17.116)$$

In particular, we consider the Gel'fand states in Eq. (17.90) and find

$$E_{12}\begin{array}{ccccc} 2 & & 1 & & 0 \\ & 2 & & 1 & \\ & & 1 & & \end{array} = \begin{array}{ccccc} 2 & & 1 & & 0 \\ & 2 & & 1 & \\ & & 2 & & \end{array} \tag{17.117}$$

It is readily seen that in terms of the field operators $E_{12} = E_{ab}$, this generator annihilates orbital b and creates a. Thus

$$\begin{aligned} &E_{ab}\left\{\frac{-[2bba - bab - abb]}{\sqrt{6}}\right\} \\ &= \frac{-[2aba + 2baa - aab - baa - aab - aba]}{\sqrt{6}} \\ &= \frac{[2aab - aba - baa]}{\sqrt{6}}, \end{aligned} \tag{17.118}$$

which is the same result as above.

Very fast algorithms for the computation of the Hamiltonian matrix elements over Gel'fand states have been worked out on the basis of these results for the unitary group. This is now the manner in which the computer codes of some correlated methods of electronic structure actually work.

17.7 LIE ALGEBRAS

As can be inferred from the preceding discussion, Lie algebras, rather than their corresponding Lie groups, arise more naturally as a practical tool for the systematic calculation of matrix elements of various quantum-mechanical operators. This means that Lie algebras and their representations and realizations are of direct importance in quantum theory.

17.7.1 Definition of a Lie Algebra

A Lie algebra can be thought of as a vector space V with a law of composition $[A, B]$ that has the following properties:

1. $[A, B] \in V$ (closure)
2. $[A, B + C] = [A, B] + [A, C]$
3. $\alpha[A, B] = [\alpha A, B] = [A, \alpha B]$
4. $[A, A] = 0$
5. $[A, [B, C]] + [B, [C, A]] + [C, [A, B]] = 0$

for all complex numbers α and all $A, B, C \in V$.

Properties 1–3 define an algebra, where the composition law is usually written AB or $A\times B$ (compare our discussions of group algebras for finite groups) rather than $[A,B]$. Different kinds of algebras are then distinguished by further defining relations: a *commutative* algebra $AB = BA$ and an *associative* algebra $A(BC) = (AB)C$. For a Lie algebra the defining relations 4 and 5 show that it is neither commutative nor associative. In fact, relation 4 implies that $[A + B, A + B] = 0$ or $[A, B] = -[B, A]$, that is, anticommutative. Relation 5 is usually referred to as the *Jacobi identity*.

17.7.2 Construction of Lie Algebras

This discussion follows closely work by B. G. Adams, J. Cizek, and J. Paldus [*Adv. Quantum Chem.* **19**, 1 (1988)]. In quantum mechanics one often encounters associative algebras of operators or matrices (e.g., group algebras). In fact, many Lie algebras (such as those we have considered already) can be constructed from such associative algebras. For example, the set of all $n \times n$ matrices with real or complex elements is an n^2-dimensional vector space, which is an associative, noncommutative algebra, whose composition law is ordinary matrix multiplication. Then, given the associative algebra with the law AB, we form the Lie product as the commutator $[A, B] = AB - BA$, which straightforwardly satisfies properties 2–5, so that only the closure property (1) needs verification in each particular case.

Since Lie algebras have an underlying vector space structure, one can choose a basis $\{E_j, j = 1, 2, \ldots, N\}$ for the Lie algebra, and since the bilinearity or distributive laws of multiplication hold, specifically

$$\left[\sum_i \alpha_i E_i, \sum_j \beta_j E_j\right] = \sum_{i,j} \alpha_i \beta_j [E_i, E_j], \tag{17.119}$$

specifying the commutators of the basis elements completely specifies the Lie algebra (as above for the X_p's):

$$[E_i, E_j] = \sum_k c_{ij}^k E_k \tag{17.120}$$

where $\{c_{ij}^k\}$ is the so-called structure constants. The basis elements $\{E_i\}$ are the generators of the Lie algebra.

Given a Lie algebra with such defining commutation relations, we can consider the generators as operators on some vector space (n-dimensional) with elements $|i\rangle$:

$$E_k|j\rangle = \sum_i |i\rangle\langle i|E_k|j\rangle, \tag{17.121}$$

where $\langle i|E_k|j\rangle$ is a matrix element of the matrix $\mathbf{E_k}$. The matrices $\{\mathbf{E_k}\}$ satisfy

$$[\mathbf{E_i}, \mathbf{E_j}] = \sum c_{ij}^k \mathbf{E_k}\,, \tag{17.122}$$

and they generate a *matrix representation* of the Lie algebra.

A particularly important role is played by the unitary irreducible representations (*unirreps*) of the Lie group, namely, by Hermitian generators.

18

EXAMPLE OF A NONCOMPACT GROUP AND ALGEBRA

18.1 REPRESENTATION THEORY OF *so*(3) AND *so*(2,1)

This chapter closely follows work by B. G. Adams, J. Cizek, and J. Paldus, *Adv. Quantum Chem.* **19**, 1 (1988). The commutation relations for the generators of *so*(3) and *so*(2,1) are

$$
\begin{aligned}
[J_1, J_2] &= i\gamma J_3 \\
[J_2, J_3] &= iJ_1 \\
[J_3, J_1] &= iJ_2
\end{aligned}
\tag{18.1}
$$

with γ=1 for *so*(3) and γ=-1 for *so*(2,1). An operator (not generally in the algebra) that commutes with all the generators is called a *Casimir operator*. Here the Casimir operator is

$$J^2 = \gamma(J_1^2 + J_2^2) + J_3^2 . \tag{18.2}$$

We see that for *so*(3) that applies for discussions of the angular momentum, it is the total angular momentum square that is the Casimir operator. Thus, this entire discussion, also for *so*(2,1), is based on much of our experience with angular momenta. For instance, in terms of the ladder operators $J_\pm = J_1 \pm iJ_2$, we can write

$$
\begin{aligned}
[J_+, J_-] &= 2\gamma J_3 \\
[J_3, J_\pm] &= \pm J_\pm ,
\end{aligned}
\tag{18.3}
$$

and $J^2 = \gamma J_+J_- + J_3^2 - J_3$. becomes the Casimir operator. The maximum number of mutually commuting operators for $so(3)$ and $so(2,1)$ is two. For $so(3)$ we can choose any of the pairs $\{J^2, J_i, i = 1, 2, 3\}$, and the representation theory would be the same in each case. However, for $so(2,1)$ each such choice of mutually commuting operators leads to a different set of inequivalent representations. We choose $\{J^2, J_3\}$, which is best suited for bound states of simple quantum systems.

We select a set of basis states $\{|k, \mu\rangle\}$ that diagonalize J^2 and J_3:

$$\begin{aligned} J^2|k,\mu\rangle &= k(k+1)|k,\mu\rangle \\ J_3|k,\mu\rangle &= \mu|k,\mu\rangle \end{aligned} \tag{18.4}$$

in formal correspondence with the familiar angular momentum theory. The defining commutation relations yield

$$\begin{aligned} J_+|k,\mu\rangle &= F_{k\mu}|k,\mu+1\rangle \\ J_-|k,\mu\rangle &= G_{k\mu}|k,\mu-1\rangle . \end{aligned} \tag{18.5}$$

From these relations we easily obtain

$$\gamma J_+J_-|k,\mu\rangle = \gamma F_{k\mu-1}G_{k\mu}|k,\mu\rangle = (k+\mu)(k-\mu+1)|k,\mu\rangle , \tag{18.6}$$

which means that $F_{k\mu-1}G_{k\mu} = \gamma(k+\mu)(k-\mu+1)$.

Similarly, we get

$$F_{k\mu}G_{k\mu+1} = \gamma(k-\mu)(k+\mu+1) . \tag{18.7}$$

We choose $F_{k\mu} = \gamma(k-\mu)$, and $G_{k\mu} = k+\mu$. Any other factorization of Eq. (18.7) yields equivalent results and leads at most to a phase change $(-)^{\mu}$ of the basis states.

Thus, one finds

$$\begin{aligned} J_+|k,\mu\rangle &= \gamma(k-\mu)|k,\mu+1\rangle \\ J_-|k,\mu\rangle &= (k+\mu)|k,\mu-1\rangle , \end{aligned} \tag{18.8}$$

and we see that the subspaces $V(K) = \{|k,\mu\rangle; K = k(k+1)\}$ are invariant under the action of J_3, and $J_\pm$. Furthermore, since $K = k(k+1)$ is invariant to the replacement $k \longrightarrow -k-1$, we obtain the further reduction $V(K) = V_k + V_{-k-1}$ into invariant subspaces $V_k = \{|k,\mu\rangle; \mu \in C\}$ where C denotes the complex number field. An exception occurs for $K = -1/4$, *i.e.* $k = -\frac{1}{2}$. We can look at the possible eigenvalue spectra $\mu = \mu_0 + m$ (m = integer and μ_0 some suitably chosen reference value) of J_3 to further classify irreps.

1. This spectrum can be unbounded, that is, neither $k - \mu = 0$ nor $k + \mu = 0$ for any $\mu (J_\pm|k,\mu\rangle \neq 0)$, which means $k \pm m_0 \neq$ *integer*. We denote the corresponding irreps $D(K, \mu_0)$.

2. The spectrum can be bounded from below where $J_-|k,\mu\rangle = 0$ for some μ but $J_+|k,\mu\rangle \neq 0$ for all μ). If we choose $\mu_0 = -k$, then the J_3 spectrum is $\mu = -k, -k+1, -k+2, \ldots$ with $2k \neq 0, 1, 2, \ldots$.[1] Irreps are denoted $D^+(k)$.
3. Similarly, the spectrum can be bounded from above ($\mu_0 = k, \mu = k, k-1, k-2, \ldots,$ with $2k \neq 0, 1, 2 \ldots$). We denote these irreps $D^-(k)$.
4. Finally, we can have a bounded spectrum $\mu = -k, -k+1, \ldots, k$, with $2k = 0, 1, 2, \ldots$ and with the finite dimensional irreps $D(k)$.

In quantum mechanics we are interested in unitary irreps of Lie groups, that is, corresponding to irreps of the Lie algebras for which the generators are Hermitian. If $J_i^\dagger = J_i$ with respect to the scalar product of our basis states, then $J_+^\dagger = J_-$, and $J_-^\dagger = J_+$, leading to conclusions or conjecture that (1) $K = k(k+1)$ and μ must be real and that (2)

$$\langle k,\mu|J_+J_-|k,\mu\rangle \geq 0 \quad \text{and} \quad \langle k,\mu|J_-J_+|k,\mu\rangle \geq 0\,. \tag{18.9}$$

This leads to $\gamma[k(k+1) - \mu(\mu \pm 1)] = \gamma[(k+\frac{1}{2})^2 - (\mu \pm \frac{1}{2})^2] \geq 0$, which must be satisfied by any unirrep of *so*(3) (γ=1) and *so*(2,1) (γ=-1).

Real $K = k(k+1)$, and μ implies that either k is real or $k = -\frac{1}{2} + i\beta$ $(\beta \neq 0)$. For unirreps of *so*(3), $k = -\frac{1}{2} + i\beta$ is not possible since the inequality $-\beta^2 - (\mu \pm 1/2)^2 \geq 0$ cannot be satisfied by real β and μ. Therefore, k must be real, and since $(k+1/2)^2 \geq (\mu \pm 1/2)^2$ can be satisfied by only a finite number of μ values for each finite k, only the finite-dimensional unirreps $D(k)$ or equivalently $D(-k-1)$ apply for *so*(3). For *so*(2,1) the situation is quite different. The inequality $(k+\frac{1}{2})^2 \leq (\mu \pm \frac{1}{2})^2$ cannot be satisfied for a bounded J_3 eigenvalue spectrum unless $k = 0$, corresponding to the trivial one-dimensional unirrep. Thus, there are no nontrivial finite-dimensional unirreps of *so*(2,1). (The corresponding Lie group is noncompact.)

Unirreps of *so*(2,1) can be summarized as follows:

$$\begin{aligned} &D^+(k)\begin{cases} q == k + m; m = 0, 1, 2, \ldots \\ \qquad k < 0 \end{cases} \\ &D^-(k)\begin{cases} q = k + m; m = 0, -1, -2, \ldots \\ \qquad k < 0 \end{cases} \end{aligned} \tag{18.10}$$

and also $D^\pm(-k-1)$, and unbounded unirreps are possible. In applications of the representation theory of *so*(2,1) to physical problems, we need a realization of the *so*(2,1) generators in coordinate or momentum space.

[1] As we can see below for a completely bounded spectrum, the greatest value of μ, $\mu_>$ and the smallest value $\mu_<$, must satisfy $\mu_> = -\mu_< = k$, and since we can proceed from one to the other in integer steps, their difference must be an integer: k — (-k) = $2k$ = integer. So when the spectrum is not bounded from both below and above, it must hold that $2k \neq$ *integer*.

Example 18.1 Radial distance and momenta in three dimensions are r and p_r, respectively. In units with $\hbar = 1$ we have $[r, p_r] = i$, with $r = [x^2 + y^2 + z^2]^{1/2}$, and $p_r = -ir^{-1}(\partial/\partial r)r$, where the radial momentum has been chosen in a form that is Hermitian with respect to the scalar product

$$\langle f|g\rangle = \int_o^\infty f^*(r)g(r)r^2 dr\,, \tag{18.11}$$

which arises from the radial part of the wavefunction in the three-dimensional spherical polar coordinate system. It is straightforward to show that

$$\langle f|p_r f\rangle - \langle f|p_r f\rangle^* = -i\int_o^\infty \frac{d}{dr}|rf(r)|^2 dr \tag{18.12}$$

and thus p_r is Hermitian for the class of functions $f(r)$ that satisfy

$$\lim_{r\to 0} rf(r) = 0 \text{ and } \lim_{r\to\infty} rf(r) = 0\,. \tag{18.13}$$

The three operators r, rp_r, and rp_r^2 have the mutual commutation relations:

$$\begin{aligned} [r, rp_r] &= ir\,, \\ [rp_r, rp_r^2] &= irp_r^2\,, \\ [rp_r^2, r] &= -2irp_r\,, \end{aligned} \tag{18.14}$$

and form a three-dimensional Lie algebra. The linear combinations

$$T_1 = \frac{rp_r^2 - r}{2}\,, \quad T_2 = rp_r, \quad T_3 = \frac{rp_r^2 + r}{2} \tag{18.15}$$

satisfy

$$[T_1, T_2] = -iT_3, \quad [T_2, T_3] = iT_1, \quad [T_3, T_1] = iT_2\,, \tag{18.16}$$

and thus give a realization of the *so*(2,1) generators. It is an excellent exercise to show the relations in Eqs. (18.14) and (18.15). J. Cizek and J. Paldus [*Int. J. Quantum Chem.* **12**, 875 (1977)] have published a pedagogical approach to the manipulation of generators of this type for a general coordinate space in N dimensions.

These generators are not sufficiently general for our purpose. The general three-dimensional momentum square is $p^2 = p_r^2 + L^2/r^2$, where L^2 is the square of the total orbital angular momentum, an operator independent of r.

In terms of the radial motion, L^2 can be replaced by a constant ξ. We form the generators

$$T_1 = \frac{rp_r^2 + \xi/r - r}{2},$$

$$T_2 = rp_r, \tag{18.17}$$

$$T_3 = \frac{rp_r^2 + \xi/r + r}{2},$$

which satisfy the commutation relations in Eq. (18.16). Again, it is a worthwhile exercise to derive the commutation relations for these generators. Even more general realizations are

$$T_1 = \frac{w^{-2}r^{2-w}p_r^2 + \xi r^{-w} - r^w}{2}$$

$$T_2 = w^{-1}\left[rp_r - i\frac{(w-1)}{2}\right] \tag{18.18}$$

$$T_3 = \frac{w^{-2}r^{2-w}p_r^2 + \xi r^{-w} + r^w}{2}$$

for w=1, 2, 3,.... These operators are Hermitian with respect to the scalar product

$$\langle f|g\rangle = \int_0^\infty f^*(r)g(r)r^w dr, \tag{18.19}$$

making, say, T_2 Hermitian, since

$$\langle f|T_2f\rangle - \langle f|T_2f\rangle^* = -\frac{i}{w}\int \frac{d}{dr}|r^{(w+1)/2}f(r)|^2 dr, \tag{18.20}$$

which vanishes for suitable boundary conditions on the functions $f(r)$. Only the generator T_3 has square integrable eigenfunctions, which is why we choose to diagonalize T^2 and T_3 in the study of the representations of $so(2,1)$. The Casimir operator [see Eq. (18.2)]

$$\begin{aligned}T^2 &= T_3^2 - T_1^2 - T_2^2 = (T_3 - T_1)(T_3 + T_1) - [T_3,T_1] - T_2^2\\ &= r^w(w^{-2}r^{2-w}p_r^2 + \xi r^{-w}) - iw^{-1}rp_r - \frac{w-1}{2w}\\ &\quad - w^{-2}rp_rrp_r + \frac{i(w-1)rp_r}{w^2} + \frac{(w-1)^2}{4w^2}\\ &= \xi + \frac{1-w^2}{4w^2} = \xi + W(W-1), \quad W = \frac{(w+1)}{2w},\end{aligned} \tag{18.21}$$

is a constant.

For this example we need the unirreps of $so(2,1)$ that have the eigenvalue spectrum of T_3 bounded from below. We consider the unirreps $D^+(-k-1)$, for which we have

$$\begin{aligned} T^2|k,\mu\rangle &= k(k+1)|k,\mu\rangle \\ T_3|k,\mu\rangle &= \mu|k,\mu\rangle \\ k(k+1) &= \xi + W(W-1); \quad k > -1 \\ \mu &= k+1+m; \quad m = 0,1,2,\ldots. \end{aligned} \tag{18.22}$$

The ladder operators

$$T_\pm = T_1 \pm iT_2 = T_3 \pm W - r^w \pm \frac{r}{w}\frac{\partial}{\partial r} \tag{18.23}$$

satisfy

$$\begin{aligned} T_+|k,\mu\rangle &= [-(k-\mu)(k+\mu+1)]^{1/2}|k,\mu+1\rangle, \\ T_-|k,\mu\rangle &= [-(k+\mu)(k-\mu+1)]^{1/2}|k,\mu-1\rangle, \end{aligned} \tag{18.24}$$

where the states $|k,\mu\rangle$ are orthonormal with respect to some suitable scalar product.

The preceding relations for the generators can be used with advantage to calculate matrix elements for operators that are expressible in terms of the generators, and this can be done without having to obtain explicit solutions for $|k,\mu\rangle$.

18.2 THE HYDROGEN ATOM AND THE ISOTROPIC LINEAR HARMONIC OSCILLATOR

The generator T_3 can be expressed as the second-order differential operator:

$$T_3 = -\left(\frac{1}{2}w^2\right)r^{2-w}\left(\frac{\partial^2}{\partial r^2} + \frac{2}{r}\frac{\partial}{\partial r}\right) + \frac{\xi}{2r^w} + \frac{r^w}{2} \tag{18.25}$$

that we can see is related to the operators that arise from various radial eigenvalue problems involving some Hamiltonian. We consider the second-order differential equation

$$\left[\frac{w^{-2}p_r^2}{2} + \frac{\xi r^{-2}}{2} - \eta r^{w-2} - \tau r^{2w-2}\right]R(r) = 0, \tag{18.26}$$

which for suitable values of the parameters ξ, η, and τ and for $w = 1$ has the form of the radial nonrelativistic hydrogenic Hamiltonian, while for

$w = 2$ it can be made to correspond to the three-dimensional isotropic harmonic oscillator radial equation.

This equation multiplied with $2r^{2-w}$ can be expressed as

$$[(1 + 2\tau)T_1 + (1 - 2\tau)T_3 - 2\eta]R(r) = 0. \tag{18.27}$$

In order to eliminate T_1 so that we can use the eigenvalue properties of T_3, we need to consider the general question of scaling transformations. An operator S affects a scaling transformation on a function $f(r)$ when $Sf(r) = f(\lambda r)$ with $\lambda > 0$. Normally one reparametrizes so that $\lambda = e^{\alpha}$, and thus α can be any real number. A scaling transformation of an operator X is defined by $\bar{X} = SXS^{-1}$.

These scaling transformations can be defined in terms of the $so(2,1)$ generator T_2. In fact, we can use the results from our discussion of finite transformations:

$$\exp\left(\alpha r\frac{d}{dr}\right)f(r) = f(e^{\alpha}r), \tag{18.28}$$

which essentially comes from expanding $f(e^{\alpha}r)$ in a Taylor series around $\alpha = 0$ and using $d/d\alpha = rd/dr$.[2] From this result and the expression for

$$T_2 = w^{-1}\left[rp_r - \frac{i(w-1)}{2}\right] = w^{-1}\left[-ird/dr - \frac{i(w+1)}{2}\right], \tag{18.29}$$

we can write

$$\exp(i\alpha T_2)f(r) = e^{\alpha W}e^{(\alpha/w)rd/dr}f(r) = e^{\alpha W}f(e^{\alpha/w}r);$$
$$W = \frac{1+w}{2w}. \tag{18.30}$$

The equality in Eq. (18.29) means that when operating on a radial function, it holds that $rp_r = -ir(d/dr) - i$. Now since

$$e^{-B}Ae^{B} = A + [A,B] + \frac{[[A,B],B]}{2!} + \frac{[[[A,B],B],B]}{3!} + \dots, \tag{18.31}$$

we get $\bar{T}_3 = \exp(i\alpha T_2)T_3\exp(-i\alpha T_2) = T_3\cosh\alpha - T_1\sinh\alpha$.

If we now reparametrize in the preceding differential equation by putting $(1 + 2\tau)/(-8\tau)^{1/2} = \sinh\alpha$, and $(1 - 2\tau)/(-8\tau)^{1/2} = \cosh\alpha$, we get

$$\left\{T_1\sinh\alpha + T_3\cosh\alpha - \frac{2\eta}{(-8\tau)^{1/2}}\right\}R(r) = 0. \tag{18.32}$$

[2] Using the chain rule two ways, we can write the relation $d/du = (re^{\alpha})^{-1}d/d\alpha = e^{-\alpha}d/dr$ for $u = e^{\alpha}r$, from which the result follows.

Now expressing T_3 in terms of $\bar{T}_3$ and T_1 yields

$$\exp(-i\alpha T_2)\left[T_3 - \frac{2\eta}{(-8\tau)^{1/2}}\right]\exp(i\alpha T_2)R(r) = 0. \tag{18.33}$$

The T_3 eigenvalue problem then becomes

$$\left[T_3 - \frac{\eta}{(-2\tau)^{1/2}}\right]\bar{R}(r) = 0, \tag{18.34}$$

with the scaling parameter $\lambda = e^{\alpha} = (-2\tau)^{-1/2}$. The T_3 eigenvalues are given by $\mu = \eta\lambda$, and the parameters of the original differential equation are $\tau = -\frac{1}{2}\eta^2\mu^{-2}$ and $\eta = \mu\lambda^{-1}$.

When we now compare this result to the nonrelativistic hydrogenic Hamiltonian

$$\left[\frac{p_r^2}{2} + \frac{l(l+1)}{2r^2} - \frac{Z}{r} - E\right]R(r) = 0, \tag{18.35}$$

we find with $w = 1$, $\xi = l(l+1)$, $\eta = Z$, and $\tau = E$, that

$$E_\mu = -\frac{Z^2\mu^{-2}}{2}, \tag{18.36}$$

where $\mu = \mu_0 + m$ $(m = 0, 1, 2, \ldots)$ and

$$\mu_0 = k + 1 = \frac{1 \pm [4\xi + (2W-1)^2]^{1/2}}{2} = l + 1, \tag{18.37}$$

which is obtained from the previous result that $k(k+1) = \xi + W(W-1)$.

Comparing the general results with the radial equation of the isotropic harmonic oscillator, we find

$$\left[\frac{p_r^2}{2} + \frac{l(l+1)}{2r^2} + \frac{\omega^2 r^2}{2} - E\right]R(r) = 0, \tag{18.38}$$

with $w = 2$, $\xi = l(l+1)/4$, $\eta = E/4$, $\tau = -\omega^2/8$, and $\lambda = 2/\omega$. Now $\mu_0 = k + 1 = (l + \frac{3}{2})/2$, $\mu = (n + \frac{3}{2})/2$ with $n = l + 2m$ $(m = 0, 1, 2, \ldots)$, and $E_n = 4\eta = 4\mu\lambda^{-1} = 2\mu\omega = \omega(n + \frac{3}{2})$.

Altough the details of the preceding brief account of the noncompact Lie algebra *so*(2,1) is a bit complicated, it should be sufficient to show the power of group theory. Not only are the angular properties of the electron states in the nonrelativistic hydrogen atom completely determined by group theory, but the radial part of the problem, which includes the energy spectrum, is also completely determined by symmetry. Obviously, this situation does not extend to the case of many-electron atoms.

This algebraic approach to the hydrogen atom problem introduced by Pauli in 1926 is just as old as the differential equation approach by Schrödinger from the same year. Nevertheless, the differential equation approach dominated the presentations in both chemistry and physics for a long time. The significance of the symmetry concepts became evident with elementary particle physics, where the lack of explicit forms of Hamiltonians forces the study of descriptions of the physical systems based on symmetry assumptions. The obvious success of so-called noncompact dynamical groups [such as $SO(2,1)$] has inspired the search for other symmetries in particle physics.

In quantum chemistry and also in nuclear physics, approximate treatments of the many-body problem have utilized the unitary group with great success. In this text we have tried to provide the briefest of introductions to this rich area of research.

Obviously, a deeper understanding of the role of continuous groups in the study of many-particle problems in electronic structure, and perhaps even more importantly, in molecular reaction dynamics, must be achieved by a much more thorough treatment of groups and algebras than is attempted in this text. There is little doubt that such knowledge will facilitate the future development of successful theoretical treatments of the complex many-body systems of chemistry.

REFERENCES AND ADDITIONAL READING

Group theory is presented in numerous texts at different levels. There are sufficient variations of style that allow a reader to be able to find more than one book to his/her own liking. Interesting books on group theory include

Adams, B. G., Cizek, J., and Paldus, J., in *Adv. Quantum Chem.*, **19**, 1 (1988) (the application of Lie algebra methods has been presented in interesting ways).

Boerner, H., *Representations of Groups*, North Holland, Amsterdam, 1969.

Coleman, A. J., "The Symmetric Group Made Easy," *Adv. Quantum Chem.*, **4**, 83 (1968) (interesting and useful reading).

Cotton, F. A., *Chemical Applications of Group Theory*, Wiley-Interscience, New York, 1971.

Gilmore, R., *Lie Groups, Lie Algebras, and Some of Their Applications*, Wiley, New York, 1974.

Griffith, J. S., *The Theory of Transition-Metal Ions*, Cambridge University Press, 1961 (spin–orbit coupling is treated in some detail).

Hall, G. G., *Applied Group Theory*, American Elsevier, New York, 1967.

Kotani, M., "Coupling of Identical Particles in a Field of Given Symmetry," *J. Phys. Soc.* (Japan) **19**, 215 (1964).

Ledermann, W., *Introduction to the Theory of Finite Groups*, Oliver & Boyd, London 1957.

Löwdin, P.-O., "Group Algebra and Convolution Algebra," *Rev. Modern Phys.* **39**, 259 (1967) (a detailed account of ideas on this subject).

Miller, W., *Symmetry Groups and Their Applications*, Academic Press, New York, 1972.

Pauncz, R., *Spin Eigenfunctions*, Plenum Press, New York, 1979.

Weyl, H., *Group Theory and Quantum Mechanics*, Dover, New York, 1950 (reprint).

Wigner, E. P., *Group Theory and its Application to The Quantum Mechanics of Atomic Spectra*, Academic Press, New York, 1959 (a classic treatment of symmetry).

PROBLEMS

1. Prove that the powers g^n of an element of a finite group form a cyclic subgroup, which has the same order as that of the element g.

2. Find all the elements of the group generated by the matrices

$$\mathbf{A} = \begin{bmatrix} 0 & 1 \\ 1 & 0 \end{bmatrix} \quad \text{and} \quad \mathbf{B} = \begin{bmatrix} 0 & i \\ -i & 0 \end{bmatrix}$$

under matrix multiplication. Construct the multiplication table of the group.

3. Consider the group of order 8 with the elements $\pm E$, $\pm J_1$, $\pm J_2$, $\pm J_3$, with E the identity, and the generating relations

$$J_i J_i = -E, J_1 J_2 = J_3, J_2 J_3 = J_1, J_3 J_1 = J_2 .$$

(a) Determine the conjugate classes of the group.
(b) Determine the dimensions of the irreps.
(c) Construct the character table.
(d) Determine the matrices of all the irreps.

4. Show that the ordinary three-dimensional vectors do not form a linear associative algebra when the vector product is the rule of multiplication. Indicate which defining properties of a linear associative algebra are not satisfied.

5. Show that the convolution product of functions over the group is associative.

6. Show that the convolution product of functions over the group is not, in general, commutative.

7. Show that the convolution product of characters is commutative.

8. Construct the multiplication table of the point group C_{4v}.

9. Find all the subgroups of the point group C_{4v}.

10. Construct the normalizers $N(C_4)$, and $N(\sigma_d)$ of the point group C_{4v}.

11. Determine the matrices of the regular representation of the point group D_{2h}.

12. Consider the irrep $[2^3]$ of the symmetric group S_6. Calculate the dimension of this irrep. Also determine the character of the irrep at the conjugate class (1,2,3).

13. Determine the orders of the 11 classes of the symmetric group S_6.

14. Consider the symmetric group S_7 and determine the dimension of the irrep $[3^2,1]$, and the order of the conjugate class $(1,2^3)$.
    (a) Calculate the character of $[3^2,1]$ at $(1,2^3)$.

15. Consider the symmetric group S_5.
    (a) Determine the matrices of the permutations (23) and (34) in the orthogonal irrep $[2,1^3]$.
    (b) Do the same for the permutation (1234).

16. Consider the point group D_{6h}. Determine the irreps of the inner direct product group $D_{6h}[\times]D_{6h}$ contained in the following representations formed as direct products of the irreps of D_{6h}:

$$A_{1g} \otimes B_{1g};\ A_{1u} \otimes A_{1u};\ B_{2u} \otimes E_{1g};\ E_{1g} \otimes E_{2u};$$

$$E_{1g} \otimes B_{2g} \otimes A_{2u} \otimes E_{1u};$$

17. Prove that if Γ is an irrep and Δ a one-dimensional rep of a finite group G, then the inner direct product $\Gamma[\times]\Delta$ is an irrep of G.

18. Consider the NH_3 molecule in its ground-state equilibrium geometry. Determine the reducible representation matrices spanned by the $2p_x, 2p_y, 2p_z$ atomic orbitals on nitrogen. Determine the irreps and the corresponding matrices contained in this representation.

19. By considering the group $SO(3)$ and the character of the irrep Γ^l of dimension $2l + 1$ at the class of rotations $C(\alpha)$

$$\chi^l(C(\alpha)) = 1 + 2\sum_{m=1}^{l} \cos m\alpha\,,$$

determine

(a) The atomic spin doublet terms obtained from the electron configuration d^3

(b) The atomic spin triplet terms obtained from the electron configuration f^4

20. Consider an octahedral molecule (point group O_h) . By considering the group $SO(3)$ and the character of the irrep Γ^l, for $l = 2$, determine which irreps of $\Gamma^{l=2} \downarrow O_h$ are spanned by the the d orbitals on the central atom.

(a) Distort the molecule to D_{4h} symmetry. How can this be done?

(b) Show how the octahedral molecular orbital energy levels that belong to the d orbitals on the central atom split when the molecule undergoes this distortion.

21. In a molecule with octahedral symmetry, the molecular orbitals are labeled by the irreps of the point group O. When a perturbation is applied, the molecule assumes a lower symmetry:

(a) T

(b) D_3

(c) C_4

In each case determine how the possible molecular orbital energy levels of the system are split by the perturbation.

22. Consider the subgroup D_3 of the point group D_{6h}. Construct the representation matrices of the irrep E of D_3. This problem is the same as the construction of the orthogonal irrep [2,1] of the symmetric group S_3, which is isomorphic to D_3.

(a) Construct the induced representation $E \uparrow D_{6h}$.

23. Consider the H_2S_2 molecule with fixed bond distances and SSH bond angles $\approx 91°$, where the only geometric parameter that changes is the dihedral angle ϕ. Use only the $1s$ orbitals on the hydrogens and the $3p$ orbitals on the sulfurs to obtain appropriate symmetry-adapted orbitals and label each according to the appropriate irrep of the full point group for the following three cases:

(a) Planar moleculer with $\phi = 0$, that is, the cis conformation.

(b) Planar molecule with the dihedral angle $\phi = \pi$, that is, the trans conformation.

(c) Dihedral angle $0 < \phi < \pi$.

Let the S—S bond lie along the x axis, and let the z axis be a symmetry axis.

24. Consider the C_2H_4 molecule in its planar ground-state equilibrium geometry. After having determined the full point group of the molecule:

(a) Determine the vibrational symmetry modes.

(b) Find the C—H and the C—C stretch symmetry modes and display them graphically.
(c) Likewise, determine the in-plane bends.
(d) Finally, determine the out-of-plane bends.
(e) Label the Raman active, IR active, and inactive symmetry modes.

25. Consider the C_{3v} molecule PH_3. It has Raman and IR active vibrational modes at 2421, 2327, 1121, and 991 cm^{-1}. Suggest an assignment of these frequencies and draw a sketch of each symmetry (normal) mode.

26. Show that the commutator of an operator A with an integer power of another operator B satisfies the relation

$$[A, B^n] = \sum_{k=0}^{n-1} B^k [A, B] B^{n-k-1}.$$

INDEX

Printed and bound by CPI Group (UK) Ltd, Croydon, CR0 4YY
16/09/2026
14963442-0001